8086/8088, 80286, 80386, AND 80486 ASSEMBLY LANGUAGE PROGRAMMING

BARRY B. BREY
DeVry Institute of Technology

Merrill, an imprint of
Macmillan Publishing Company
New York

Maxwell Macmillan Canada
Toronto

Maxwell Macmillan International
New York   Oxford   Singapore   Sydney

Cover photo: Richard Wahlstrom Photography, Inc. By permission of Intel Corporation.
Editor: Dave Garza
Production Editor: Louise N. Sette
Art Coordinator: Lorraine Woost
Cover Designer: Thomas Mack
Production Buyer: Patricia A. Tonneman
Illustrations: Academy ArtWorks, Inc.

This book was set in Times Roman by Publication Services and was printed and bound by
Book Press, Inc., a Quebecor America Book Group Company. The cover was printed by
Phoenix Color Corp.

The Publisher offers discounts on this book when ordered in bulk quantities. For more
information, write to: Special Sales Department, Macmillan Publishing Company, 445
Hutchinson Avenue, Columbus, OH 43235, or call 1-800-228-7854

Macmillan Publishing Company
866 Third Avenue
New York, NY 10022

Macmillan Publishing Company is part of the Maxwell Communication Group of
Companies.

Maxwell Macmillan Canada, Inc.
1200 Eglinton Avenue East, Suite 200
Don Mills, Ontario M3C 3N1

Library of Congress Cataloging-in-Publication Data

Brey, Barry B.
  8086/8088, 80286, 80386, and 80486 assembly language programming /
Barry B. Brey.
        p.   cm.
  Includes index.
  ISBN 0-02-314247-2
    1. Intel 80xxx series microprocessors–Programming.
  2. Microprocessors–Programming. 3. Assembler language (Computer
program language) 4. Microsoft Macro assembler.   I. Title.
  QA76.8.I2674B73   1993
  005.265–dc20
                                                       93-93
                                                       CIP

Printing: 2 3 4 5 6 7 8 9   Year: 4 5 6 7 8

MERRILL'S INTERNATIONAL SERIES IN ENGINEERING TECHNOLOGY

INTRODUCTION TO ENGINEERING TECHNOLOGY

Pond, *Introduction to Engineering Technology, 2nd Edition*, 0-02-396031-0

ELECTRONICS TECHNOLOGY

Electronics Reference

Adamson, *The Electronics Dictionary for Technicians*, 0-02-300820-2

Berlin, *The Illustrated Electronics Dictionary*, 0-675-20451-8

Reis, *Becoming an Electronics Technician: Securing Your High-Tech Future*, 0-02-399231-X

DC/AC Circuits

Boylestad, *DC/AC: The Basics*, 0-675-20918-8

Boylestad, *Introductory Circuit Analysis, 7th Edition*, 0-02-313161-6

Ciccarelli, *Circuit Modeling: Exercises and Software, 2nd Edition*, 0-02-322455-X

Floyd, *Electric Circuits Fundamentals, 2nd Edition*, 0-675-21408-4

Floyd, *Electronics Fundamentals: Circuits, Devices, and Applications, 2nd Edition*, 0-675-21310-X

Floyd, *Principles of Electric Circuits, 4th Edition*, 0-02-338531-6

Floyd, *Principles of Electric Circuits: Electron Flow Version, 3rd Edition*, 0-02-338501-4

Keown, *PSpice and Circuit Analysis, 2nd Edition*, 0-02-363526-6

Monssen, *PSpice with Circuit Analysis*, 0-675-21376-2

Murphy, *DC Circuit Tutor: A Software Tutorial Using Animated Hypertext*, 0-02-385141-4

Murphy, *AC Circuit Tutor: A Software Tutorial Using Animated Hypertext*, 0-02-385144-9

Tocci, *Introduction to Electric Circuit Analysis, 2nd Edition*, 0-675-20002-4

Devices and Linear Circuits

Berlin & Getz, *Fundamentals of Operational Amplifiers and Linear Integrated Circuits*, 0-675-21002-X

Berube, *Electronic Devices and Circuits Using MICRO-CAP II*, 0-02-309160-6

Berube, *Electronic Devices and Circuits Using MICRO-CAP III*, 0-02-309151-7

Bogart, *Electronic Devices and Circuits, 3rd Edition*, 0-02-311701-X

Bogart, *Linear Electronics*, 0-02-311601-3

Floyd, *Basic Operational Amplifiers and Linear Integrated Circuits*, 0-02-338641-X

Floyd, *Electronic Devices, 3rd Edition*, 0-675-22170-6

FLoyd, *Electronic Devices: Electron Flow Version*, 0-02-338540-5

Floyd, *Fundamentals of Linear Circuits*, 0-02-338481-6

Schwartz, *Survey of Electronics, 3rd Edition*, 0-675-20162-4

Stanley, *Operational Amplifiers with Linear Integrated Circuits, 3rd Edition*, 0-02-415556-X

Tocci, *Electronic Devices: Conventional Flow Version, 3rd Edition*, 0-675-21150-6

Tocci & Oliver, *Fundamentals of Electronic Devices, 4th Edition*, 0-675-21259-6

Digital Electronics

Floyd, *Digital Fundamentals, 5th Edition*, 0-02-338502-2

Foster, *Sequential Logic Tutor: A Software Tutorial Using Animated Hypertext*, 0-02-338731-9

Foster, *Combinational Logic Tutor: A Software Tutorial Using Animated Hypertext*, 0-02-338735-1

McCalla, *Digital Logic and Computer Design*, 0-675-21170-0

Reis, *Digital Electronics through Project Analysis*, 0-675-21141-7

Tocci, *Fundamentals of Pulse and Digital Circuits, 3rd Edition*, 0-675-20033-4

Microprocessor Technology

Antonakos, *The 68000 Microprocessor: Hardware and Software Principles and Applications, 2nd Edition*, 0-02-303603-6

Antonakos, *An Introduction to the Intel Family of Microprocessors: A Hands-On Approach Utilizing the 8088 Microprocessor*, 0-675-22173-0

Brey, *8086/8066, 80286, 80386, and 80486 Assembly Language Programming*, 0-02-314247-2

Brey, *The Advanced Intel Microprocessor*, 0-02-314245-6

Brey, *The Intel Microprocessors: 8086/8088, 80186, 80286, 80386, and 80486: Architecture, Programming, and Interfacing, 3rd Edition*, 0-02-314250-2

Brey, *Microprocessors and Peripherals: Hardware, Software, Interfacing, and Applications, 2nd Edition*, 0-675-20884-X

Driscoll, Coughlin, & Villanucci, *Data Acquisition and Process Control with the MC68HC11 Microcontroller*, 0-02-330555-X

Gaonkar, *Microprocessor Architecture, Programming, and Applications with the 8085/8080A, 2nd Edition*, 0-675-20675-6

Gaonkar, *The Z80 Microprocessor: Architecture, Interfacing, Programming, and Design, 2nd Edition*, 0-02-340484-1

Goody, *Programming and Interfacing the 8086/8088 Microprocessor: A Product-Development Laboratory Process*, 0-675-21312-6

MacKenzie, *The 8051 Microcontroller*, 0-02-373650-X

Miller, *The 68000 Family of Microprocessors: Architecture, Programming, and Applications, 2nd Edition*, 0-02-381560-4

Quinn, *The 6800 Microprocessor*, 0-675-20515-8

Subbarao, *16/32 Bit Microprocessors: 68000/68010/68020 Software, Hardware, and Design Applications*, 0-675-21119-0

Electronic Communications

Monaco, *Introduction to Microwave Technology*, 0-675-21030-5

Monaco, *Preparing for the FCC Radio-Telephone Operator's License Examination*, 0-675-21313-4

Schoenbeck, *Electronic Communications: Modulation and Transmission, 2nd Edition*, 0-675-21311-8

Young, *Electronic Communication Techniques, 3rd Edition*, 0-02-431201-0

Zanger & Zanger, *Fiber Optics: Communication and Other Applications*, 0-675-20944-7

Microcomputer Servicing

Adamson, *Microcomputer Repair*, 0-02-300825-3

Asser, Stigliano, & Bahrenburg, *Microcomputer Servicing: Practical Systems and Troubleshooting, 2nd Edition*, 0-02-304241-9

Asser, Stigliano, & Bahrenburg, *Microcomputer Theory and Servicing, 2nd Edition*, 0-02-304231-1

Programming

Adamson, *Applied Pascal for Technology*, 0-675-20771-1

Adamson, *Structured BASIC Applied to Technology, 2nd Edition*, 0-02-300827-X

Adamson, *Structured C for Technology*, 0-675-20993-5

Adamson, *Structured C for Technology (with disk)*, 0-675-21289-8

Nashelsky & Boylestad, *BASIC Applied to Circuit Analysis*, 0-675-20161-6

Instrumentation and Measurement

Berlin & Getz, *Principles of Electronic Instrumentation and Measurement*, 0-675-20449-6

Buchla & McLachlan, *Applied Electronic Instrumentation and Measurement*, 0-675-21162-X

Gillies, *Instrumentation and Measurements for Electronic Technicians, 2nd Edition*, 0-02-343051-6

Transform Analysis

Kulathinal, *Transform Analysis and Electronic Networks with Applications*, 0-675-20765-7

Biomedical Equipment Technology

Aston, *Principles of Biomedical Instrumentation and Measurement*, 0-675-20943-9

Mathematics

Monaco, *Essential Mathematics for Electronics Technicians*, 0-675-21172-7

Davis, *Technical Mathematics*, 0-675-20338-4

Davis, *Technical Mathematics with Calculus*, 0-675-20965-X

INDUSTRIAL ELECTRONICS/INDUSTRIAL TECHNOLOGY

Bateson, *Introduction to Control System Technology, 4th Edition*, 0-02-306463-3

Fuller, *Robotics: Introduction, Programming, and Projects*, 0-675-21078-X

Goetsch, *Industrial Safety and Health: In the Age of High Technology*, 0-02-344207-7

Goetsch, *Industrial Supervision: In the Age of High Technology*, 0-675-22137-4

Geotsch, *Introduction to Total Quality: Quality, Productivity, and Competitiveness*, 0-02-344221-2

Horath, *Computer Numerical Control Programming of Machines*, 0-02-357201-9

Hubert, *Electric Machines: Theory, Operation, Applications, Adjustment, and Control*, 0-675-20765-7

Humphries, *Motors and Controls*, 0-675-20235-3

Hutchins, *Introduction to Quality: Management, Assurance, and Control*, 0-675-20896-3

Laviana, *Basic Computer Numerical Control Programming*, 0-675-21298-7

Pond, *Fundamentals of Statistical Quality Control*

Reis, *Electronic Project Design and Fabrication, 2nd Edition*, 0-02-399230-1

Rosenblatt & Friedman, *Direct and Alternating Current Machinery, 2nd Edition*, 0-675-20160-8

Smith, *Statistical Process Control and Quality Improvement*, 0-675-21160-3

Webb, *Programmable Logic Controllers: Principles and Applications, 2nd Edition*, 0-02-424970-X

Webb & Greshock, *Industrial Control Electronics, 2nd Edition*, 0-02-424864-9

MECHANICAL/CIVIL TECHNOLOGY

Dalton, *The Technology of Metallurgy*, 0-02-326900-6

Keyser, *Materials Science in Engineering, 4th Edition*, 0-675-20401-1

Kokernak, *Fluid Power Technology*, 0-02-305705-X

Kraut, *Fluid Mechanics for Technicians*, 0-675-21330-4

Mott, *Applied Fluid Mechanics, 4th Edition*, 0-02-384231-8

Mott, *Machine Elements in Mechanical Design, 2nd Edition*, 0-675-22289-3

Rolle, *Thermodynamics and Heat Power, 4th Edition*, 0-02-403201-8

Spiegel & Limbrunner, *Applied Statics and Strength of Materials, 2nd Edition*, 0-02-414961-6

Spiegel & Limbrunner, *Applied Strength of Materials*, 0-02-414970-5

Wolansky & Akers, *Modern Hydraulics: The Basics at Work*, 0-675-20987-0

Wolf, *Statics and Strength of Materials: A Parallel Approach to Understanding Structures*, 0-675-20622-7

DRAFTING TECHNOLOGY

Cooper, *Introduction to VersaCAD*, 0-675-21164-6

Ethier, *AutoCAD in 3 Dimensions*, 0-02-334232-3

Goetsch & Rickman, *Computer-Aided Drafting with AutoCAD*, 0-675-20915-3

Kirkpatrick & Kirkpatrick, *AutoCAD for Interior Design and Space Planning*, 0-02-364455-9

Kirkpatrick, *The AutoCAD Book: Drawing, Modeling, and Applications, 2nd Edition*, 0-675-22288-5

Kirkpatrick, *The AutoCAD Book: Drawing, Modeling, and Applications, Including Release 12, 3rd Edition*, 0-02-364440-0

Lamit & Lloyd, *Drafting for Electronics, 2nd Edition*, 0-02-367342-7

Lamit & Paige, *Computer-Aided Design and Drafting*, 0-675-20475-5

Maruggi, *Technical Graphics: Electronics Worktext, 2nd Edition*, 0-675-21378-9

Maruggi, *The Technology of Drafting*, 0-675-20762-2

Sell, *Basic Technical Drawing*, 0-675-21001-1

TECHNICAL WRITING

Croft, *Getting a Job: Resume Writing, Job Application Letters, and Interview Strategies*, 0-675-20917-X

Panares, *A Handbook of English for Technical Students*, 0-675-20650-2

Pfeiffer, *Proposal Writing: The Art of Friendly Persuausion*, 0-675-20988-9

Pfeiffer, *Technical Writing: A Practical Approach, 2nd Edition*, 0-02-395111-7

Roze, *Technical Communications: The Practical Craft, 2nd Edition*, 0-02-404171-8

Weisman, *Basic Technical Writing, 6th Edition*, 0-675-21256-1

This book is dedicated to all who seethe with the passion to unravel the interior workings of the microprocessor-based personal computer.

"We work in the dark—we do what we can—we give what we have. Our doubt is our passion, and our passion is our task."

Henry James
The Middle Years (1893)

PREFACE

This text is written for the student in a course of study that seeks to impart a thorough knowledge of programming the Intel family of microprocessors. It is a practical reference text for anyone interested in all programming aspects of this important microprocessor family. Today, anyone seeking to excel in a field of study that uses computers must understand assembly language programming of this important family of microprocessors. Intel microprocessors have gained wide applications in many areas of electronics, communications, control systems, and particularly in desktop computer systems.

Organization and Coverage

To try to cultivate a comprehensive approach to learning, each chapter of the text begins with a set of objectives that specify the outcomes of mastering the chapter. This is followed by the body of the chapter, which includes many programming applications to illustrate the main topics of the chapter. At the end of each chapter, a summary, which doubles as a study guide, reviews the information presented in the chapter. An end-of-the-chapter glossary defines each new key term. Finally, questions and problems are provided to allow practice in applying the concepts presented in the chapter.

This text contains many example programs, using the Microsoft Macro Assembler program, to demonstrate how to program the Intel microprocessors. The programming environment covered includes the linker, the library, macros, DOS functions, and BIOS functions.

Approach

Because the Intel family of microprocessors is quite diverse, this text initially concentrates on real mode programming, which is compatible with all versions of the Intel microprocessors. Instructions for each microprocessor, including the 80386 and 80486, are compared and contrasted with those for the 8086/8088. This entire series of microprocessors is very similar, which promotes learning the advanced versions once the basic 8086/8088 is understood.

In addition to fully explaining the programming and operation of the microprocessor, this text also covers the programming and operation of the numeric coprocessor (8087/80287/80387/80486/7). The numeric coprocessor provides access to floating-point calculations that are important in applications such as control systems, video graphics, and computer-aided design (CAD). The numeric coprocessor allows a program to access complex arithmetic operations that are otherwise difficult to achieve with normal microprocessor programming.

Through this approach, stressing the operation of the microprocessor and programming with the advanced family members, a working and practical background is attainable. On completion of a course of study based on this text, you should be able to:

1. Develop control software to control an application interface to the 8086/8088, 80286, 80386, or 80486 microprocessor. Generally, the software developed will function on all versions of the microprocessor. This software also includes DOS-based applications.
2. Program using DOS function calls to control the keyboard, video display system, and disk memory in assembly language.
3. Use the BIOS functions to control the keyboard, display, and various other components in the computer system.
4. Develop software that uses interrupt hooks and hot-keys to gain access to terminate and stay resident software.
5. Program the numeric coprocessor (80287/80387) to solve complex equations.
6. Explain the differences between the family members and highlight the features of each member.
7. Describe and use real and protected mode operation of the 80286, 80386, and 80486 microprocessors.

Content Overview

Chapter 1 introduces the Intel family of microprocessors, with emphasis on the microprocessor-based computer system. This first chapter serves to introduce the microprocessor, its history, its operation, and the methods used to store data in a microprocessor-based system.

Chapter 2 begins by exploring the programming model of the microprocessor and system architecture. Both real and protected mode operation is explained in this introductory chapter. Once an understanding of the basic machine is grasped, Chapters 3–6 explain how each instruction functions with the Intel family of microprocessors. As instructions are explained, simple applications are presented to illustrate their operation and develop basic programming concepts.

Once the basis for programming is developed, Chapters 6–8 provide applications using the assembler program. These applications include programming using DOS and BIOS function calls. Disk files are explained, as well as keyboard and video operation on a personal computer system. This chapter provides the tools required to develop virtually any program on a personal computer system.

Chapter 9 explains some advanced programming techniques used in everyday software. These techniques include interrupt hooks, terminate and stay resident software, and hot-key programs.

Chapter 10 details the operation, instruction set, and programming of the 80X87 numeric coprocessor family. Applications for the coprocessor include problems that solve complex equations and allow data to be displayed in floating-point format on the video screen.

Appendices are included to supplement the text coverage. They are comprised of the following:

Appendix A. A complete listing of the DOS INT 21H function calls. Appendix A also details the use of the assembler program and many of the BIOS function calls.

Appendix B. Complete listing of all 8086/8088/80286/80386/80486 instructions, including many example instructions and machine coding in hexadecimal.

Appendix C. Answers for all the even-numbered questions and problems in the text.

I am grateful to the following reviewers for their helpful suggestions: Robert L. Douglas (Memphis State University, Memphis, TN), James DeLoach (DeVry Institute of Technology–Atlanta), Omer Farook (Purdue University–Calumet, Hammond, IN), Chandra Sekhar (Purdue University–Calumet, Hammond, IN).

CONTENTS

CHAPTER 1

Introduction to the Microprocessor and Computer

INTRODUCTION

This chapter presents an overview of the microprocessor and its applications. Included is a discussion of the history of computers, as well as a description of the function of the microprocessor-based computer. The terms and jargon of the computer field are also introduced so that computerese can be understood and used when discussing microprocessors and computers.

Also introduced is the block diagram of the computer system and a description of the function of each block. The chapter also details the memory and input/output system of the personal computer.

Finally, the way that data are stored in the memory is provided so each data type can be used with software development. Data are stored as integers, floating-point and binary-coded decimal (BCD) numbers, and as ASCII (American Standard Code for Information Interchange) codes to represent alphanumeric data.

1–1 CHAPTER OBJECTIVES

Upon completion of this chapter, you will be able to:

1. Converse using the appropriate computer terms, including *bit, byte, data, memory, expanded memory, extended memory, DOS, I/O,* and so forth.
2. Briefly describe the history of the computer and list some applications performed by the computer.
3. Draw the block diagram of a computer system and explain the purpose of each block.
4. Describe the function of the microprocessor and detail its basic operation.
5. Detail the contents of the memory system in a personal computer.
6. Define and represent numeric and alphabetic information as integers, floating-point, BCD, and ASCII data.

1

1-2   A HISTORICAL BACKGROUND

This first section of the text outlines the historical events that led to the development of the microprocessor and, specifically, the extremely powerful and current 80386 and 80486 microprocessors. Although history is not essential to understanding the microprocessor, it furnishes interesting reading. It also provides a historical retrospective of the evolution of the computer.

The Mechanical Age

The idea of a computing system is not new—it has been around long before electronic devices. The idea of calculating with a machine dates to 500 B.C., when the Babylonians invented the *abacus,* a mechanical calculator. This first mechanical calculator, which was used extensively and is still in use, was not improved until 1642, when Blaise Pascal, the mathematician, invented a calculator constructed of gears and wheels. Each gear contained 10 teeth that, when moved one complete revolution, advanced the next gear one place. This is the same as the odometer mechanism in your automobile and is the basis of all mechanical calculators. Incidentally, the Pascal programming language was named in honor of Blaise Pascal and his pioneering work with mathematics and his mechanical calculator.

The idea of automatically computing information dates to at least the early 1800s and probably much earlier. Bear in mind that this is long before Edison invented the light bulb. This is even before much was known about electricity. In this dawn of the computer age, humans dreamed of mechanical machines that could compute numerical facts using a program and not merely calculate facts as with a calculator.

The discovery in 1937 of his plans and journals brought the fact to light that one of the early pioneers of the computing machine was Charles Babbage. Babbage was commissioned in 1823 by the Royal Astronomical Society of Great Britain to produce a programmable calculating machine. He accepted the challenge and began to create what he called the *"Analytical Engine."* His engine was a mechanical computer that was capable of storing 1,000 twenty-digit decimal numbers. It could also store a program that could be used to perform various calculations. Input to his machine was through punched cards, much as computers in the 1950s and 1960s used punched cards.

After years of work, his dream began to fade when he realized that the machinists of his day were unable to create the mechanical parts needed to complete his work. The Analytical Engine called for more than 50,000 machined parts, which could not be made with enough precision to allow his engine to function reliably.

The Electrical Age

The late 1800s saw the advent of the electric motor; with it came a multitude of electric adding machines, all based on the mechanical calculator developed by Pascal. These mechanical calculators were common pieces of office equipment until well into the early 1970s, when the electronic calculator, first introduced by Bomar, appeared.

In 1889 *Herman Hollerith* developed the punched card for storing data. He also developed a mechanical machine—driven by one of the new electric motors—that counted, sorted, and collated information stored on punched cards. The idea of calculating by machinery intrigued the government so much that Hollerith was commissioned to use his punched-card system to store and tabulate information for the 1890 census.

In 1896, Hollerith formed a company called the *Tabulating Machine Company.* This company developed a line of machinery that used punched cards for tabulations. After a number of mergers, the Tabulating Machine Company was formed into the *International Business Machines Corporation,* now more commonly referred to as IBM. We often refer to the punched cards used in computer systems as *Hollerith cards* in honor of Herman Hollerith. The 12-bit code used on a punched card is called the *Hollerith code*.

Mechanical machines driven by electric motors continued to dominate the information-processing world until the advent of the first electronic calculating machine in 1942 by a German inventor named Konrad Zuse. His calculating computer, the Z3, was used in aircraft and missile design during World War II as part of the German war effort.

The first electronic computer was placed into operation in 1943 to break secret German military codes. This first electronic computer system, which used vacuum tubes, was invented by Alan Turing. Turing called his machine *Colossus,* most likely because of its size. A problem with Colossus was that its design allowed it to break German codes, but it could not solve any other problems. It was not programmable; it was a fixed-program computer system, of a type which we often call a *special-purpose computer.*

The first general-purpose, programmable electronic **computer** system was developed in 1946 at the University of Pennsylvania. This first modern computer was called the ENIAC (*Electronic Numerical Integrator and Calculator*). The ENIAC was a huge machine containing over 17,000 vacuum tubes and over 500 miles of wires. This massive machine weighed over 30 tons, yet performed only about 100,000 operations per second. Still and all, the ENIAC thrust the world into the age of electronic computers.

Breakthroughs that followed included the development of the transistor in 1948, followed by the invention of the integrated circuit in 1958 by Jack Kilby. The integrated circuit was to lead to the development of the first microprocessor in 1971 at Intel Corporation. At this time, Intel and one of its engineers, Marcian E. Hoff, developed the 4004 microprocessor—the device that started the microprocessor revolution that continues unabated today.

The Microprocessor Age

The world's first microprocessor, the Intel 4004, is a 4-bit microprocessor—a programmable controller on a chip—that is meager by today's standards. It addressed a mere 4,096 four-bit memory locations. (A **bit** is a binary digit with a value of one or zero. A 4-bit memory location is often called a **nibble.**) The 4004 instruction set contained only 45 instructions. It was fabricated with P-channel MOSFET technology that only allowed it to execute instructions at the slow rate of 50 KIPS (*kilo-instructions per second*). This was slow when compared to the 100,000 instructions executed each second by the 30-ton ENIAC computer in 1946. The main difference was that the 4004 weighed less than an ounce.

At first, applications abounded for this device. The 4-bit microprocessor debuted in early video game systems and small microprocessor-based control systems. One early video game, a shuffleboard game, was produced by Bally. The main problems with this early microprocessor were its speed, data size, and memory size. The evolution of the 4-bit microprocessor ended when Intel released the 4040, an updated version of the earlier 4004. The 4040 operated at a higher speed, although it lacked improvements in word and memory size. Other companies, in particular Texas Instruments, also produced 4-bit microprocessors (the TMS-1000). The 4-bit microprocessor still survives in low-end applications such as microwave ovens and small control systems and is still available from some microprocessor manufacturers.

Later, in 1971, realizing that the microprocessor was a commercially viable product, Intel Corporation released the 8008—an extended 8-bit version of the 4004 microprocessor. The 8008 contained an expanded memory size (16K bytes) and additional instructions (a total of 48) that provided an opportunity for its application in more advanced systems. (A **byte** is generally an 8-bit-wide binary number and a computer **K** is 1,024. Often, memory size is rated in *K bytes*.)

As engineers developed more demanding uses for the 8008 microprocessor, they discovered that its somewhat small memory size, slow speed, and instruction set limited its usefulness. Intel recognized these limitations and, in 1973, introduced the 8080 microprocessor—the first of the modern 8-bit microprocessors. The flood gates opened as the 8080 ushered in the *age of the microprocessor.* Soon, many other companies began to introduce their own versions of the 8-bit microprocessor. Table 1–1 lists several of these early microprocessors and their manufacturers. Of these early microprocessor producers, only Intel and Motorola continue to enjoy success with newer and improved versions of the microprocessor.

What Was Special about the 8080? Not only could the 8080 address more memory and execute more instructions, but it executed them 10 times faster than the 8008. An addition, which took 20 μs (50,000 instructions per second) on an 8008-based

TABLE 1–1 Early 8-bit microprocessors

Manufacturer	Part number
Fairchild	F-8
Intel	8080
MOS Technology	6502
Motorola	MC6800
National Semiconductor	IMP-8
Rockwell International	PPS-8

system, required only 2.0 μs (500,000 instructions per second) on an 8080-based system. Also, the 8080 was compatible with TTL (transistor–transistor logic) and the 8008 was not directly compatible. This made interfacing much easier and less expensive. The 8080 could also address four times more memory (64K bytes) than the 8008 (16K bytes). These improvements are responsible for ushering in the era of the 8080, and the continuing era of the microprocessor.

The 8085 Microprocessor. In 1977, Intel Corporation introduced an updated version of the 8080—the 8085. This was to be the last 8-bit general-purpose microprocessor developed by Intel. Although only slightly more advanced than an 8080, the 8085 executes software at an even higher speed. An addition, which took 2.0 μs (500,000 instructions per second) on the 8080, requires only 1.3 μs (769,230 instructions per second) on the 8085. The main advantages of the 8085 are its internal clock generator, internal system controller, and higher clock frequency. This higher level of component integration reduced its cost and increased the usefulness of the 8085 microprocessor. Intel has sold well over 100 million copies of the 8085 microprocessor, its most successful 8-bit microprocessor. Because the 8085 is also manufactured (*second-sourced*) by many other companies, there are well over 200 million of these microprocessors in existence. Applications that contain the 8085 are still being used and designed, and it will likely continue to be popular well into the future.

The Modern Microprocessor

In 1978, Intel released the 8086 microprocessor and a year or so later the 8088. Both devices are 16-bit microprocessors, which execute instructions in as little as 400 ns (2.5 **MIPS,** or 2.5 *millions of instructions per second*). This represents a major improvement over the execution speed of the 8085. In addition, the 8086 and 8088 address 1M byte of memory, 16 times more memory than the 8085. (A *1M byte memory* contains 1,024K byte-sized memory locations, or 1,048,576 bytes.) These higher execution speeds and larger memory sizes allow the 8086 and 8088 to replace smaller minicomputers in many applications.

The increase in memory size and additional instructions of the 8086 and 8088 have led to many sophisticated applications for microprocessors. Improvements to the instruction set included a multiply and divide instruction, which is missing from all prior versions of the microprocessor. Also the number of instructions increased from 45 on the 4004, to 246 on the 8085, to well over 20,000 variations on the 8086 and 8088 microprocessors. These additional instructions ease the task of developing efficient and sophisticated applications even though their sheer number is at first overwhelming. The 16-bit microprocessor also provides more internal register storage space than the 8-bit microprocessor. These additional registers allow software to be written more efficiently.

The 16-bit microprocessor evolved because of the need for larger memory systems. Applications such as spreadsheets, word processors, spelling checkers, and computer-based thesauruses are memory intensive and require more than the 64K bytes of memory found in 8-bit microprocessors to execute efficiently. The 16-bit 8086 and 8088 provide 1M byte of memory for these applications. Soon, even 1M byte of memory proved limiting for large spreadsheets and other applications. This led to the introduction of the 80286 microprocessor by Intel in 1983.

The 80286 Microprocessor.   The 80286 microprocessor is almost identical to the 8086 and 8088 except it addresses a 16M byte memory system instead of 1M byte. The instruction set of the 80286 is also almost identical to the 8086 and 8088 except for a few additional instructions that manage the extra 15M bytes of memory. The clock speed of the 80286 is increased so it executes some instructions in as little as 250 ns (4.0 MIPS) with the original 8.0 MHz version.

The 32-bit Microprocessor.   Applications began to demand faster microprocessor speeds, more memory, and wider data widths. This led to the introduction of the 80386 in 1986 by Intel Corporation. The 80386 represented a major overhaul of the 16-bit microprocessor's architecture. The 80386 is a full 32-bit microprocessor that contains a 32-bit data bus and a 32-bit memory address. The 80386 addresses up to 4G bytes of memory. (***One G*** of memory contains 1,024M locations, or 1,073,741,824 locations.) A 4G-byte memory can store an astounding 1 million typewritten double-spaced pages of data. The 80386 is also available in a few modified versions such as the 80386SX, which addresses 16M bytes of memory through a 16-bit data and 24-bit address bus, and the 80386SL/80386SLC, which addresses 32M bytes of memory through a 16-bit data and 25-bit address bus. The 80386SLC version contains an internal cache memory to allow it to process data at even higher rates.

Applications that require higher microprocessor speeds and large memory systems include software systems that use a ***GUI,*** or *graphical user interface*. Modern graphical displays often contain 256,000 or more picture elements (***pixels*** or ***pels***). The least sophisticated VGA (*variable graphics array*) video display has a resolution of 640 pixels per scanning line with 480 scanning lines. In order to display one screen of information, each picture element must be changed. This requires a high-speed microprocessor. Many new software packages use this type of video interface. These GUI-based packages require high microprocessor speeds for quick and efficient manipulation of video text and graphical data. The most striking system, which requires high-speed computing for its graphical display interface, is Microsoft Corporation's Windows*.

The 32-bit microprocessor is needed because of the size of its data bus, which transfers real (single-precision floating-point) numbers that require 32-bit wide memory. In order to efficiently process 32-bit real numbers, the microprocessor must efficiently pass them between itself and memory. If they pass through an 8-bit data bus, it takes four read or write cycles, but when they pass through a 32-bit data bus only one read or write cycle is required. This significantly increases the speed of any program that manipulates real numbers. Most high-level languages, spreadsheets, and database management systems use real numbers for data storage. Real numbers are also used in graphic design packages that use vectors to plot images on the video screen.

Besides providing higher clocking speeds, the 80386 includes a memory management unit that allows memory resources to be allocated and managed by the

---

*Windows is a registered trademark of Microsoft Corporation.

operating system. Earlier microprocessors left memory management completely to the software. The 80386 includes hardware circuitry for memory management and memory assignment, which improves its efficiency.

The 80486 Microprocessor. In 1989, Intel released the 80486 microprocessor, which incorporated an 80386-like microprocessor, an 80387-like numeric coprocessor, and an 8K-byte cache memory system into one integrated package. Although the 80486 is not radically different from the 80386, it does include one major change. The internal structure of the 80486 is modified from the 80386 so about half its instructions execute in 1 clock instead of 2 clocks. Because the 80486 is available in a 50-MHz version, about half the instructions execute in 25 ns (50 MIPS). The average speed improvement for a typical mix of instructions is about 50 percent over the 80386 operated at the same clock speed. Newer versions of the 80486 execute instructions at even higher speeds with a 66-MHz and hopefully a 100-MHz clock rate.

Table 1–2 lists many microprocessors produced by Intel and Motorola with information about their data widths and memory sizes. Other companies produce

TABLE 1–2 Many modern microprocessors

Manufacturer	Part number	Data bus width	Memory size
Intel	8048	8	2K internal
	8051	8	8K internal
	8085A	8	64K
	8086	16	1M
	8088	8	1M
	8096	16	8K internal
	80186	16	1M
	80188	8	1M
	80286	16	16M
	80386DX	32	4G
	80386SL	16	32M
	80386SLC	16	32M
	80386SX	16	16M
	80486DX	32	4G
	80486SX	32	4G
Motorola	6800	8	64K
	6805	8	2K
	6809	8	64K
	68000	16	16M
	68008Q	8	1M
	68008D	8	4M
	68010	16	16M
	68020	32	4G
	68030	32	4G
	68040	32	4G

microprocessors, but none have had the success of Intel and, to a lesser degree, Motorola.

1–3   THE MICROPROCESSOR-BASED COMPUTER SYSTEM

The computer system has undergone many changes in recent history. Machines that once filled rooms have been reduced to small desktop computer systems because of the microprocessor. Even though these desktop computers are compact, they possess power only dreamed of a few years ago. Million-dollar mainframe computer systems developed in the early 1980s are not as powerful as the 80386/80486 microprocessor of today.

This section of the chapter discusses the structure of the microprocessor-based computer system. This discussion includes information about the memory and operating system used in many microprocessor-based computer systems.

Refer to Figure 1–1 for the block diagram of a computer system. Notice that the block diagram is composed of three blocks that are interconnected with buses. (A *bus* is a set of common connections that carry the same type of information. For example, the address bus, which contains 20 to 32 connections, is a bus that conveys the memory address to the memory.) These blocks and their function in a personal computer are outlined next.

The Memory and I/O System

The memory structure of all Intel 8086–80486 based computer systems is very similar. This similarity marks the first personal computer systems based on the 8088 that were introduced in 1981 by IBM and the most powerful, high-speed versions of today based on the 80486 microprocessor. Figure 1–2 illustrates the memory map of a personal computer system. This map applies to any IBM personal computer or any of the host of clones that are in existence.

The memory system is divided into three main parts: TPA (transient program area), system area, and extended memory. The type of microprocessor in your

FIGURE 1–1   The block diagram of a microprocessor-based computer system.

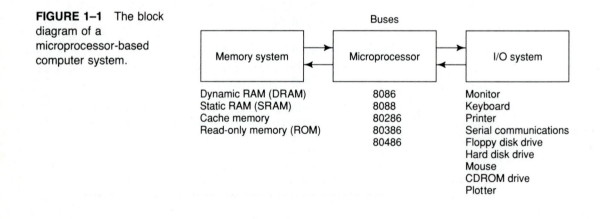

Buses		
Memory system	Microprocessor	I/O system

Dynamic RAM (DRAM)	8086	Monitor
Static RAM (SRAM)	8088	Keyboard
Cache memory	80286	Printer
Read-only memory (ROM)	80386	Serial communications
	80486	Floppy disk drive
		Hard disk drive
		Mouse
		CDROM drive
		Plotter

FIGURE 1–2 The memory map of a personal computer system.

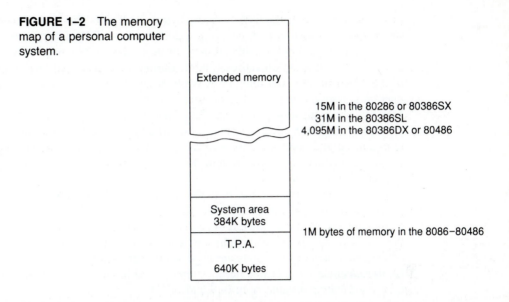

Extended memory

15M in the 80286 or 80386SX
31M in the 80386SL
4,095M in the 80386DX or 80486

System area
384K bytes

T.P.A.

640K bytes

1M bytes of memory in the 8086–80486

computer determines whether you have extended memory. If the computer is based on an 8086 or 8088 (a **PC**** or **XT**†), the TPA and system areas exist, but there is no extended memory area. The PC and XT contain 640K bytes of TPA and 384K bytes of system memory for a total memory size of 1M bytes.

Computer systems based on the 80286, 80386, and 80486 also contain the TPA (640K bytes) and system area (384K bytes). In addition, they contain extended memory. These machines are often called **AT**‡ class machines. The PS/2, produced by IBM, is another version of the same basic memory design. Sometimes these machines are also referred to as **ISA** (*industry standard architecture*) machines. The PS/2 is referred to as a *microchannel architecture* system or ISA system, depending on the model number. **Extended memory** contains 15M bytes in the 80286-based computer and 4,095M bytes in the 80386/80486-based computer in addition to the first 1M byte. The ISA machine is available with an 8-bit bus used to interface 8-bit devices to the computer in the 8086/8088-based PC or XT computer system. The AT class machine, also an ISA machine, uses a 16-bit bus for interface and may contain an 80286–80486 microprocessor. The **EISA** (extended ISA) bus is a 32-bit interface bus found in 80386- and 80486-based systems. Note that each of these buses are compatible with the earlier versions. That is, an 8-bit interface card functions in the 16-bit ISA or 32-bit EISA bus standards; likewise, a 16-bit interface card functions in the 32-bit EISA standard.

The TPA. The transient program area (**TPA**) holds the operating system and other programs that control the computer system. It also holds any currently active application

---

*PC is a trademark of IBM Corporation for a personal computer.
†XT is a trademark of IBM Corporation for the extended technology personal computer.
‡AT is a trademark of IBM Corporation used to denote an advanced class computer system.

program. The size of the TPA is 640K bytes. To repeat, this area of memory holds the operating system, which occupies a portion of the TPA. In practice, the amount of memory remaining for application software is about 628K bytes if MSDOS* version 5.0 is used as an operating system. Earlier versions of MSDOS required more of the TPA and often left only 530K bytes or less for application programs. Another operating system often found in personal computers is PCDOS†. Both PCDOS and MSDOS are compatible, so either functions just as the other does for application programs. The **DOS** (disk operating system) controls the way that the disk memory is organized and controlled, as well as determining the function and control of the I/O devices connected to the system. Figure 1–3 shows the organization of the TPA in a computer system.

The memory map shows how the many areas of the TPA are used for system programs, data, and drivers. It also shows a large area of memory available for application programs. To the left of each area are hexadecimal numbers that represent the memory addresses that begin and end each data area. Hexadecimal memory addresses or memory locations are used to number each byte of the memory system. (A **hexadecimal** number is a number represented in radix 16, or base 16, with each digit representing a value from 0–9 and A–F.)

Interrupt vectors are used to access various features of the DOS, BIOS (basic I/O system), and applications. The **BIOS** is a collection of programs that operate the

FIGURE 1–3 The memory map for the DOS TPA. This map varies from one version of DOS to another and also changes with different drivers and configurations.

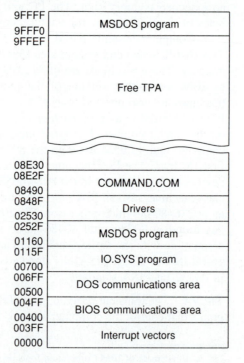

*MSDOS (Microsoft Disk Operating System) is a trademark of Microsoft Corporation.
†PCDOS (Personal Computer Disk Operating System) is a trademark of IBM Corporation.

I/O devices connected to your computer system. These programs are stored in the system area defined later in this section of the chapter.

The BIOS and DOS communications areas contain transient data used by programs to access I/O devices and the internal features of the computer system. These are stored in the TPA so they can be changed as the system operates. Note that the TPA contains read/write memory so it can change as a program executes.

The IO.SYS is a program that loads into the TPA from the disk whenever an MSDOS or PCDOS system is started. The IO.SYS contains programs that allow DOS to use the keyboard, video display, printer, and other I/O devices often found in the computer system.

The MSDOS (PCDOS) program occupies two areas of memory. One area is 16 bytes in length and is located at the top of the TPA and the other, which is much larger, is located near the bottom of the TPA. The MSDOS program controls the operation of the computer system. The size of the MSDOS area depends on the version of MSDOS installed in the computer memory and how the MSDOS program is installed.

The size of the *driver* area changes from one computer to another. Drivers are programs that control installable I/O devices such as a mouse or CDROM memory or installable programs. *Installable drivers* are programs that control or drive devices or programs that are added to the computer system. Because few computer systems are identical, this area will vary in size and contain different numbers and types of drivers.

The COMMAND.COM program—command processor—controls the operation of the computer from the keyboard. It processes the DOS commands as they are typed on the keyboard. For example, if DIR is typed, the COMMAND.COM program displays a directory of the disk files in the current disk directory.

The free TPA area is used to hold applications programs as they are executed. These applications programs include word processors, spreadsheet programs, CAD (computer-aided design) programs, and so forth. The TPA also holds *TSR* (terminate and stay resident) programs that remain in memory in an inactive state until activated by a hot-key sequence or other event such as an interrupt. A calculator program is an example of a TSR that is activated by an ALT-C key as a hot-key. A *hot-key* is a combination of keys on the keyboard that activates a TSR program. A TSR program is often called a *pop-up* program, because, when activated, it appears inside of another program.

The System Area.   The system area is smaller than the TPA, but is just as important. The system area contains programs on a read-only memory (*ROM*) and also areas of read/write (*RAM*) memory for data storage. Figure 1–4 shows the system area of a typical computer system. As with the map of the TPA, this map also includes the hexadecimal memory addresses of the various areas.

The first area of the system space contains video display RAM and video control programs on ROM (video BIOS). This area generally starts at location A0000H and extends to location C7FFFH. The size and amount of memory used depends on the type of video display adapter attached to the system. Display adapters often attached to a computer include: CGA (*color graphics adapter*), EGA (*enhanced graphics adapter*), or a form of VGA (*variable graphics array*). Generally, the video RAM located between memory locations A0000H and AFFFFH stores graphic data,

FIGURE 1–4 The system area of a typical computer system.

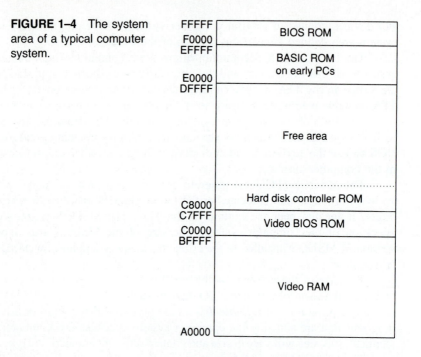

whereas memory between B0000H and BFFFFH stores text data. The video BIOS, located on a ROM, is located between C0000H and C7FFFH and contains programs that control the video display.

If a hard-disk memory is attached to the computer, the interface card might contain a ROM. The ROM holds low-level format software for the hard-disk memory, at location C8000H. The size, location, and presence of the ROM depends on the hard-disk adapter attached to the computer.

The area between locations C8000H and DFFFFH is often open or free. This area is used as *expanded memory* in a PC or XT or *upper memory* in an AT system. Its use depends on the system and its configuration. The expanded memory system allows a 64K-byte page of memory to be used by application programs. This 64K-byte page frame (usually location D0000H through DFFFFH) is expandable by switching in other 64K-byte pages of memory. Most application programs that state they are LIM 4.0 driver compatible can use expanded memory. The LIM 4.0 memory-management driver is the result of Lotus, Intel, and Microsoft having standardized access to expanded memory systems.

Memory locations E0000H–EFFFFH contain the BASIC language on ROM found in early personal IBM computer systems. This area is often open or free in newer computer systems.

Finally, the system BIOS ROM is located in the top 64K bytes of the system area. This ROM controls the operation of the basic I/O devices connected to the computer system. It does not control the operation of the video system, which has its own BIOS ROM at location C0000H.

I/O Space. The I/O (*input/output*) space in a computer system extends from I/O port 0000H to port FFFFH. (A *port address* is similar to a memory address except instead of addressing memory, it addresses an I/O device.) The **I/O devices** allow the microprocessor to communicate between itself and the outside world. The I/O space allows the computer to access up to 64K different I/O devices. A great number of these locations are available for expansion in most computer systems. Figure 1–5 shows the I/O map found in many 8086–80486-based personal computer systems.

The I/O area contains two major sections. The area below I/O location 0400H is considered reserved for system devices, with many of them depicted in Figure 1–5. The remaining area is available I/O space for expansion that extends from I/O port 0400H through FFFFH.

Various I/O devices that control the operation of the system are usually not directly addressed. The BIOS addresses these basic devices, which can vary slightly in location from one computer to the next. Access to I/O should always be made through DOS or BIOS to maintain compatibility from one computer system to another. This map is provided as a guide to illustrate the I/O space in the system.

The Operating System

The operating system is the program that operates the computer. This text assumes that the operating system is either MSDOS or PCDOS, which are by far the most common operating systems found in personal computers. The operating system is stored on a disk that is either placed in one of the floppy disk drives or found on a hard disk drive that is either resident to the computer or to a local area network (LAN). Each time the computer is powered up or reset, the operating system is read from the disk. We call this operation *booting* the system. Once DOS is installed in memory by the boot, it controls the operation of the computer system, its I/O devices, and application programs. In addition to the DOS operating system, other operating systems are sometimes used to control or operate the computer. These other systems include OS/2, Unix, and many others.

The first task of the DOS operating system, after loading into memory, is to use a file called the CONFIG.SYS file. This file specifies various drivers that load into the memory, setting up or *configuring* the machine for operation. Example 1–1 illustrates a sample CONFIG.SYS file. Note that the statements in this file vary from machine to machine and the one illustrated is just an example.

EXAMPLE 1–1

```
FILES=30
BUFFERS=30
STACKS=64,128
FCBS=48
SHELL=C:\DOS\COMMAND.COM C:\DOS\ /E:256 /P
DEVICE=C:\DOS\HIMEM.SYS
DOS=HIGH,UMB
DEVICE=C:\DOS\EMM386.EXE I=C800-EFFF NOEMS
DEVICEHIGH SIZE=1EB0 C:\LASERLIB\SONY_CDU.SYS /D:SONY_001 /B:340 /Q:* /T:* /M:H
```

FIGURE 1–5 The I/O map of a personal computer, illustrating many of the fixed I/O areas.

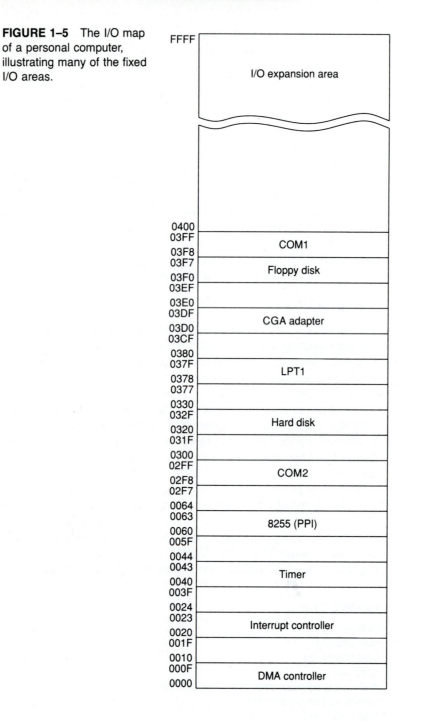

```
DEVICEHIGH SIZE=0190 C:\DOS\SETVER.EXE
DEVICEHIGH SIZE=3150 C:\MOUSE1\MOUSE.SYS
LASTDRIVE = F
```

The first four statements in this CONFIG.SYS file set up the number of files, buffers, stacks, and file control blocks required to execute various programs. These settings should be adequate for just about any program loaded into memory using DOS. In general, if a program requires more buffers, etc., the documentation indicates that the CONFIG.SYS file must be changed to reflect an increased need. Some programs automatically adjust the CONFIG.SYS file when installed by changing these parameters or by adding additional statements.

The SHELL command specifies the command processor used with DOS. In this example, the COMMAND.COM file is the command processor (also selected by default) using the E:256 /P switches. The E:256 switch sets the environment size to 256 bytes. The /P switch tells the command processor to run the AUTOEXEC.BAT file and make COMMAND.COM permanent.

The first DEVICE (driver) loaded into memory in this example is a program called HIMEM.SYS (high memory driver). A **driver** is usually a program that controls an I/O device or a program that must remain in the computer system memory. The HIMEM program allows a 64K-byte section of extended memory, just above the first 1M byte of memory, to be used for programs in an 80286-, 80386-, or 80486-based system. (This extra 64K memory space is supported by most computer systems, but not all.) This driver allows a DOS-based system access to 1M plus 64K bytes of memory. This extra 64K-byte section of memory can hold most of the MSDOS version 5.0 program, freeing additional space in the TPA. The next command (DOS=HIGH,UMB) tells the computer to load DOS into this high part of memory and also to use upper memory blocks.

In order to use upper memory blocks, available only in an 80386- or 80486-based system, we load the EMM386.EXE program. This is a driver that emulates expanded memory in extended memory on an 80386- or 80486-based system. *Expanded memory* fills free areas of memory within the system area so programs can be loaded into this area and accessed directly by DOS applications. The I=C800-EFFF switch tells EMM386.EXE to use memory area C8000H–EFFFFH for this expanded memory area. DOS treats this area as **upper memory,** or *upper memory blocks (UMB)*. Drivers and programs can be loaded into upper memory freeing even more area in the TPA for application programs. Before using the I=C800-EFFF switch, make sure that your computer does not contain any system ROM/RAM in this area of the memory. Note that NOEMS tells EMM386.EXE to exclude expanded memory. Expanded memory can be installed by replacing NOEMS with a number that indicates how much extended memory to allocate to LIM 4.0 expanded memory.

The DEVICEHIGH command loads drivers and programs into upper memory blocks allocated by the EMM386.EXE driver. In the CONFIG.SYS file illustrated, three drivers are loaded into upper memory blocks beginning at location C8000H. The first is a program that operates a SONY CDROM drive; the second loads a program called SETVER; and the third loads the MOUSE driver.

The last statement in the CONFIG.SYS file illustrated shows the LASTDRIVE statement. This tells DOS which is the last disk drive connected to your computer system. Other drivers may also be loaded into memory using the CONFIG.SYS file such as a PRINT.SYS driver, ANSI.SYS driver, or any other program that functions as a driver. Driver programs normally contain the DOS extension .SYS used to indicate a system file. Be very careful when changing the CONFIG.SYS file, because an error locks up the computer system. Once the computer is locked up by a CONFIG.SYS error, the only way to recover is to boot off a DOS disk that contains the operating system with a functioning CONFIG.SYS file.

Once the operating system completes its configuration as dictated by CONFIG.SYS, the AUTOEXEC.BAT (automatic execution batch) file is executed by the computer. If none exists, the computer asks for the TIME and DATE. Example 1–2 shows a typical AUTOEXEC.BAT file. This is only an example, and often variations occur from system to system. The AUTOEXEC.BAT file contains commands that are executed at the start of system operation when power is first applied to the computer. These are the same commands that could be typed from the keyboard, but the AUTOEXEC.BAT saves us from doing so each time the computer is powered up.

EXAMPLE 1–2

```
PATH C:\DOS;C:\;C:\MASM\BIN;C:\MASM\BINB\;C:\UTILITY;C:\WS;C:\LASERLIB
SET BLASTER=A220 I7 D1 T3
SET INCLUDE=C:\MASM\INCLUDE\
SET HELPFILES=C:\MASM\HELP\*.HLP
SET INIT=C:\MASM\INIT\
SET ASMEX=C:\MASM\SAMPLES\
SET TMP=C:\MASM\TMP
SET SOUND=C:\SB
LOADHIGH C:\LASERLIB\MSCDEX.EXE /D:SONY_001 /L:F /M:8
LOADHIGH C:\LASERLIB\LLTSR.EXE ALT-Q
LOADHIGH C:\DOS\FASTOPEN C:=256
LOADHIGH C:\DOS\DOSKEY /BUFSIZE=1024
LOADHIGH C:\LASERLIB\PRINTF.COM
DOSKEY GO=DOSSHELL
DOSSHELL
```

The PATH statement specifies the search paths whenever a program name is typed at the command line. The order of the path search is the same as the order of the paths in the path statement. For example, if PROG is typed at the command line, the machine first searches C:\DOS, then the root directory C:\, then C:\MASM\BIN and so forth until the program named PROG is found. If it isn't found, the command interpreter (COMMAND.COM) informs the user that the program is not found.

The SET statement sets a variable name to a path. This allows names to be associated with paths for batch programs. It also can be used to set command strings (environments) for various programs. The first SET command sets the environment for the sound blaster card. The second SET command sets INCLUDE to the path C:\MASM\INCLUDE\. Note that the SET statements are stored in the DOS environment space that was reserved in the CONFIG.SYS file using the SHELL statement. If the environment becomes too large, you must change the SHELL statement to allow more space.

LOADHIGH places programs into upper memory blocks defined by the EMM386.EXE program. LOADHIGH is used at any DOS command prompt for loading a program into the high (upper) memory area as long as the computer is an 80386 or 80486. The last command in this AUTOEXEC.BAT file is DOSSHELL. The DOSSHELL program is a menu program included with MSDOS version 5.0. This last command is replaced by WIN or WIN/R to run Microsoft Windows in place of DOSSHELL.

Once the CONFIG.SYS and AUTOEXEC.BAT files are executed, the program name last shown in the AUTOEXEC.BAT file is executed. In this example, the system operates from the DOSSHELL program.

The Microprocessor

At the heart of the microprocessor-based computer system is the microprocessor integrated circuit. The **microprocessor** is the controlling element in a computer system and is sometimes referred to as the **CPU** (*central processing unit*). The microprocessor controls memory and I/O through a series of connections called *buses*. The buses are used to select an I/O or memory device, transfer data between an I/O device or memory and the microprocessor, and control the I/O and memory system. Memory and I/O are controlled through instructions that are stored in the memory and executed by the microprocessor.

The microprocessor performs three main tasks for the computer system: (1) data transfer between itself and the memory or I/O systems, (2) simple arithmetic and logic operations, and (3) program flow through simple decisions. Albeit these are simple tasks, through them the microprocessor can perform virtually any series of operations.

The power of the microprocessor lies in its ability to execute millions of instructions per second from a *program* (group of instructions) stored in the memory system. This *stored program concept* has made the microprocessor and computer system a very powerful device. Recall that Babbage also wanted to use the stored program concept in his Analytical Engine.

Table 1–3 shows the arithmetic and logic operations that the 8086–80486 microprocessors execute. These operations are very basic, but through them very complex problems are solved. Data from the memory system or internal registers is operated upon. Data widths are variable; they include a **byte** (8 bits), a **word** (16 bits), and a **double word** (32 bits). Note that only the 80386 and 80486 directly manipulate 8-, 16-, and 32-bit numbers. The earlier 8086–80286 directly manipulate 8- and 16-bit numbers, but not 32-bit numbers.

Another feature that makes the microprocessor powerful is its ability to make simple decisions. These decisions are based upon numerical facts. For example, a microprocessor can decide if a number is zero, if it is positive, and so forth. These simple decisions allow the microprocessor to modify the program flow so programs

TABLE 1–3 Simple arithmetic and logic operations

Operation	Comment
Addition	
Subtraction	
Multiplication	
Division	
AND	Logical multiplication
OR	Logical addition
NOT	Logical inversion
NEG	Arithmetic negation
Shift	
Rotate	

appear to think—simply by making these little decisions. Table 1–4 lists the decision making abilities of the 8086–80486 microprocessors.

Buses.   A *bus* is a group of wires that interconnect components in a computer system. The buses that interconnect the sections of a computer system transfer address, data, and control information between the microprocessor and its memory and I/O systems. In the microprocessor-based computer system three buses exist for this transfer of information: address, data, and control. Figure 1–6 shows how these buses connect to various system components such as read/write memory (RAM), read-only memory (ROM), and a few I/O devices.

The ***address bus*** is used by the microprocessor to request a memory location from the memory or an I/O location from the I/O devices. If I/O is addressed, the address bus contains a 16-bit I/O address of 0000H through FFFFH. The 16-bit I/O address or port number selects one of 64K different I/O devices. If memory is addressed, the address bus contains a memory address. The memory address varies in width with the different versions of the 8086–80486 microprocessor. The 8086 and 8088 address 1M byte of memory using a 20-bit address that selects locations 00000H–FFFFFH. The 80286 and 80386SX address 16M bytes of memory using a 24-bit address that selects locations 000000H–FFFFFFH. The 80386SL/80386SLC address 32M bytes of memory using a 25-bit address that selects locations 0000000H–1FFFFFFH. The 80386DX, 80486SX, and 80486DX address 4G bytes of memory using a 32-bit address that selects locations 00000000H–FFFFFFFFH.

TABLE 1–4 Decisions made by an 8086–80486 microprocessor

Decision	Comment
Zero	Test a number for zero or not-zero
Sign	Test a number for positive or negative
Carry	Test for a carry after an addition or a borrow after a subtraction
Parity	Test a number for the count of the number of ones expressed as even or odd

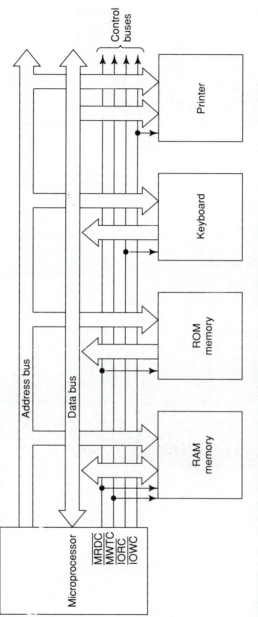

FIGURE 1–6 The block diagram of a computer system showing the address, data, and control buses. Note that decoders are not shown in this illustration.

19

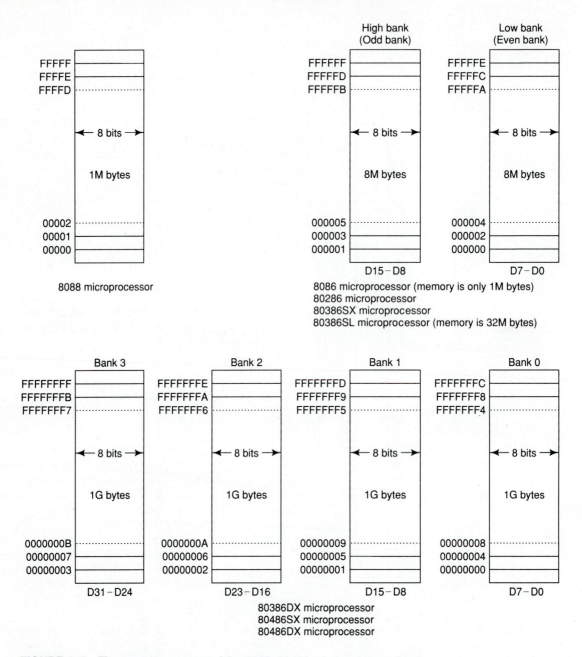

FIGURE 1–7 The memory systems of the 8086–80486 microprocessor family.

005.265

The **data bus** transfers information between the microprocessor and its memory and I/O address space. Data transfers vary in size from 8 bits wide to 32 bits wide in various members of the 8086–80486 family. The 8088 contains an 8-bit data bus that transfers 8 bits of data at a time. The 8086, 80286, 80386SL, and 80386SX transfer 16-bit data through their data buses. The 80386DX, 80486SX, and 80486DX transfer 32-bit data. The advantage of a wider data bus is speed in applications that use wide data. For example, if a 32-bit number is stored in memory, it takes the 8088 microprocessor 4 transfer operations to complete because its data bus is only 8 bits wide. The 80486 accomplishes the same task with one transfer because its data bus is 32 bits wide. Figure 1–7 shows the memory widths and sizes of the 8086–80486 microprocessors. Notice how the memory sizes and organizations differ between various members of the 8086–80486 microprocessor family. In all family members the memory is numbered by byte.

The **control bus** contains lines that select the memory or I/O and cause them to perform a read or a write operation. In most computer systems, there are four control bus connections: $\overline{\text{MRDC}}$ (memory read), $\overline{\text{MWTC}}$ (memory write), $\overline{\text{IORC}}$ (I/O read), and $\overline{\text{IOWC}}$ (I/O write).

The microprocessor reads the contents of a memory location by sending the memory an address through the address bus. Next it sends the memory the memory read control signal to cause it to read data. Finally the data read from the memory is passed to the microprocessor through the data bus.

1–4   COMPUTER DATA FORMATS

Successful programming depends on a clear understanding of data formats. In this section, we describe the common data formats used with the 8086–80486 microprocessors. Data appear as ASCII, BCD, signed and unsigned integers, and floating-point numbers (real numbers).

ASCII Data

ASCII (*American Standard Code for Information Interchange*) data (see Table 1–5) represent alphanumeric characters in the memory of a computer system. The *ASCII code* is a 7-bit code with the eighth and most significant bit used to hold parity in some systems. If ASCII data are used with a printer, the most significant bit is a 0 for alphanumeric printing, and a 1 for graphics printing. In the personal computer, extended characters are selected by placing a logic 1 in the leftmost bit. Table 1–6 shows the extended characters using code 80H–FFH. The extended characters store foreign letters and punctuation, Greek characters, mathematical characters, box drawing characters, and other special characters. Note that extended characters can vary from one printer to another.

The ASCII control characters, also listed in Table 1–5, perform control functions in a computer system. These include: clear screen (FF), backspace (BS), line feed, and the like. To enter the control codes through the computer keyboard, the

TABLE 1–5 The ASCII Code

							Second									
First	X0	X1	X2	X3	X4	X5	X6	X7	X8	X9	XA	XB	XC	XD	XE	XF
0X	NUL	SOH	STX	ETX	EOT	ENQ	ACK	BEL	BS	HT	LF	VT	FF	CR	SO	SI
1X	DLE	DC1	DC2	DC3	DC4	NAK	SYN	ETB	CAN	EM	SUB	ESC	FS	GS	RS	US
2X	SP	!	"	#	$	%	&	'	(	)	*	+	,	-	.	/
3X	0	1	2	3	4	5	6	7	8	9	:	;	<	=	>	?
4X	@	A	B	C	D	E	F	G	H	I	J	K	L	M	N	O
5X	P	Q	R	S	T	U	V	W	X	Y	Z	[	\	]	^	_
6X	`	a	b	c	d	e	f	g	h	i	j	k	l	m	n	o
7X	p	q	r	s	t	u	v	w	x	y	z	{	¦	}	~	▦

control key is held down while typing a letter. To obtain the control code 01H, type a control-A; a 02H is obtained by typing control-B, and so on.

To use Table 1–5 or 1–6 for converting alphanumeric or control characters into ASCII characters, first locate the alphanumeric code for conversion. Next find the first digit (row) of the hexadecimal ASCII code. Then find the second digit (column). For example, the letter *A* is ASCII code 41H. The letter *a* is ASCII code 61H.

BCD (Binary-Coded Decimal) Data

Binary-coded decimal (BCD) information is stored in either packed or unpacked forms. Packed BCD data are stored as two digits per byte and unpacked BCD data are stored as one digit per byte. The range of BCD data extends from 0000 to 1001, or 0–9 decimal.

Table 1–7 shows some decimal numbers converted to both the packed and unpacked BCD forms. Applications that require BCD data are point-of-sale terminals, and almost any device that performs a minimum amount of simple

TABLE 1–6 The Extended ASCII Code

							Second									
First	X0	X1	X2	X3	X4	X5	X6	X7	X8	X9	XA	XB	XC	XD	XE	XF
8X																
9X																
AX	á	í	ó	ú	ñ	Ñ	ª	º	¿	⌐	¬	½	¼	¡	«	»
BX	▓	▒	▓	│	┤	╡	╢	╖	╕	╣	║	╗	╝	╜	╛	┐
CX	└	┴	┬	├	─	┼	╞	╟	╚	╔	╩	╦	╠	═	╬	╧
DX	╨	╤	╥	╙	╘	╒	╓	╫	╪	┘	┌	■	▄	▌	▐	▀
EX	α	β	Γ	π	Σ	σ	µ	τ	Φ	Θ	Ω	δ	∞	φ	ε	∩
FX	≡	±	≥	≤	⌠	⌡	÷	≈	°	•	·	√	η	2	■	

TABLE 1-7 Packed and Unpacked BCD Data

Decimal	Packed		Unpacked		
12	0001 0010		0000 0001	0000 0010	
623	0000 0110	0010 0011	0000 0110	0000 0010	0000 0011
910	0000 1001	0001 0000	0000 1001	0000 0001	0000 0000

arithmetic. If a system requires complex arithmetic, BCD data is seldom used because there is no simple and efficient method of performing complex BCD arithmetic.

Byte-Sized Data

Byte-sized data are stored as unsigned and signed integers. Figure 1-8 illustrates both the unsigned and signed forms of the byte-sized integer. The difference in these forms is the weight of the leftmost bit position. Its value is 128 for the unsigned integer and -128 for the signed integer. In the signed integer format, the leftmost bit represents the sign bit of the number as well as a weight of -128. For example, an 80H represents a value of 128 as an unsigned number, and as a signed number, it represents a value of -128. Unsigned integers range in value from 00H–FFH (0–255). Signed integers range in value from -128 to 0 to $+127$.

Although negative signed numbers are represented in this way, they are stored in the ***two's complement form.*** The method of evaluating a signed number, using the weights of each bit position, is much easier than the act of two's complementing a number to find its value. This is especially true in the world of calculators designed for programmers.

Whenever a number is two's complemented, its sign changes from negative to positive or positive to negative. For example, the number 00001000 is a $+8$. Its negative value (-8) is found by two's complementing the $+8$. To form a two's complement, we first one's complement the number. To one's complement a number, invert each bit of a number from zero to one or from one to zero. Once the one's complement is formed, the two's complement is found by adding a one to the one's complement. Example 1-3 shows how numbers are two's complemented using this technique.

FIGURE 1-8 The format of (a) an 8-bit unsigned number and (b) an 8-bit signed number.

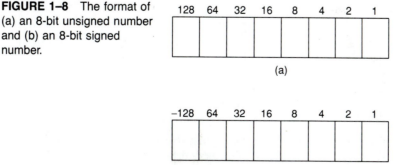

EXAMPLE 1–3

+ 8 = 00001000

        11110111 (one's complement)
+   1

- 8 = $\overline{11111000}$ (two's complement)

Word-Sized Data

A *word* (16 bits) is formed with two bytes of data. The least significant byte is always stored in the lowest numbered memory location, and the most significant byte in the highest. Figure 1–9(a) shows the weights of each bit position in a word of data, and Figure 1–9(b) shows how the number 1234H appears when stored in the memory. The only difference between a signed and an unsigned word is the leftmost bit position. In the unsigned form, the leftmost bit is unsigned and in the signed form its weight is a −32,768. As with byte-sized signed data, the signed word is in two's complement form when representing a negative number.

Double-Word-Sized Data

Double-word-sized data require four bytes to store because we are dealing with a 32-bit number. Double-word data appear as a product after a multiplication and also as a dividend before a division. In the 80386/80486 microprocessor, memory and registers are also 32-bit double words. Figure 1–10 shows the form used to store double words in memory.

    When a double word is stored in memory, its least significant byte is stored in the lowest numbered memory location and its most significant byte is stored in

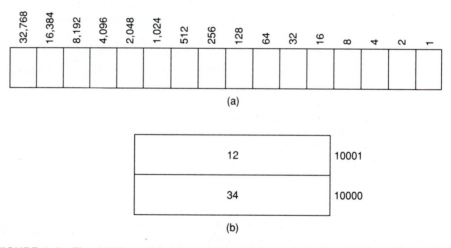

(a)

(b)

FIGURE 1–9 The 16-bit word format showing (a) the weight of each bit position and (b) a word (1234H) stored at memory byte 10000H and 10001H.

FIGURE 1–10 A 32-bit double word stored in memory locations 00100H through 00103H.

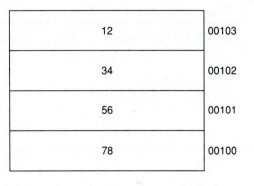

12	00103
34	00102
56	00101
78	00100

the highest numbered memory location. For example, a 12345678H stored in memory location 00100H–00103H is stored with the 78H in memory location 00100H, the 56H in location 00101H, the 34H in location 00102H, and the 12H in location 00103H.

Real Numbers

Because many high-level languages use the 8086–80486 microprocessors, real numbers are often encountered. A real number, or as it is often called, a *floating-point number,* contains two parts: a *mantissa,* or **significand,** and an *exponent.* Figure 1–11 depicts both the 4- and 8-byte forms of real numbers as they are stored in any Intel system. Note that the 4-byte real number is called *single-precision* and

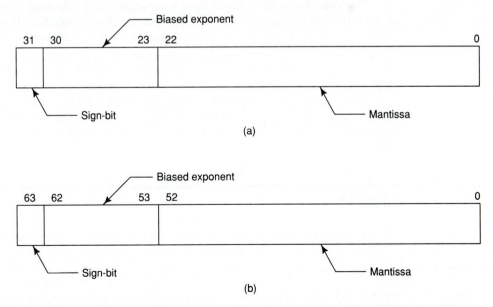

(a)

(b)

FIGURE 1–11 The (a) single-precision and (b) double-precision floating-point numbers.

TABLE 1–8   Single-precision real numbers

Decimal	Binary	Normalized	Sign	Biased exponent	Mantissa
+12	1100	1.1×2^3	0	10000010	1000000 00000000 00000000
−12	1100	-1.1×2^3	1	10000010	1000000 00000000 00000000
+100	1100100	1.1001×2^6	0	10000101	1001000 00000000 00000000
−1.75	1.11	-1.11×2^0	1	01111111	1100000 00000000 00000000
+0.25	.01	1.0×2^{-2}	0	01111101	0000000 00000000 00000000
+0.0	0	0	0	00000000	0000000 00000000 00000000

the 8-byte form is called *double-precision*. The form presented here is the same form specified by the IEEE* standard, IEEE-754, version 10.0. This standard has been adopted as the standard form of real numbers with virtually all high-level programming languages and many applications packages. Figure 1–11(a) shows the single-precision form that contains a sign-bit, an 8-bit exponent, and a 24-bit fraction (mantissa).

Simple arithmetic indicates that it should take 33 bits to store all three pieces of data. Not true—the 24-bit mantissa contains an implied (hidden) one-bit that allows the mantissa to represent 24-bits while being stored in only 23-bits. The **hidden bit** is the first bit of the normalized real number. When normalizing a number, it is adjusted so its value is at least 1, but less than 2. For example, if we convert a 12 to binary (1100) and normalize it, the result is a 1.1×2^3. The 1. is not stored in the 23-bit mantissa portion of the number. The 1. is the hidden one-bit. Table 1–8 shows the single-precision form of this number and others.

The exponent is stored as a biased exponent. With the single-precision form of the real number, the bias is 127 (7FH) with the double-precision form it is 1023 (3FFH). The **bias** adds to the exponent before it is stored into the exponent portion of the floating-point number. In the previous example, there is an exponent of 2^3, represented as a biased exponent of $127 + 3$ or 130 (82H) in the short form or as 1026 (402H) in the long form.

There are two exceptions to the rules for floating-point numbers. The number 0.0 is stored as all zeros. The number infinity is stored as all ones in the exponent and all zeros in the mantissa. The sign-bit indicates a ±0.0 or a ±∞.

SUMMARY

1. The mechanical computer age began with the advent of the abacus in 500 B.C. This first mechanical calculator remained unchanged until 1642, when Blaise Pascal improved it. The first mechanical computer system was the Analytical

---

*IEEE is the Institute of Electrical and Electronic Engineers.

Engine, developed by Charles Babbage in 1823. Unfortunately, this machine was never functional because of the inability to properly machine parts in that day.

2. The first electronic calculating machine was developed during World War II by Konrad Zuse. His computer, the Z3, was used in aircraft and missile design for the German war effort.

3. The first electronic computer, which used vacuum tubes, was placed into operation in 1943 to break secret German military codes. This first electronic computer system, the Colossus, was invented by Alan Turing. Its only problem was that the program was fixed and could not be changed.

4. The first general-purpose, programmable electronic computer system was developed in 1946 at the University of Pennsylvania. This first of the modern computers was called the ENIAC (Electronic Numerical Integrator and Calculator).

5. The world's first microprocessor, the Intel 4004, a 4-bit microprocessor—a programmable controller on a chip—was meager by today's standards. It addressed a mere 4,096 four-bit memory locations. Its instruction set contained only 45 different instructions.

6. Microprocessors that are common today include the 8086/8088, which were the first 16-bit microprocessors. Following these early 16-bit machines came the 80286, 80386, and 80486 microprocessor. With each newer version, improvements followed that increased computers' speed and performance.

7. Microprocessor-based personal computers contain memory systems that include three main areas: the TPA (transient program area), system area, and extended memory. The TPA holds application programs, the operating system, and drivers. The system area contains memory used for video display cards, disk drives, and the BIOS ROM. The extended memory area is only available to the 80286–80486 microprocessor in an AT-style personal computer system.

8. The 8086/8088 addresses 1M bytes of memory from location 00000H through location FFFFF. The 80286 and 80386SX address 16M bytes of memory from location 000000H through FFFFFFH. The 80386SL addresses 32M bytes of memory from location 0000000H through 1FFFFFFH. The 80386DX and 80486 address 4G bytes of memory from location 00000000H through FFFFFFFFH.

9. All versions of the 8086–80486 microprocessor address 64K bytes of I/O address space.

10. The operating system in many personal computers is either MSDOS (Microsoft disk operating system) or PCDOS (personal computer disk operating system). The operating system performs the task of operating or controlling the computer system along with its I/O devices.

11. The microprocessor is the controlling element in a computer system. The microprocessor performs data transfers, carries out simple arithmetic and logic operations, and makes simple decisions. The microprocessor executes programs stored in the memory in order to perform complex operations in short periods of time.

12. All computer systems contain three buses to control memory and I/O. The address bus is used to request a memory location or I/O device. The data bus transfers data between the microprocessor and its memory and I/O spaces. The control bus controls the memory and I/O and requests reading or writing of data.

13. The ASCII code is used in a computer system to store alphabetic or numeric data. The ASCII code is a 7-bit code and can have the 8th bit used to extend the character set from 128 codes to 256 codes.

14. Binary-coded decimal (BCD) data are sometimes used in a computer system to store decimal data. This data is stored as either packed (2 digits per byte) or unpacked (one digit per byte) data.

15. Binary data are stored as a byte (8 bits), word (16 bits), or double word (32 bits) in a computer system. This data may be unsigned or signed. Signed negative data are always stored in the two's complement form.

16. Floating-point data are used in a computer system to store whole, mixed, or fractional numbers. A floating-point number is composed of a sign, a mantissa, and an exponent.

1–6   GLOSSARY

Address bus   A group of connections that convey the memory and I/O address to the memory and I/O devices in the computer system.

ASCII   The American Standard Code for Information Interchange encodes alphanumeric data in a computer system. The ASCII code is a 7-bit code that represents a standard 128 characters; it is extended to 8 bits in most personal computer systems to represent 256 characters.

AT   Advanced architecture personal computer system based on the 80286, 80386, or 80486 microprocessor.

BCD   Binary-coded decimal code is sometimes used to store decimal numbers in computer memory systems.

Bias   In a floating-point number, the bias is added to the exponent before it is stored in the floating-point number format.

BIOS   The basic I/O system is a collection of programs that control the basic I/O devices attached to the computer system. The BIOS is normally stored in a read-only memory.

Bit   A single binary digit with a value of 1 or 0.

Bus   A group of wires that interconnect components in a computer system. All computer systems contain three buses: address, data, and control.

Byte   A grouping of eight binary digits.

Computer   A programmable system that contains a microprocessor, memory, and I/O devices.

Control bus   A group of connections that control the memory and I/O devices in a computer system.

CPU   The microprocessor in a computer system is often called the central processing unit or CPU.

Data bus   A group of connections that transfer data between the microprocessor and its memory and I/O system.

DOS   Disk operating system, which performs the function of controlling the computer system's access to the disk memory and most I/O devices.

Double word   A grouping of 32 binary bits.

Driver   A program loaded into a computer system to control I/O and other devices attached to the computer system.

EISA   Extended IBM standard architecture that supports 8-, 16-, and 32-bit interfaces.

Expanded memory   Memory that appears in the system area of a personal computer as 64K-byte memory pages.

Exponent   The binary power of two in a floating-point number.

Extended memory   The memory that lies above the first 1M bytes is often called extended memory in a personal computer system.

Floating-point number   Mixed, whole, and fractional numbers are stored as floating-point numbers in computer systems.

G   Used to indicate a computer billion, which is equal to 1,024M, or 1,073,741,824.

GUI   Graphical user interface.

Hexadecimal   The number system often used to represent data and memory addresses in a computer system. A hexadecimal (radix 16) number uses the digits 0–9 and A–F to represent numbers.

Hidden bit   A bit of the mantissa in a floating-point number that is always a logic one, but is never stored with the number.

High memory   Memory that exists between locations A0000H and FFFFFH in an 80386 or 80486 microprocessor that uses the driver EMM386.EXE.

Hot key   A key that is programmed to activate a TSR program.

I/O   Input/output devices are used to pass information between the computer and external machinery or humans. Examples of I/O devices are keyboards, printers, and so forth.

ISA   Used to specify industry standard architecture.

K   Used to indicate a computer thousand, which is equal to 1,024.

M   Used to indicate a computer million, which is equal to 1,024K or 1,048,576.

Mantissa   The fractional portion of a floating-point number.

Microprocessor   A programmable controller that is the heart of a computer system. The microprocessor performs data transfers, arithmetic and logic operations, and makes simple decisions.

MIPS Millions of instructions per second.

Nibble A term used to refer to a memory location that contains 4 binary bits.

PC Personal computer system based on the 8086 or 8088 microprocessor.

PEL Used to indicate a picture element (the smallest displayable dot) in a video display system.

Pixel *See* PEL.

RAM A memory device that can be written to or read by the microprocessor in a computer system. A random access memory device is a RAM.

ROM Read-only memory is a device that cannot be written to, but paramountly stores data and programs for use by the computer system.

Significand *See* Mantissa.

TPA Transient program area used to hold the application programs, operating system, and drivers in a computer system.

TSR Terminate and stay residend stay resident program is a program that remains in memory until it is activated.

Two's complement Used to store negative numbers in a computer system. The two's complement is formed by first inverting each one to a zero and each zero to a one. Next, a one is added to this inverted number to form the two's complement.

Upper memory *See* High memory.

Word A grouping of 16 binary bits.

XT Extended architecture personal computer system based on the 8086 or 8088 microprocessor.

1-7   QUESTIONS

1. Who developed the Analytical Engine?
2. The 1890 census used a new device called a punched card. Who developed the punched card?
3. Who was the founder of the IBM Corporation?
4. Who developed the first electronic calculator?
5. The first truly electronic computer system was developed for what purpose?
6. The first general-purpose computer was called the _____.
7. The world's first microprocessor was developed in 1971 by _____.
8. Which 8-bit microprocessor ushered in the age of the microprocessor?
9. The 8085 microprocessor, introduced in 1977, has sold _____ copies.
10. Which microprocessor was the first to address 1M bytes of memory?
11. The 80386SL addresses _____ bytes of memory.
12. How much memory is available to the 80486 microprocessor?
13. What is MIPS?

14. A computer K is equal to _____ bytes.
15. A computer M is equal to _____ K bytes.
16. How many typewritten pages of information are stored in a 4G-byte memory system?
17. The first 1M byte of memory in a computer system contains a _____ and a _____ area.
18. How much memory is found in the transient program area?
19. How much memory is found in the system area?
20. The 8086 microprocessor addresses _____ bytes of memory.
21. The 80286 microprocessor addresses _____ bytes of memory.
22. Which microprocessors address 4G bytes of memory?
23. Memory above the first 1M bytes is called _____ memory.
24. What is the BIOS?
25. What is DOS?
26. What is the difference between an XT and an AT computer system?
27. A driver is stored in the _____ area.
28. What is a TSR?
29. How is a TSR often accessed?
30. What is the purpose of the CONFIG.SYS file?
31. What is the purpose of the AUTOEXEC.BAT file?
32. What program processes DOS command line statements?
33. The personal computer system addresses _____ bytes of I/O space.
34. The DEVICE or DEVICEHIGH statement is found in what file?
35. Where are the upper memory blocks used by MSDOS version 5.0?
36. Draw the block diagram of a computer system.
37. What is the purpose of the microprocessor in a microprocessor-based computer system?
38. List the three buses found in all computer systems.
39. What bus transfers the memory address to the I/O device or to the memory device?
40. Which control signal causes the memory to perform a read operation?
41. What is the purpose of the $\overline{\text{IORC}}$ signal?
42. Define *byte*, *word*, and *double word*.
43. Convert the following words into ASCII character strings.
    a. FROG
    b. Arc
    c. Water.
    d. Well
44. What is the difference between an extended ASCII character and a standard ASCII character?
45. Convert the following decimal numbers into 8-bit signed binary numbers.
    a. +32
    b. −12
    c. +100
    d. −92

46. Convert the following decimal numbers into signed binary words.
    a. +1,000
    b. −120
    c. +800
    d. −3,212

47. Show how the following 16-bit hexadecimal numbers are stored in the memory system.
    a. 1234H
    b. A122H
    c. B100H

48. Convert the following decimal numbers into both packed and unpacked forms.
    a. 102
    b. 44
    c. 301
    d. 1,000

49. Convert the following binary numbers into signed decimal numbers.
    a. 10000000
    b. 00110011
    c. 10010010
    d. 10001001

50. Convert the following binary numbers into BCD numbers (assume that these are packed numbers).
    a. 10001001
    b. 00001001
    c. 00110010
    d. 00000001

51. Convert the following decimal numbers into single-precision floating-point numbers.
    a. +1.5
    b. −10.625
    c. +100.25
    d. −1,200

52. Convert the following single-precision floating-point numbers into decimal.
    a. 0   10000000   11000000000000000000000
    b. 1   01111111   00000000000000000000000
    c. 0   10000010   10010000000000000000000
    d. 1   01111110   11000000000000000000000

CHAPTER 2

The Microprocessor and Its Architecture

INTRODUCTION

This chapter presents the microprocessor as a programmable device by first looking at its internal programming model and then at how it addresses its memory space. The architecture of the entire 8086–80486 family is presented simultaneously as are the ways that the family addresses the memory system.

The addressing modes for this powerful microprocessor are described for both the real and protected modes of operation. Real mode memory exists at locations 00000H–FFFFFH, the first 1M byte of the memory system, and is present on all versions of the microprocessor. Protected mode memory exists at any location in the entire memory system, but is only available to the 80286–80486. Protected mode memory for the 80286 contains 16M bytes and for the 80386/80486 contains 4G bytes.

2–1   CHAPTER OBJECTIVES

Upon completion of this chapter, you will be able to:

1. Describe the function and purpose of each program-visible register with the 8086–80486 microprocessor.
2. Detail the flag register and the purpose of each of the flag bits.
3. Describe how memory is accessed using real mode memory-addressing techniques.
4. Describe how memory is accessed using protected mode memory-addressing techniques.
5. Describe the program-invisible registers within the 80286, 80386, and 80486 microprocessors.

2–2   ## 8086–80486 INTERNAL ARCHITECTURE

Before a program is written or any instruction investigated, the internal structure of the microprocessor must be known. This section of the chapter details the program-visible internal architecture of the 8086–80486 microprocessors. Also detailed are the function and purpose of each of these internal registers.

The Programming Model

The programming model of the 8086–80486 microprocessor is considered *program visible* because the registers in it are used during programming and are specified by the instructions. Other registers, detailed later, are considered program invisible because they are not normally used during applications programming, but may be used during system programming. Only the 80286–80486 contain program invisible registers.

Figure 2–1 illustrates the programming model of the 8086–80486 microprocessor. The early 8086, 8088, and 80286 contain 16-bit internal architectures, a subset of the registers shown in Figure 2–1. The 80386 and 80486 microprocessors contain full 32-bit internal architectures. The shaded areas in this illustration are not found in the 8086, 8088, or 80286 microprocessors and are enhancements provided on the 80386 and 80486 microprocessors.

The register structure contains 8-, 16-, and 32-bit registers. The 8-bit registers are AH, AL, BH, BL, CH, CL, DH, and DL and are referred to by these two-letter designations when an instruction is formed. The 16-bit registers are AX, BX, CX, DX, SP, BP, DI, SI, IP, FLAGS, CS, DS, ES, SS, FS, and GS. Likewise these registers are also referenced by these two-letter designations. The extended 32-bit registers are labeled EAX, EBX, ECX, EDX, ESP, EBP, EDI, ESI, EIP, and EFLAGS. These 32-bit extended registers and 16-bit registers FS and GS are only available in the 80386 or 80486 microprocessors and are referenced using the extended registers' designations and FS or GS for the two new 16-bit registers.

Some registers are *general-* or *multipurpose registers*, while some have special purposes. The general-purpose registers include EAX, EBX, ECX, EDX, EBP, EDI, and ESI. These registers hold various data sizes (bytes, words, or double words) and are used for almost any purpose.

Multipurpose Registers.

EAX   *(Accumulator)*—used as a 32-bit register (EAX), as a 16-bit register (AX), or as two 8-bit registers AH and AL. Note that if an 8- or 16-bit register is addressed, only that portion of the 32-bit register changes; the remaining bits are not affected. The accumulator is used for certain instructions such as multiplication, division, and some of the adjustment instructions. For these instructions, the accumulator has a special purpose, but it is generally considered a multipurpose register. In the 80386 and 80486 microprocessors, the EAX register is also used to address the memory system.

FIGURE 2–1 The internal program-visible register set of the 8086–80486 microprocessor.

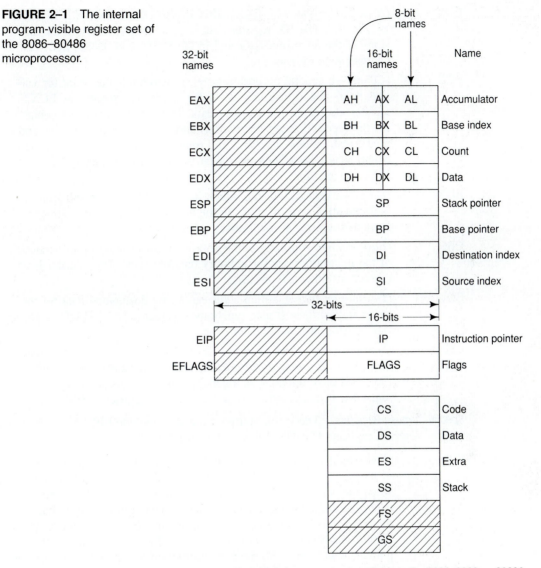

Notes: 1. The shaded areas are not available to the 8086, 8088, or 80286 microprocessors.
2. No special names are given to the FS and GS registers.

EBX   *(Base Index)*—as with EAX, EBX is also addressable as EBX, BX, BH, or BL. The BX register is used to address memory in all versions of the microprocessor and in the 80386 and 80486; EBX can address memory data.

ECX   *(Count)*—a general-purpose register, it also holds the count for various 8086–80486 instructions. In the 80386 and 80486, the ECX register also addresses memory data. Instructions that use a count are the repeated string instructions and the shift, rotate, LOOP, and repeat instructions. The shift and rotate instructions use CL as the count, while the LOOP and repeat instructions use either CX or ECX as the count.

EDX   *(Data)*—a general register, which holds a part of the result from a multiplication or a division. In the 80386 and 80486 microprocessors, this register also addresses memory data.

EBP   *(Base Pointer)*—usually points to a memory location in all versions of the microprocessor for memory data transfers. This register is addressed as either BP or EBP.

EDI   *(Destination Index)*—often addresses string destination data for the string instructions. It also functions as either a 32-bit (EDI) or 16-bit (DI) general-purpose register.

ESI   *(Source Index)*—used as either ESI or SI. The source index register often addresses source string data for the string instructions. Like EDI, ESI also functions as a general-purpose register. As a 16-bit register it is addressed as SI and as a 32-bit register as ESI.

Special-Purpose Registers.   The special-purpose registers include EIP, ESP, EFLAGS, and the segment registers CS, DS, ES, SS, FS, and GS.

EIP   *(Instruction Pointer)*—addresses the next instruction in a section of memory defined as a code segment. This register is IP (16-bits) when the 8086–80486 operate in the real mode and EIP (32-bits) when the 80386/80486 operate in the protected mode. Note that the 8086, 8088, and 80286 do not contain EIP and only the 80286, 80386, and 80486 operate in the protected mode. The instruction pointer, which points to the next instruction in a program, is used by the microprocessor to find the next sequential instruction in a program located within the code segment.

ESP   *(Stack Pointer)*—addresses an area of memory called the stack. The stack memory stores data through this pointer. This technique is explained later in the text with instructions that address stack data.

EFLAGS   *(Flags)*—indicate the condition of the microprocessor as well as control its operation. Figure 2–2 shows the flag registers of all versions of the 8086–80486 microprocessor. Note that the flags are upwardly compatible from the 8086/8088 to the 80486 microprocessor. The

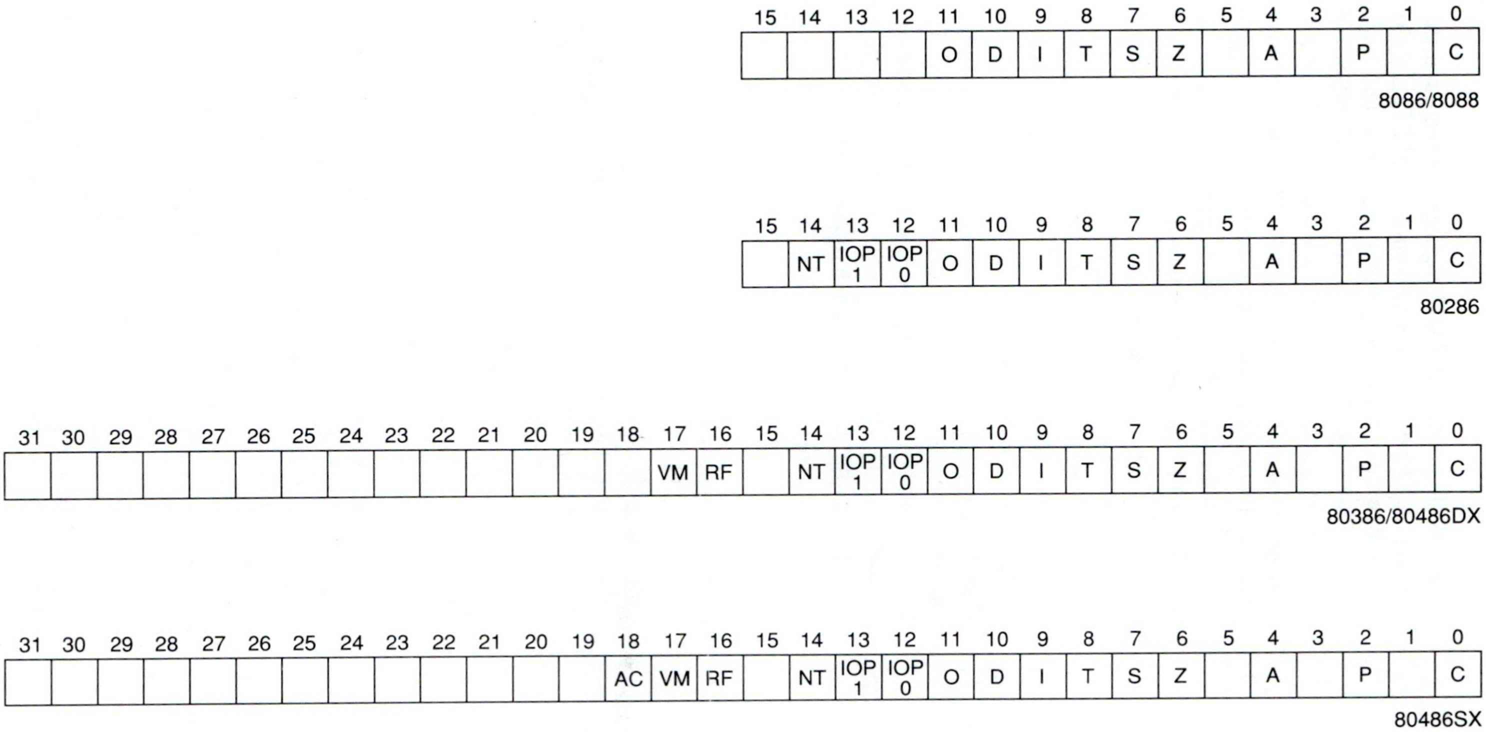

8086/8088

15	14	13	12	11	10	9	8	7	6	5	4	3	2	1	0
				O	D	I	T	S	Z		A		P		C

80286

15	14	13	12	11	10	9	8	7	6	5	4	3	2	1	0
	NT	IOP 1	IOP 0	O	D	I	T	S	Z		A		P		C

80386/80486DX

31	30	29	28	27	26	25	24	23	22	21	20	19	18	17	16	15	14	13	12	11	10	9	8	7	6	5	4	3	2	1	0
														VM	RF		NT	IOP 1	IOP 0	O	D	I	T	S	Z		A		P		C

80486SX

31	30	29	28	27	26	25	24	23	22	21	20	19	18	17	16	15	14	13	12	11	10	9	8	7	6	5	4	3	2	1	0
													AC	VM	RF		NT	IOP 1	IOP 0	O	D	I	T	S	Z		A		P		C

Note: The flag registers of all family members. Note how they are all upwardly compatible.

FIGURE 2–2 The flag registers of all Intel family members. Note how they are all upwardly compatible.

8086–80286 contain a FLAG register (16 bits) and the 80386–80486 contain an EFLAG register (32-bit extended flag register).

The flag bits change after many arithmetic and logic instructions execute. Some of the flags are also used to control features found in the microprocessor. Following is a list of each flag bit with a brief description of its function. As instructions are introduced in subsequent chapters, additional detail on the flag bits is provided.

C   (*Carry*)—indicates a carry after addition or a borrow after subtraction. The carry flag also indicates error conditions in some programs and procedures.

P   (*Parity*)—is a logic 0 for odd parity and a logic 1 for even parity. Parity is a count of ones in a number expressed as even or odd. For example, if a number contains 3 binary one bits, it has odd parity. If a number contains zero one bits it is considered to have even parity.

A   (*Auxiliary Carry*)—holds a carry after addition or a borrow after subtraction between bit positions 3 and 4 of the result. This highly specialized flag bit is tested by the DAA and DAS instructions to adjust the value of AL after a BCD addition or subtraction. Otherwise, the A flag bit is not used by the microprocessor.

Z   (*Zero*)—indicates that the result of an arithmetic or logic operation is zero. If Z = 1, the result is zero, and if Z = 0, the result is not zero.

S   (*Sign*)—indicates the arithmetic sign of the result after an addition or subtraction. If S = 1, the sign is set, or negative, and if S = 0, the sign is cleared, or positive.

T   (*Trap*)—when the trap flag is set, it enables trapping through the on-chip debugging feature. More detail of this debugging feature is provided later in the text.

I   (*Interrupt*)—an **interrupt** controls the operation of the INTR (interrupt request) input pin. If I = 1, the INTR pin is enabled, and if I = 0, the INTR pin is disabled. The state of the I flag bit is controlled by the STI (set I flag) and CLI (clear I flag) instructions.

D   (*Direction*)—controls the selection of increment or decrement for the DI and/or SI registers during string instructions. If D = 1, the registers are automatically decremented, and if D = 0, the registers are automatically incremented. The D flag is set with the STD (set direction) and cleared with the CLD (clear direction) instructions.

O   (*Overflow*)—is a condition that occurs when signed numbers are added or subtracted. An **overflow** indicates that the result has exceeded the capacity of the machine. For example, if a 7FH (+127) is added to a 01H (+1), the result is 80H (−128). This result represents an overflow condition indicated by the overflow flag for signed addition. For unsigned operations, we ignore the overflow flag.

IOPL   (*Input/Output Privilege Level*)—used in protected mode operation to select the privilege level for I/O devices. If the current privilege level is higher or more trusted than the IOPL, then I/O executes without hindrance. If the IOPL is lower than the current privilege level, an interrupt occurs causing execution to be suspended. Note that an IOPL of 00 is the highest or most trusted and an IOPL of 11 is the lowest or least trusted.

NT   (*Nested Task*)—indicates that the current task is nested within another task in protected mode operation. This flag is set when the task is nested by software.

RF   (*Resume*)—used with debugging to control resuming execution after the next instruction.

VM   (*Virtual Mode*)—selects virtual mode operation in a protected mode system. A virtual mode system allows multiple DOS memory partitions.

AC   (*Alignment Check*)—if a word or double word is addressed on a non-word or non-double-word boundary, this flag bit is set. Only the 80486SX microprocessor contains the alignment check bit that is primarily used with its companion numeric coprocessor.

Segment Registers.   Additional registers, called **segment registers**, generate memory addresses along with other registers in the microprocessor. There are either four or six segment registers in various versions of the 8086–80486 microprocessors. A segment register functions differently in the real mode when compared to protected mode operation of the microprocessor. Detail on their function in real and protected mode is provided later in this chapter. Following is a list of each segment register along with its function in the system.

CS   (*Code*)—the **code segment** is a section of memory that holds programs and procedures used by programs. The code segment register defines the starting address of the section of memory that is holding code. In real mode operation it defines the start of a 64K-byte section of memory and in protected mode it selects a descriptor that describes the starting address and length of a section of memory holding code. The code segment is limited to 64K bytes in length in the 8088–80286 and 4G bytes in the 80386/80486.

DS   (*Data*)—the **data segment** is a section of memory that contains most data used by a program. Data are accessed in the data segment by an offset address or by the contents of other registers that hold the offset address.

ES   (*Extra*)—the **extra segment** is an additional data segment that is used by some of the string instructions.

SS   (*Stack*)—the **stack segment** defines the area of memory used for stack. The location of the current entry point in the stack

	determined by the stack pointer (SP) register. The BP register also addresses data within the stack segment.
FS and GS	These supplemental segment registers are available in the 80386 and 80486 microprocessors to allow two additional memory segments for access by programs.

2–3   REAL MODE MEMORY ADDRESSING

The 80286–80486 microprocessor operates in either real or protected mode. The 8086/8088 only operate in the real mode. This section of the text details the operation of the microprocessor in the real mode. **Real mode** operation allows the microprocessor to address only the first 1M byte of memory space even if it is an 80486 microprocessor. Both the MSDOS or PCDOS operating systems assume that the microprocessor is operated in the real mode at all times. Real mode operation allows application software written for the 8086/8088, which only contain 1M byte of memory, to function with the 80286, 80386, and 80486 microprocessors. In all cases, each of these microprocessors begins operation in the real mode by default whenever power is applied or the microprocessor is reset.

Segments and Offsets

The combination of a segment address and an offset address accesses a memory location in the real mode. All real mode memory addresses consist of a segment address plus an offset address. The **segment address**, located within one of the segment registers, defines the beginning address of any 64K-byte memory segment. The **offset address** selects a location within that 64K-byte memory segment. Figure 2–3 shows how the segment plus offset addressing scheme selects a memory location. This illustration shows a memory **segment** that begins at location 10000H and ends at location 1FFFFH; it is 64K bytes in length. It also shows how an offset, sometimes called a *displacement*, of F000H selects location 1F000H in the memory system. Note that the offset or displacement is the distance above the start of the segment.

The segment register in Figure 2–3 contains a 1000H, yet it addresses a starting segment location of 10000H. In the real mode, each segment register is internally appended with a *0H* on its rightmost end to form a 20-bit memory address allowing it to access the start of the segment at any location within the first 1M byte of memory. For example, if a segment register contains a 1200H, it addresses a 64K-byte memory segment beginning at location 12000H. Likewise, if a segment register contains a 1201H, it addresses a memory segment beginning at location 12010H. Because of the internally appended 0H, segments can begin at any 16-byte boundary in the memory system. We often call this 16-byte boundary a *paragraph*.

Because a real mode segment of memory is 64K in length, once the beginning address is known, the ending address is found by adding an FFFFH to the starting address. For example, if a segment register contains 3000H, the first address of the

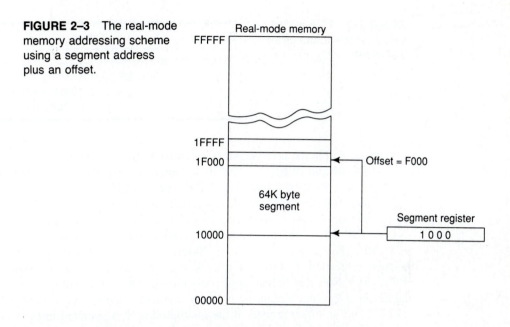

FIGURE 2–3 The real-mode memory addressing scheme using a segment address plus an offset.

segment is 30000H and the last address is 30000H + FFFFH, or 3FFFFH. Table 2–1 shows several examples of segment register contents and the starting and ending addresses of the memory segments selected by each segment address.

The offset address is added to the start of the segment to address a memory location in the memory segment. For example, if the segment address is 1000H and the offset address is 2000H, the microprocessor addresses memory location 12000H. The segment and offset address is sometimes written as *1000:2000* for a segment address of 1000H with an offset of 2000H.

Default Segment and Offset Registers

The microprocessor has a set of rules that apply whenever memory is addressed. These rules, which apply in either the real or protected mode, define the segment register and offset register combination used by certain addressing modes. For example, the code segment register is always used with the instruction pointer to address the next instruction in a program. This combination is CS:IP or CS:EIP depending upon the microprocessor and mode of operation. The code segment register defines the start of the code segment and the instruction pointer points to the next

TABLE 2–1 Example segment addresses

Segment Register	Starting Address	Ending Address
2000H	20000H	2FFFFH
2100H	21000H	30FFFH
AB00H	AB000H	BAFFFH
1234H	12340H	2233FH

TABLE 2–2 8086–80286 default segment and offset addresses

Segment	Offset
CS	IP
SS	SP or BP
DS	BX, DI, SI, or a 16-bit number
ES	DI for string instructions

instruction within the code segment executed by the microprocessor. For example, if CS = 1400H and IP/EIP = 1200H, the microprocessor fetches its next instruction from memory location 14000H + 1200H, or 15200H.

Another default is the stack. Stack data are referenced through the stack segment at the memory location addressed by either the stack pointer (SP/ESP) or the base pointer (BP/EBP). These combinations are referred to as SS:SP (SS:ESP) or SS:BP (SS:EBP). For example, if SS = 2000H and BP = 3000H, the microprocessor addresses location 23000H for a stack segment memory location addressed by the BP register. Note that in real mode, only the rightmost 16 bits of the extended register addresses a location within the memory segment. Never place a number larger than FFFFH into an offset register if the microprocessor is operated in the real mode.

Other defaults are shown in Table 2–2 for addressing memory using the 8086–80286 microprocessor. Table 2–3 shows the defaults assumed in the 80386 and 80486 microprocessors. Note that the 80386 and 80486 microprocessors have a far greater selection of segment/offset address combinations than the 8086–80286 microprocessors.

The 8086–80286 allow four memory segments and the 80386 and 80486 allow six memory segments. Figure 2–4 shows a system that contains four memory segments. Note that a memory segment can touch or even overlap if 64K bytes of memory are not required for a segment. Think of segments as windows that can be moved over any area of memory to access data or code.

Suppose that an application program requires 1000H bytes of memory for its code, 190H bytes of memory for its data, and 200H bytes of memory for its stack. This application does not require an extra segment. When this program is placed in the memory system by DOS, it is loaded in the TPA at the first available area of memory above the drivers and other TPA programs. Figure 2–5 shows how this

TABLE 2–3 80386 and 80486 default segment and offset addresses

Segment	Offset
CS	EIP
SS	ESP or EBP
DS	EAX, EBX, ECX, EDX, EDI, ESI, an 8-bit number, or 32-bit number
ES	EDI for string instructions
FS	no default
GS	no default

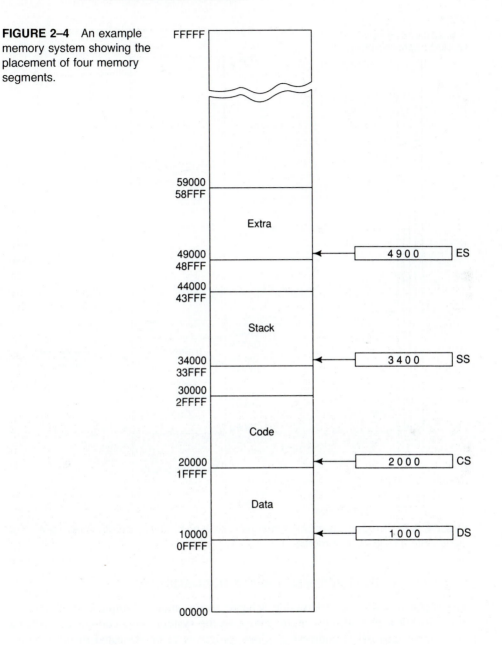

FIGURE 2–4 An example memory system showing the placement of four memory segments.

application is stored in the memory system. The segments show an overlap because the amount of data in them does not require 64K bytes of memory. The side view of the segments clearly shows the overlap and how segments can be slid to any area of memory. Fortunately for us, DOS calculates and assigns segment addresses. This is explained in a later chapter that details the operation of the assembler, BIOS, and DOS for an assembly language program.

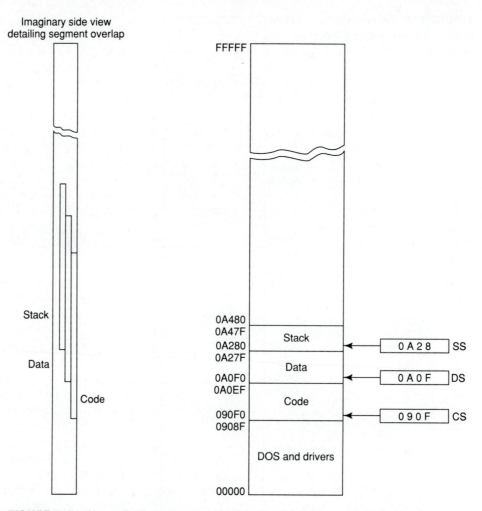

FIGURE 2–5   An application program containing a code, data, and stack segment loaded into a DOS system memory.

Segment and Offset Addressing Allows Relocation

The segment and offset addressing scheme seems unduly complicated. It is complicated, but it also affords an advantage to the system. This complicated scheme of segment plus offset addressing allows programs to be relocated in the memory system. A *relocatable program* is one that can be placed into any area of memory and executed without change. *Relocatable data* is data that can be placed in any area of memory and used without any change to the program. The segment and offset addressing scheme allows both programs and data to be relocated without changing a thing in a program or data. This is perfect for use in a general-purpose computer system where not all machines contain the same memory areas. The personal computer memory structure being different from machine to machine requires relocatable software and data.

Because memory is addressed within a segment by an offset address, the memory segment can be moved without changing any of the offset addresses. This is accomplished by moving the entire program, as a block, to a new area and then changing only the contents of the segment registers. If an instruction is 4 bytes above the start of the segment, its offset address is 4. If the entire program is moved to a new area of memory, this offset address of 4 still points to a location 4 bytes above the start of the segment. The only thing that must be changed are the contents of the segment register to address the program in the new area of memory. Relocation has made the personal computer based upon Intel microprocessors very powerful and very common. Without this feature, a program would have to be rewritten or altered before it is moved. This would take additional time or require many versions of a program for the many different configurations of computer systems.

2–4   PROTECTED MODE MEMORY ADDRESSING

Protected mode memory addressing (using 80286, 80386, and 80486 only) allows access to data and programs located above the first 1M byte of memory. Addressing this extended section of the memory system requires a change in the segment plus offset addressing scheme used with real mode memory addressing. When data and programs are addressed in extended memory, the offset address is still used to access information located within the segment. The segment address, discussed with real mode memory addressing, is no longer present in the protected mode. In place of the segment address, the segment register contains a *selector* that selects a *descriptor*. The descriptor describes the memory segment's location, length, and access rights. Because the segment register and offset address still access memory, protected mode instructions look just like real mode instructions. In fact, most programs written to function in the real mode will function without any changes in the protected mode. The difference between modes lies in the way that the segment register accesses the memory segment.

Selectors and Descriptors

The *selector*, located in the segment register, selects one of 8,192 descriptors from a table of descriptors. The *descriptor* describes the location, length, and access rights of a segment of memory. Indirectly, the segment register still selects a memory segment, but not directly as in the real mode.

There are two *descriptor tables* used with the segment registers: one contains global descriptors and the other contains local descriptors. The *global descriptors* contain segments that apply to all programs, while the *local descriptors* are usually unique to an application. Each descriptor table contains 8,192 descriptors, so a total of 16,384 descriptors are available at any time. Because the descriptor describes a memory segment, this allows up to 16,384 memory segments to be described for each application.

Figure 2–6 shows the format of a descriptor for the 80286 and 80386/80486 microprocessors. Note that each descriptor is 8 bytes in length; therefore, the global and local descriptor tables are each a maximum of 64K bytes in length. Descriptors for the 80286 and the 80386/80486 differ slightly, but the 80286 descriptor is upwardly compatible with the 80386 and 80486 microprocessors.

The **base address** portion of the descriptor is used to indicate the starting location of the memory segment. For the 80286 microprocessor, the base address is a 24-bit address, so segments can begin at any of its 16M bytes of memory. The 80386/80486 uses a 32-bit base address that allows segments to begin at any of its 4G-byte memory locations. Notice how the 80286 descriptor's base address is upwardly compatible to the 80386/80486 descriptor.

The segment **limit** contains the last offset address found in a segment. For example, if a segment begins at memory location F00000H and ends at location F000FFH, the base address if F00000H and the limit is FFH. For the 80286 microprocessor, the base address is F00000H and the limit is 00FFH. For the 80386/80486 microprocessors, the base address is 00F00000H and the limit is 000FFH. Notice that the limit for the 80286 is a 16-bit limit and the limit for the 80386/80486 is 20 bits. The 80286 accesses memory segments that are from 1 byte to 64K bytes in length. The 80386/80486 accesses memory segments that are from 1 byte to 1M byte or from 4K bytes to 4G bytes in length.

There is another feature found in the 80386/80486 descriptor that is not found in the 80286 descriptor: the G bit, or **granularity bit.** If G = 0, the limit specifies a segment limit of from 1 to 1M bytes in length. If G = 1, the value of the limit is multiplied by 4K bytes. If G = 1, the limit can be any multiple of 4K bytes. This allows a segment length of 4K bytes to 4G bytes in steps of 4K bytes. The reason that the segment length is 64K bytes in the 80286 is that the offset address is always 16 bits, while the offset address in the protected mode operation of the 80386/80486 is 32 bits. This 32-bit offset address allows segment lengths of 4G bytes, whereas the 16-bit offset address allows segment lengths of 64K bytes.

The AV bit in the 80386/80486 descriptor is used by the operating system and indicates that the segment is available (AV = 1) or not available (AV = 0). The D bit indicates how the 80386/80486 instructions access register and memory data in the protected mode. If D = 0, the 80386/80486 assumes that the instructions are 16-bit

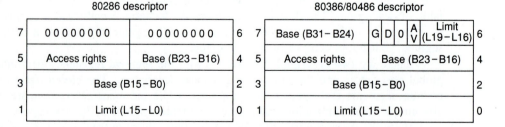

FIGURE 2–6   The descriptor formats for the 80286 and 80386/80486 microprocessors.

instructions compatible with the 8086–80286 microprocessors. This means that the instructions use 16-bit offset addresses and 16-bit registers. This mode is often called the *16-bit instruction mode*. If D = 1, the 80386/80486 assumes that the instructions are 32-bit instructions. The *32-bit instruction mode* assumes all offset addresses are 32-bits as well as all registers. The MSDOS or PCDOS operating system requires that the instructions are always used in the 16-bit instruction mode. More detail on these modes and their application to the instruction set appears in Chapters 3 and 4.

The ***access rights*** byte, illustrated in Figure 2–7, controls access to the memory segment. This byte describes how the segment functions in the system. Notice how complete the control is over the segment. If the segment is a data segment, the direction of growth can be specified. If the segment grows beyond its limit, the microprocessor's program is interrupted. You can even specify if a data segment can be written to or is write-protected. The code segment is also controlled in a similar fashion and can have reading inhibited.

Descriptors are chosen from the descriptor table by the segment register. Figure 2–8 shows how the segment register functions in the protected mode system. The segment register contains a 13-bit selector field, a table selector bit, and a requested privilege level field. The 13-bit selector chooses one of the 8,192 descriptors from the descriptor table. The TI bit selects either the global descriptor table (TI = 0) or

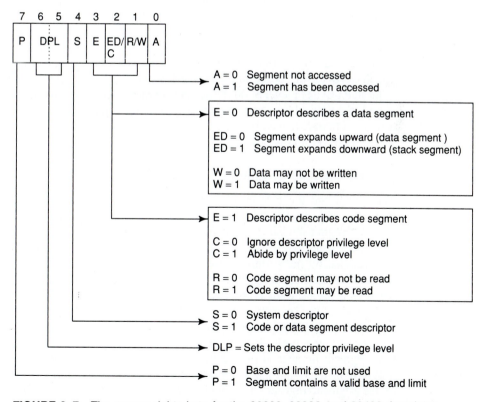

FIGURE 2–7 The access rights byte for the 80286, 80386, and 80486 descriptor.

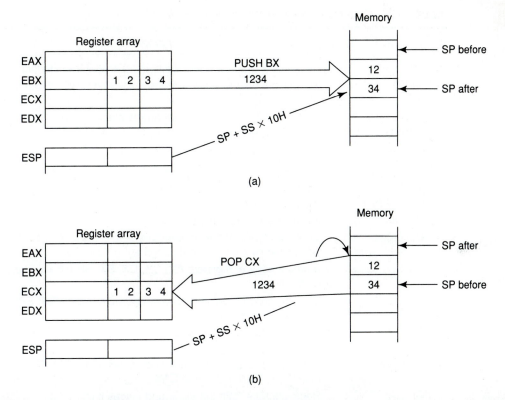

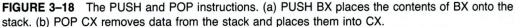

FIGURE 3–18 The PUSH and POP instructions. (a) PUSH BX places the contents of BX onto the stack. (b) POP CX removes data from the stack and places them into CX.

Whenever data are popped from the stack (see Figure 3-18b), the low-order 8 bits are removed from the location addressed by SP. The high-order 8 bits are removed from the location addressed by SP+1. The SP register is then incremented by 2. Table 3–11 lists some of the PUSH and POP instructions available in the 8086–80486 microprocessor. Note that PUSH and POP always store or retrieve *words* of data—never bytes—in the 8086–80286 microprocessor. The 80386/80486 allow words or double words to be transferred to and from the stack. Data may be pushed onto the stack from any 16-bit register or segment register and in the 80386/80486 from any 32-bit extended register. Data may be popped off the stack into any 16-bit register or any segment register except CS. The reason that we may not pop data from the stack into CS is that this changes part of the address of the next instruction.

The PUSHA and POPA instructions either push or pop all of the registers, except the segment registers, on the stack. These instructions are not available on the early 8086/8088 microprocessor. The push immediate instruction is also new to the 80286–80486 microprocessor. Note the examples in Table 3–11 that show the order of the registers transferred by the PUSHA and POPA instructions. The 80386/80486 also allow extended registers to be pushed or popped.

TABLE 3–11   Example PUSH and POP instructions

Assembly Language	Operation
POPF	Removes a word from the stack and places it into the flags
POPFD	Removes a double word from the stack and places it into the EFLAGS
PUSHF	Stores a copy of the flag word on the stack
PUSHFD	Stores a copy of the EFLAGS on the stack
PUSH AX	Stores a copy of AX on the stack
POP BX	Removes a word from the stack and places it into BX
PUSH DS	Stores a copy of DS on the stack
PUSH 1234H	Stores a 1234H on the stack
POP CS	*Not allowed* (illegal instruction)
PUSH [BX]	Stores a copy of the word contents of the memory location addressed by BX in the data segment on the stack
PUSHA	Stores a copy of registers AX, CX, DX, BX, SP, BP, DI, and SI on the stack
PUSHAD	Stores a copy of registers EAX, ECX, EDX, EBX, ESP, EBP, EDI, and ESI on the stack
POPA	Removes data from the stack and places it into SI, DI, BP, SP, BX, DX, CX, and AX
POPAD	Removes data from the stack and places it into ESI, EDI, EBP, ESP, EBX, EDX, ECX, and EAX
POP EAX	Removes data from the stack and places it into EAX
PUSH EDI	Stores a copy of EDI onto the stack

3–5   SUMMARY

1. The data-addressing modes include: register, immediate, direct, register indirect, base-plus-index, register relative, and base relative-plus-index addressing. In the 80386/80486 microprocessor an additional addressing mode, called scaled-index addressing, exists.
2. The program memory-addressing modes include: direct, relative, and indirect addressing.
3. Table 3–12 lists all real mode data-addressing modes available to the 8086–80286 microprocessor. In the protected mode, the function of the segment register is to address a descriptor that contains the base address of the memory segment.
4. The 80386 and 80486 microprocessors have additional addressing modes that allow the extended registers EAX, EBX, ECX, EDX, EBP, EDI, and ESI to address memory. These addressing modes are too numerous to list in tabular form, but in general any of these registers function in the same way as those

TABLE 3–12 8086–80286 real mode data-addressing modes

Assembly Language	Address Generation
MOV AL,BL	8-bit register addressing
MOV DI,BP	16-bit register addressing
MOV DS,BX	Segment register addressing
MOV AL,LIST	(DS × 10H) + LIST
MOV CH,DATA1	(DS × 10H) + DATA1
MOV DS,DATA2	(DS × 10H) + DATA2
MOV AL,12	Immediate data of 12 decimal
MOV AL,[BP]	(SS × 10H) + BP
MOV AL,[BX]	(DS × 10H) + BX
MOV AL,[DI]	(DS × 10H) + DI
MOV AL,[SI]	(DS × 10H) + SI
MOV AL,[BP+2]	(SS × 10H) + BP + 2
MOV AL,[BX−4]	(DS × 10H) + BX − 4
MOV AL,[DI+1000H]	(DS × 10H) + DI + 1000H
MOV AL,[SI+300H]	(DS × 10H) + SI + 0300H
MOV AL,LIST[BP]	(SS × 10H) + BP + LIST
MOV AL,LIST[BX]	(DS × 10H) + BX + LIST
MOV AL,LIST[DI]	(DS × 10H) + DI + LIST
MOV AL,LIST[SI]	(DS × 10H) + SI + LIST
MOV AL,LIST[BP+2]	(SS × 10H) + BP + LIST + 2
MOV AL,LIST[BX−6]	(DS × 10H) + BX + LIST − 6
MOV AL,LIST[DI+100H]	(DS × 10H) + DI + LIST + 100H
MOV AL,LIST[SI+20H]	(DS × 10H) + SI + LIST + 20H
MOV AL,[BP+DI]	(SS × 10H) + BP + DI
MOV AL,[BP+SI]	(SS × 10H) + BP + SI
MOV AL,[BX+DI]	(DS × 10H) + BX + DI
MOV AL,[BX+SI]	(DS × 10H) + BX + SI
MOV AL,[BP+DI+2]	(SS × 10H) + BP + DI + 2
MOV AL,[BX+SI−4]	(SS × 10H) + BP + SI − 4
MOV AL,[BX+DI+30H]	(DS × 10H) + BX + DI + 30H
MOV AL,[BX+SI+10H]	(DS × 10H) + BX + SI + 10H
MOV AL,LIST[BP+DI]	(SS × 10H) + BP + DI + LIST
MOV AL,LIST[BP+SI]	(SS × 10H) + BP + SI + LIST
MOV AL,LIST[BX+DI]	(DS × 10H) + BX + DI + LIST
MOV AL,LIST[BX+SI]	(DS × 10H) + BX + SI + LIST
MOV AL,LIST[BP+DI+2]	(SS × 10H) + BP + DI + LIST + 2
MOV AL,LIST[BP+SI−7]	(SS × 10H) + BP + SI + LIST − 7
MOV AL,LIST[BX+DI−10H]	(DS × 10H) + BX + DI + LIST − 10H
MOV AL,LIST[BX+SI+1AFH]	(DS × 10H) + BX + SI + LIST + 1AFH

listed in Table 3–12. For example, the MOV AL,TABLE[EBX+2*ECX+10H] is a valid addressing mode for the 80386/80486 microprocessor.

5. The MOV instruction copies the contents of the source operand into the destination operand. The source never changes for any instruction.

6. Register addressing specifies any 8-bit register (AH, AL, BH, BL, CH, CL, DH, or DL) or any 16-bit register (AX, BX, CX, DX, SP, BP, SI, or DI). The segment registers (CS, DS, ES, or SS) are also addressable for moving data between a segment register and a 16-bit register/memory location or for PUSH and POP. In the 80386/80486 microprocessor, the extended registers also are used for register addressing; they consist of: EAX, EBX, ECX, EDX, ESP, EBP, EDI, and ESI. Also available to the 80386/80486 are the FS and GS segment registers.

7. A MOV instruction that uses immediate addressing transfers the byte or word immediately following the opcode into a register or a memory location. Immediate addressing manipulates constant data in a program. In the 80386/80486 a double-word of immediate data may also be loaded into a 32-bit register or memory location.

8. Direct addressing takes two forms in the microprocessor: (a) direct addressing and (b) displacement addressing. Both forms of addressing are identical except direct addressing is used to transfer data between either AX or AL and memory, while displacement addressing is used with any register–memory transfer. Direct addressing requires 3 bytes of memory, while displacement addressing requires 4 bytes.

9. Register indirect addressing allows data to be addressed at the memory location pointed to by either a base (BP and BX) or index register (DI and SI). In the 80386/80486, extended registers EAX, EBX, ECX, EDX, EBP, EDI, and ESI are used to address memory data.

10. Base-plus-index addressing often addresses data in an array. The memory address for this mode is formed by adding a base register, index register, and the contents of a segment register times 10H. In the 80386/80486 the base and index registers may be any 32-bit register except EIP and ESP.

11. Register relative addressing uses either a base or index register plus a displacement to access memory data.

12. Base relative-plus-index addressing is useful for addressing a two-dimensional memory array. The address is formed by adding a base register, an index register, a displacement, and the contents of a segment register times 10H.

13. Scaled-index addressing is unique to the 80386/80486 microprocessor. The second of two registers (the index) is scaled by a factor of 2X, 4X, or 8X to access words, double words, or quad words in memory arrays.

14. Direct program memory addressing is allowed with the JMP and CALL instructions to any location in the memory system. With this addressing mode, the offset address and segment address are stored with the instruction.

15. Relative program addressing allows a JMP or CALL instruction to branch forward or backward in the current code segment by ±32K bytes. In the 80386/80486 the 32-bit displacement allows a branch to any location in the current code segment using a displacement value of ±2G bytes.

16. Indirect program addressing allows the JMP or CALL instructions to address another portion of the program or subroutine indirectly through a register or memory location.

17. The PUSH and POP instructions transfer a word between the stack and a register or memory location. A PUSH immediate instruction is available to place immediate data on the stack. The PUSHA and POPA instructions transfer AX, CX, DX, BX, BP, SP, SI, and DI between the stack and these registers. In the 80386/80486, the extended register and extended flags can also be transferred between registers and the stack. A PUSHFD stores the EFLAGS, while a PUSHF stores the FLAGS.

3–6   GLOSSARY

Absolute   A fixed memory location such as address [1000H].

Base-plus-index   A form of addressing memory wherein two registers are added to form the indirect memory location in either the data or stack segment.

Destination   In an instruction, the destination is the place where the data is moved. If two operands appear in an instruction, the destination is to the left.

Displacement   A distance used with data memory addressing and also program memory addressing modes.

Far   A label is said to be a far label if it is outside of the current code segment.

Immediate   Constant data that immediately follow the hexadecimal opcode in the memory is called immediate data.

Intersegment   A term used to indicate that the instruction or data are outside of the current code or data segment.

Intrasegment   A term used to indicate that the instruction or data are within the current code or data segment.

Label   A symbolic name for a memory location.

Near   A label is said to be near if it lies within the current segment.

Opcode   The hexadecimal or mnemonic code that directs the computer to perform an operation. Example opcodes are MOV, ADD, and so on.

Operand   The data operated upon by the opcode is called an operand.

Relative   A relative address is an address that is in relation to the instruction pointer.

Scaled-index   An index register that is scaled or multiplied by a 2, 4, or 8 in the 80386 or 80486 microprocessor.

Short   A label is said to be short if it lies within +127 to −128 bytes from the next instruction.

Source   In an instruction, the source is the place where the data originate. If the instruction has two operands, the source is to the right.

Stack   An area of memory address in the stack segment located by the stack pointer (SP/ESP).

Symbolic   A memory location referred to using a label such as DATA, TABLE, and so forth.

3–7   **QUESTIONS**

1. What do the following MOV instructions accomplish?
   a. MOV AX,BX
   b. MOV BX,AX
   c. MOV BL,CH
   d. MOV ESP,EBP
   e. MOV AX,CS
2. List the 8-bit registers that are used for register addressing.
3. List the 16-bit registers that are used for register addressing.
4. List the 32-bit registers that are used for register addressing in the 80386 and 80486 microprocessors.
5. List the 16-bit segment registers used with register addressing by MOV, PUSH, and POP.
6. What is wrong with the MOV BL,CX instruction?
7. What is wrong with the MOV DS,SS instruction?
8. Select an instruction for each of the following tasks:
   a. copy EBX into EDX
   b. copy BL into CL
   c. copy SI into BX
   d. copy DS into AX
   e. copy AL into AH
9. Select an instruction for each of the following tasks:
   a. move a 12H into AL
   b. move a 123AH into AX
   c. move a 0CDH into CL
   d. move a 1000H into SI
   e. move a 1200A2H into EBX
10. What special symbol is sometimes used to denote immediate data?
11. What is a displacement? How does it determine the memory address in a MOV [2000H],AL instruction?
12. What do the symbols [] indicate?
13. Suppose that DS = 0200H, BX = 0300H, and DI = 400H. Determine the memory address accessed by each of the following instructions, assuming real mode operation:
    a. MOV AL,[1234H]
    b. MOV EAX,[BX]
    c. MOV [DI],AL

14. What is wrong with a MOV [BX],[DI] instruction?
15. Choose an instruction that requires BYTE PTR.
16. Choose an instruction that requires WORD PTR.
17. Choose an instruction that requires DWORD PTR.
18. Explain the difference between the MOV BX,DATA instruction and the MOV BX,OFFSET DATA instruction.
19. Given that DS = 1000H, SS = 2000H, BP = 1000H, and DI = 0100H, determine the memory address accessed by each of the following, assuming real mode operation:
    a. MOV AL,[BP+DI]
    b. MOV CX,[DI]
    c. MOV EDX,[BP]
20. Given that DS = 1200H, BX = 0100H, and SI = 0250H, determine the address accessed by each of the following instructions, assuming real mode operation:
    a. MOV [100H],DL
    b. MOV [SI+100H],EAX
    c. MOV DL,[BX+100H]
21. Given that DS = 1100H, BX = 0200H, LIST = 0250H, and SI = 0500H, determine the address accessed by each of the following instructions, assuming real mode operation:
    a. MOV LIST[SI],EDX
    b. MOV CL,LIST[BX+SI]
    c. MOV CH,[BX+SI]
22. Given that DS = 1300H, SS = 1400H, BP = 1500H, and SI = 0100H, determine the address accessed by each of the following instructions, assuming real mode operation:
    a. MOV EAX,[BP+200H]
    b. MOV AL,[BP+SI−200H]
    c. MOV AL,[SI−0100H]
23. Which base register addresses data in the stack segment?
24. Given that EAX = 00001000H, EBX = 00002000H, and DS = 0010H, determine the addresses accessed by the following instructions, assuming real mode operation:
    a. MOV ECX,[EAX+EBX]
    b. MOV [EAX+2*EBX],CL
    c. MOV DH,[EBX+4*EAX+1000H]
25. List all three program memory-addressing modes.
26. How many bytes of memory store a far direct jump instruction? What is stored in each of the bytes?
27. What is the difference between an intersegment and intrasegment jump?
28. If a near jump uses a signed 16-bit displacement, how can it jump to any memory location within the current code segment?
29. What is a far jump?
30. If a JMP instruction is stored at memory location 100H within the current code segment, it cannot be a _____ jump if it is jumping to memory location 200H within the current code segment.

31. Show which JMP instruction assembles (short, near, or far) if the JMP THERE instruction is stored at memory address 10000H and the address of THERE is:
    a. 10020H
    b. 11000H
    c. 0FFFEH
    d. 30000H
32. Form a JMP instruction that jumps to the address pointed to by the BX register.
33. Select a JMP instruction that jumps to the location stored in memory at location TABLE. Assume that it is a near JMP.
34. How many bytes are stored on the stack by PUSH instructions?
35. Explain how the PUSH [DI] instruction functions.
36. What registers are placed on the stack by the PUSHA instruction and in what order?
37. What does the 80386/80486 instruction PUSHAD accomplish?

CHAPTER 4

Data Movement Instructions

INTRODUCTION

In this chapter, we explain the 8086–80486 data movement instructions. The data movement instructions include MOV, MOVSX, MOVZX, PUSH, POP, BSWAP, XCHG, XLAT, IN, OUT, LEA, LDS, LES, LFS, LGS, LSS, LAHF, SAHF, and the string instructions, MOVS, LODS, STOS, INS, and OUTS. We present data movement instructions first because they are used more often and are easy to understand.

The microprocessor requires an assembler program, which generates machine language, because machine language instructions are too complex to generate by hand. This chapter describes the assembly language syntax and some of its directives. This text assumes that the user is developing software on an IBM personal computer or clone using the Microsoft MACRO assembler (MASM), Intel Assembler (ASM), Borland Turbo assembler (TASM), or similar software. This text presents information that functions with the Microsoft MASM assembler, but most programs function without change using other assemblers. Appendix A explains the Microsoft assembler and provides detail on the linker program.

4–1 CHAPTER OBJECTIVES

Upon completion of this chapter, you will be able to:

1. Explain the operation of each data movement instruction with applicable addressing modes.
2. Explain the purposes of the assembly language pseudo-operations and key words such as: ALIGN, ASSUME, DB, DD, DW, END, ENDS, ENDP, EQU, OFFSET, ORG, PROC, PTR, USE16, USE32, and SEGMENT.

3. Given a specific data movement task, select the appropriate assembly language instructions to accomplish it.
4. Given a hexadecimal machine language instruction, determine the symbolic opcode, source, destination, and addressing mode.
5. Use the assembler to set up a data segment, stack segment, and a code segment.
6. Show how to set up a procedure using PROC and ENDP.
7. Explain the difference between memory models and full segment definitions for the MASM assembler.

4–2   MOV REVISITED

We used the MOV instruction in Chapter 3 to explain the diverse 8086–80486 addressing modes. In this chapter, we use MOV to introduce the machine language instructions available to the programmer for various addressing modes and instructions. Machine code is introduced because it may occasionally be necessary to interpret machine language programs generated by an assembler. Interpretation of the machine's language allows debugging or modification at the machine language level. We also show how to convert between machine and assembly language instructions using Appendix B.

Machine Language

Machine language is the native binary code that the microprocessor understands and uses as the instructions that control its operation. Machine language instructions, for the 8086–80486, vary in length from 1 to as many as 13 bytes. Although machine language appears complex, there is order to this microprocessor's machine language. There are over 20,000 variations of machine language instructions for the 8086–80486 microprocessor, which means there is no complete list of these variations. Because of this, some binary bits in a machine language instruction are given, and the remainder must be determined for each variation of the instruction.

Instructions for the 8086–80286 are 16-bit mode instructions that take the form found in Figure 4–1a. These 16-bit mode instructions are compatible with the 80386 and 80486 microprocessor if they are set to use 16-bit mode instruction formats. The 80386/80486 assumes that all instructions are 16-bit mode instructions when the machine is operated in the real mode. In the protected mode, the upper byte of the 80386/80486 descriptor contains the D-bit that selects either the 16- or 32-bit instruction mode. The 80386 and 80486 also use 32-bit mode instructions in the form shown in Figure 4–1b. These instructions can occur in the 16-bit instruction mode, by the use of prefixes.

The first 2 bytes of the 32-bit instruction mode format are called override prefixes because they are not always present. The first modifies the size of the address used by the instruction and the second modifies the register size. If the 80386/80486 is operating as a 16-bit instruction mode machine (real or

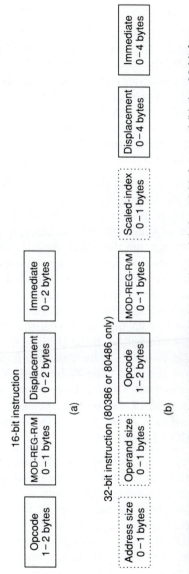

16-bit instruction

| Opcode 1 – 2 bytes | MOD-REG-R/M 0 – 1 bytes | Displacement 0 – 2 bytes | Immediate 0 – 2 bytes |

(a)

32-bit instruction (80386 or 80486 only)

| Address size 0 – 1 bytes | Operand size 0 – 1 bytes | Opcode 1 – 2 bytes | MOD-REG-R/M 0 – 1 bytes | Scaled-index 0 – 1 bytes | Displacement 0 – 4 bytes | Immediate 0 – 4 bytes |

(b)

FIGURE 4–1 The formats of the 8086–80486 instructions: (a) the 16-bit form and (b) the 32-bit form.

FIGURE 4–2 Byte one of many machine language instructions, showing the position of the D and W bits.

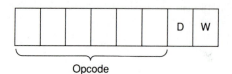

Opcode

protected mode) and a 32-bit register is used, the ***operand-size prefix*** (66H) is appended to the front of the instruction. If the 80386/80486 is operating as a 32-bit instruction mode machine (protected mode only) and a 32-bit register is used, the operand-size prefix is not present. If a 16-bit register appears in an instruction, in the 32-bit instruction mode, the ***register-size prefix*** is present to select a 16-bit register. The address size-prefix is used in a similar fashion as explained later. The prefixes toggle the size of the register and operand from 16-bit to 32-bit or 32-bit to 16-bit for the prefixed instruction.

The Opcode. The *opcode* selects the operation (addition, subtraction, move, and the like) performed by the microprocessor. The opcode is either one or two bytes in length for machine language instructions. Figure 4–2 illustrates the general form of the first opcode byte of many, but not all, machine language instructions. Here the first 6 bits of the first byte are the binary opcode. The remaining 2 bits indicate the *direction* (D) of the data flow and whether the data are a *byte* or a *word* (W). In the 80386/80486, double words and words are both specified when W = 1. The instruction mode and operand-size prefix select whether W represents a word or a double word in the 80386/80486 microprocessor.

If the D-bit = 1, data flow to the register (REG) field from the R/M field in the next byte of the instruction. If the D-bit = 0, data flow to the R/M field from the REG field. If the W-bit = 1, the data size is a word/double word, and if the W-bit = 0, data size is a byte. The W-bit appears in most instructions while the D-bit mainly appears with the MOV and a few other instructions. Refer to Figure 4–3 for the binary bit pattern of the second opcode byte (reg-mod-r/m) of many instructions. This illustration shows the location of the REG (*register*), R/M (*register/memory*), and MOD (*mode*) fields.

MOD Field. The *MOD field* specifies the addressing mode (MOD) for the selected instruction. The MOD field selects the type of addressing and whether a ***displacement*** is present with the selected type. Table 4–1 lists the operand forms available to the MOD field for 16-bit instruction mode in the 8086–80486 microprocessors. If the MOD field contains a 11, it selects the register addressing mode. Register addressing uses the R/M field to specify a register instead of a memory location. If the MOD field contains a 00, 01, or 10, the R/M field selects one of the data memory-addressing modes. When MOD selects a memory-addressing mode, it indi-

FIGURE 4–3 The second byte of many machine language instructions showing the position of the MOD, REG, and R/M fields.

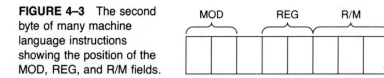

TABLE 4–1 MOD field specifications for the 16-bit instruction mode

MOD	Function
00	No displacement
01	8-bit sign-extended displacement
10	16-bit displacement
11	R/M is a register

cates that the addressing mode contains no displacement (00), an 8-bit sign-extended displacement (01), or a 16-bit displacement (10). The MOV AL,[DI] instruction is an example that uses no displacement, a MOV AL,[DI+2] instruction uses an 8-bit displacement (+ 2), and a MOV AL,[DI+1000] instruction uses a 16-bit displacement (+ 1000).

All 8-bit displacements are *sign extended* into 16-bit displacements when the microprocessor executes the instruction. If the 8-bit displacement is 00H–7FH (positive), it is sign extended to 0000H–007FH before being added to a segment address. If the 8-bit displacement is 80H–FFH (negative), it is sign extended to FF80H–FFFFH. To sign extend a number, its sign-bit is copied to the next higher order byte, which generates a 00H or an FFH in the higher order byte.

In the 80386/80486 microprocessor, the MOD field may be the same as in Table 4–1 or, if the instruction mode is 32 bits, it is as appears in Table 4–2. The MOD field is interpreted as selected by the address override prefix or the operating mode of the microprocessor. This change in the interpretation of the MOD field and instruction supports the numerous additional addressing modes allowed in the 80386/80486 microprocessor. The main difference occurs when the MOD field is a 10. This causes the 16-bit displacement to become a 32-bit displacement to allow any protected mode memory location (4G bytes) to be accessed. The 80386/80486 only allows an 8- or 32-bit displacement when operated in the 32-bit instruction mode.

Register Assignments. Table 4–3 lists the register assignments for the REG field and the R/M field (MOD = 11). This table contains three lists of register assignments: one is used when W-bit = 0 (bytes), and the other two are employed when the W-bit = 1 (words or double words). Note that double-word registers are only available to the 80386/80486.

Suppose that a 2-byte instruction, 8BECH, appears in a machine language program. Because neither a 66H (operand-size override prefix) nor 67H (register-size override prefix) appears as the first byte, the first byte is the opcode. Assuming that the microprocessor is operated in 16-bit instruction mode, this instruction is converted to binary and placed in the instruction format of bytes 1 and 2 as illustrated in Figure 4–4.

TABLE 4–2 MOD field specifications for 32-bit instruction mode (80386/80486 only)

MOD	Function
00	No displacement
01	8-bit sign-extended displacement
10	32-bit displacement
11	R/M is a register

TABLE 4–3 REG and R/M (when MOD = 11) assignments

Code	W = 0 (Byte)	W = 1 (Word)	W = 1 (Double word)
000	AL	AX	EAX
001	CL	CX	ECX
010	DL	DX	EDX
011	BL	BX	EBX
100	AH	SP	ESP
101	CH	BP	EBP
110	DH	SI	ESI
111	BH	DI	EDI

The opcode is 100010. If you refer to Appendix B, which lists the machine language instructions, you will find that this is the opcode for a MOV instruction. Also notice that both the D- and W-bits are a logic 1, which means that a word moves into the register specified in the REG field. The REG field contains a 101, indicating register BP, so the MOV instruction moves data into register BP. Because the MOD field contains a 11, the R/M field also indicates a register. Here, R/M = 100 (SP); therefore, this instruction moves data from SP into BP and is written in symbolic form as a MOV BP,SP instruction.

    Suppose that a 668BE8H instruction appears in an 80386/80486 operated in the 16-bit instruction mode. The first byte (66H) is the operand-size override prefix that selects 32-bit operands for the 16-bit instruction mode. The remainder of the instruction indicates that the opcode is that of the MOV with a source operand of EAX and a destination operand of EBP. This instruction is a MOV EBP,EAX. The same instruction becomes a MOV BP,AX instruction in the 80386/80486 microprocessor if it is originally operating in the 32-bit instruction mode and then overridden. Luckily the assembler program keeps track of the operand- and address-size prefixes.

R/M Memory Addressing. If the MOD field contains a 00, 01, or 10, the R/M field takes on a new meaning. Table 4–4 lists the memory-addressing modes for the R/M field when MOD is a 00, 01, or 10 for the 16-bit instruction mode.

Opcode	D	W		MOD	REG	R/M
1 0 0 0 1 0	1	1		1 1	1 0 1	1 0 0

Opcode = MOV
D = Transfer to REG
W = Word
MOD = R/M is a register
REG = BP
R/M = SP

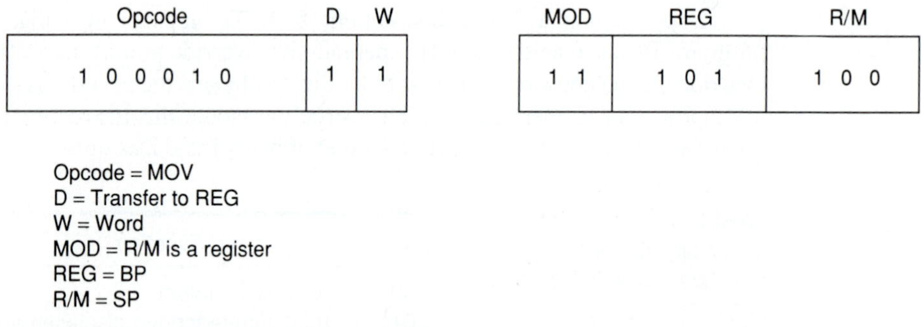

FIGURE 4–4 The BBEC instruction placed into byte 1 and 2 formats from Figures 4–2 and 4–3. This instruction is a MOV BP,SP.

TABLE 4-4 The 16-bit R/M memory-addressing modes

Code	Function
000	DS:[BX+SI]
001	DS:[BX+DI]
010	SS:[BP+SI]
011	SS:[BP+DI]
100	DS:[SI]
101	DS:[DI]
110	SS:[BP]*
111	DS:[BX]

Note: See text under Special Addressing Mode.

All of the 16-bit addressing modes represented in Chapter 3 appear in Table 4-4. The displacement, discussed in Chapter 3, is defined by the MOD field. If MOD = 00 and R/M = 101, the addressing mode is [DI]. If MOD = 01 or 10, the addressing mode is [DI+33H] or LIST [DI+22H] for the 16-bit instruction mode. (This example uses LIST, 33H, and 22H as arbitrary values for the displacement.)

Figure 4-5 illustrates the machine language version of the 16-bit instruction MOV DL,[DI] or instruction (8A15H). This instruction is 2 bytes long and has an opcode 100010, D = 1 (*to REG from R/M*), W = 0 (*byte*), MOD = 00 (*no displacement*), REG = 010 (*DL*), and R/M = 101 (*[DI]*). If the instruction changes to MOV DL,[DI+1], the MOD field changes to 01, for an 8-bit displacement, but the first 2 bytes of the instruction otherwise remain the same. The instruction now becomes 8A5501H instead of 8A15H. Notice that the 8-bit displacement appends to the first 2 bytes of the instruction to form a 3-byte instruction instead of 2 bytes. If the instruction is again changed to a MOV DL,[DI+1000H], the machine language form becomes an 8A750010H. Here the 16-bit displacement of 1000H (coded as 0010H) appends the opcode.

Special Addressing Mode. There is a special addressing mode (for 16-bit instructions) that does not appear in Tables 4-2, 4-3, or 4-4 that occurs whenever memory data

Opcode	D	W		MOD	REG	R/M
1 0 0 0 1 0	1	0		0 0	0 1 0	1 0 1

Opcode = MOV
D = Transfer to REG
W = Byte
MOD = No displacement
REG = DL
R/M = DS:[DI]

FIGURE 4-5 A MOV DL,[DI] instruction converted to its machine language form.

are referenced by only the displacement mode of addressing for 16-bit instructions. Examples are the MOV [1000H],DL and MOV NUMB,DL instructions. The first instruction moves the contents of register DL into data segment memory location 1000H. The second instruction moves register DL into symbolic data segment memory location NUMB.

Whenever an instruction has only a displacement, the MOD field is always a 00 and the R/M field is always a 110. This combination normally shows that the instruction contains no displacement and uses addressing mode [BP]. You cannot actually use addressing mode [BP] without a displacement in machine language. The assembler takes care of this by using an 8-bit displacement (MOD = 01) of 00H whenever the [BP] addressing mode appears in an instruction. This means that the [BP] addressing mode assembles as a [BP+0] even though we use [BP].

Figure 4–6 shows the binary bit pattern required to encode the MOV [1000H],DL instruction in machine language. If the individual translating this symbolic instruction into machine language does not know about the special addressing mode, it would incorrectly translate to a MOV [BP],DL instruction. Figure 4–7 shows the actual form of the MOV [BP],DL instruction. Notice that this is a 3-byte instruction with a displacement of 00H.

32-Bit Addressing Modes.   The 32-bit addressing modes found in the 80386/80486 microprocessor are obtained by running these machines in the 32-bit instruction mode or in the 16-bit instruction mode using the address-size prefix 67H. Table 4–5 shows the coding for R/M used to specify the 32-bit addressing modes. Notice that when R/M = 100, an additional byte appears in the instruction called a *scaled-index byte*. The scaled-index byte indicates the additional forms of scaled-index addressing that do not appear in Table 4–5. The scaled-index byte is mainly used when two registers are added to specify the memory address in an instruction.

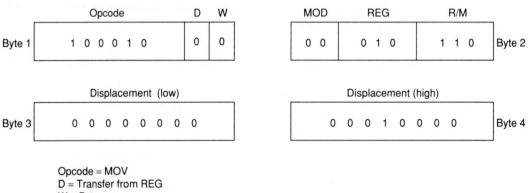

Opcode = MOV
D = Transfer from REG
W = Byte
MOD = Because R/M is [BP] (special addressing)
REG = DL
R/M = DS: [BP]
Displacement = 1000H

FIGURE 4–6   The MOV [1000H],DL instruction uses the special addressing mode.

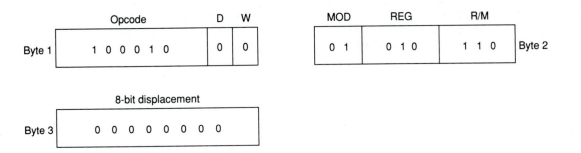

Opcode = MOV
D = Transfer from REG
W = Byte
MOD = Because R/M is [BP] (special addressing)
REG = DL
R/M = DS: [BP]
Displacement = 00H

FIGURE 4–7 The MOV [BP],DL instruction converted to binary machine language.

Figure 4–8 shows the format of the *scaled-index byte* as selected by a value of 100 in the R/M field of an instruction when the 80386/80486 uses a 32-bit address. The leftmost 2 bits select a scaling factor (multiplier) of 1X, 2X, 4X, or 8X. Note that a scaling factor of 1X is implicit if none is used in an instruction that contains two 32-bit indirect address registers. The index and base fields both contain register numbers as indicated in Table 4–3 for 32-bit registers.

The instruction MOV EAX,[EBX+4*ECX] is encoded as 67668B048BH. Notice that both the address size (67H) and register size (66H) override prefixes appear for this instruction. This means that the instruction is 67668B048BH when operated with the 80386/80486 in the 16-bit instruction mode. If the 80386/80486 is operated in the 32-bit instruction mode both prefixes disappear so the instruction becomes an

TABLE 4–5 32-bit addressing modes selected by R/M

Code	Function
000	DS:[EAX]
001	DS:[ECX]
010	DS:[EDX]
011	DS:[EBX]
100	uses scaled-index byte
101	SS:[EBP]*
110	DS:[ESI]
111	DS:[EDI]

*Note: If the MOD bits are 00, this addressing mode uses a 32-bit displacement without register EBP. This is similar to the special addressing mode for the 16-bit instruction mode.

FIGURE 4–8 The scaled-index byte of some 32-bit addressing modes.

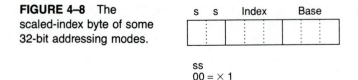

ss
00 = × 1
01 = × 2
10 = × 4
11 = × 8

8B048BH instruction. The use of the prefixes depend on the mode of operation for the 80386 and 80486 microprocessor.

An Immediate Instruction. Suppose we choose the MOV WORD PTR [BX + 1000H],1234H instruction as an example of a 16-bit instruction using immediate addressing. This instruction moves a 1234H into the word-sized memory location addressed by the sum of 1000H, BX, and DS × 10H. This 6-byte instruction uses 2 bytes for the opcode, D, W, MOD, REG, and R/M fields. Two of the 6 bytes are the data of 1234H. Two of the 6 bytes are the displacement of 1000H. Figure 4–9 shows the binary bit pattern for each byte of this instruction.

This instruction, in symbolic form, includes WORD PTR. The WORD PTR directive indicates to the assembler that the instruction uses a word-sized memory pointer. If the instruction moves a byte of immediate data, then BYTE PTR replaces

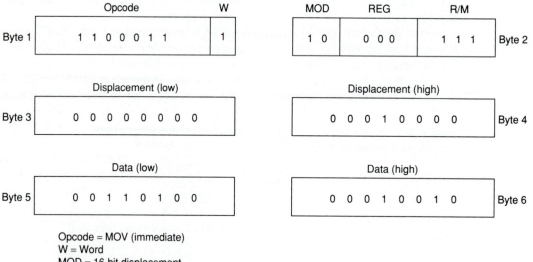

Opcode = MOV (immediate)
W = Word
MOD = 16-bit displacement
REG = 000 (not used in immediate addressing)
R/M = DS: [BX]
Displacement = 1000H
Data = 1234H

FIGURE 4–9   A MOV WORD PTR [BX+1000H],1234H instruction converted to binary machine language.

TABLE 4–6 Segment register selection bits

Code	Segment Register
000	ES
001	CS*
010	SS
011	DS
100	FS
101	GS

Note: MOV CS,?? and POP CS are not allowed by the microprocessor. The FS and GS segments are only available to the 80386/80486 microprocessor.

WORD PTR in the instruction. Likewise if the instruction uses a double-word of immediate data, the DWORD PTR directive replaces BYTE PTR. Most instructions that refer to memory through a pointer do not need the BYTE PTR, WORD PTR, or DWORD PTR directives. These are only necessary when it is not clear if the operation is a byte or a word. The MOV [BX],AL instruction is clearly a byte move, while the MOV [BX],1 instruction is not exact and could therefore be a byte-, word-, or double-word sized move. Here the instruction must be coded as MOV BYTE PTR [BX],1, MOV WORD PTR [BX],1, or MOV DWORD PTR [BX],1. If not, the assembler flags it as an error because it cannot determine the intent of this instruction.

Segment MOV Instructions. If the contents of a segment register are moved by the MOV, PUSH, or POP instruction, a special set of register bits (REG field) selects the segment register (see Table 4–6).

Figure 4–10 shows a MOV BX,CS instruction converted to binary. The opcode for this type of MOV instruction is different from that of the prior MOV instructions. Segment registers can be moved between any 16-bit register or 16-bit memory location. For example, the MOV [DI],DS instruction stores the contents of DS into the memory location addressed by DI in the data segment.

Although this has not been a complete coverage of machine language coding, it should give you a good start in machine language programming. Remem-

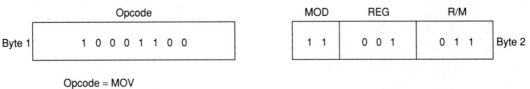

Opcode = MOV
MOD = R/M is a register
REG = CS
R/M = BX

FIGURE 4–10 A MOV BX.CS instruction converted to binary machine language.

ber a program written in symbolic assembly language (assembly language), is rarely assembled by hand into binary machine language. An *assembler* converts symbolic assembly language into machine language. With the microprocessor and its over 20,000 instruction variations, let us hope that an assembler is available for the conversion, because the process is very time-consuming, although not impossible.

4–3   PUSH/POP

The PUSH and POP instructions are important instructions that store and retrieve data from the LIFO (last-in, first-out) stack memory. The 8086–80486 microprocessor has six forms of the PUSH and POP instructions: register, memory, immediate, segment register, flags, and all registers. The PUSH and POP immediate and the PUSHA and POPA (affecting all registers) forms are not available in the earlier 8086/8088 microprocessor, but are available to the 80286, 80386, and 80486.

Register addressing allows the contents of any 16-bit register to be transferred to or from the stack. In the 80386/80486, the 32-bit extended registers and flags (EFLAGS) can also be pushed or popped from the stack. Memory addressing stores the contents of a 16-bit memory location (32 bits in the 80386/80486) on the stack or stores stack data into a memory location. Immediate addressing allows immediate data to be pushed onto the stack, but not popped off the stack. Segment register addressing allows the contents of any segment register to be pushed onto the stack or removed from the stack (CS may be pushed, but data from the stack may never be popped into CS). The flags may be pushed or popped from the stack, and the contents of all the registers may be pushed or popped.

PUSH

The 8086–80286 PUSH instruction always transfers *2 bytes* of data to the stack, and the 80386/80486 transfers 2 or 4 bytes depending on the register or size of the memory location. The source of the data may be any internal 16-bit or 32-bit register, immediate data, any segment register, or any 2 bytes of memory data. There is also a PUSHA instruction that copies the contents of the internal register set, except the segment registers, to the stack. The PUSHA (*push all*) instruction copies the registers to the stack in the following order: AX, CX, DX, BX, SP, BP, SI, and DI. The value of SP pushed to the stack is whatever it was before the PUSHA instruction executes. The PUSHF (*push flags*) instruction copies the contents of the flag register to the stack. The PUSHAD and POPAD instructions push and pop the contents of the 32-bit register set found in the 80386 and 80486 microprocessors.

Whenever data are pushed onto the stack, the first (most significant) data byte moves into the stack segment memory location addressed by SP − 1. The second (least significant) data byte moves into the stack segment memory location addressed by SP − 2. After the data are stored by a PUSH, the contents of the SP register decrement by 2. The same is true for a double word push except 4 bytes are moved to the stack memory (most significant byte first) then the stack pointer decrements by 4. Figure 4–11 shows the operation of the PUSH AX instruction. This instruction copies the contents of AX onto the stack where address SS:[SP − 1] = AH, SS:[SP − 2] = AL, and afterwards SP = SP − 2.

The PUSHA instruction pushes all the internal 16-bit registers onto the stack as illustrated in Figure 4–12. This instruction requires 16 bytes of stack memory space to store all eight 16-bit registers. After all registers are pushed, the contents of the SP register are decremented by 16. The PUSHA instruction is very useful when the entire register set (*microprocessor environment*) of the 80286–80486 must be saved during a task.

The PUSH immediate data instruction has two different opcodes, but in both cases a 16-bit immediate number moves onto the stack or, if PUSHD is used, a 32-bit immediate datum is pushed. If the value of the immediate data are 00H–FFH, the opcode is a 6AH, and if the data are 0100H–FFFFH, the opcode is 68H. The PUSH 8 instruction assembles as a 6A08H, and the PUSH 1000H instruction assembles as 680010H.

Table 4–7 lists the forms of the PUSH instruction, including PUSHA and PUSHF. Notice how the instruction set is used to specify different data sizes with the assembler.

POP

The POP instruction performs the inverse operation of a PUSH instruction. The POP instruction removes data from the stack and places it into the target 16-bit register,

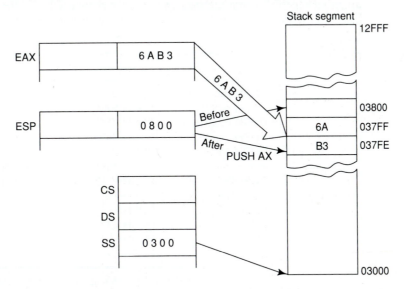

FIGURE 4–11 The effect of the PUSH AX instruction on ESP and stack memory locations 37FFH and 37FEH.

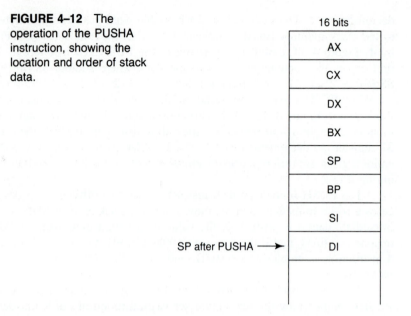

FIGURE 4–12 The operation of the PUSHA instruction, showing the location and order of stack data.

segment register, or a 16-bit memory location. In the 80386/80486 a POP can also remove 32-bit data from the stack and use a 32-bit address. The POP instruction is not available as an immediate POP. The POPF (*pop flags*) instruction removes a 16-bit number from the stack and places it into the flag register, and the POPFD removes a 32-bit number from the stack and places it into the extended flag register. The POPA (*pop all*) instruction removes 16 bytes of data from the stack and places them into the following registers in the order shown: DI, SI, BP, SP, BX, DX, CX, and AX. This is the reverse order from the way they are placed on the stack by

TABLE 4–7 The PUSH instructions

Symbolic	Example	Note
PUSH reg16	PUSH BX	16-bit register
PUSH reg32	PUSH EAX	32-bit register
PUSH mem16	PUSH [BX]	16-bit addressing mode
PUSH mem32	PUSH [EAX]	32-bit addressing mode
PUSH seg	PUSH DS	any segment register
PUSH imm8	PUSH 12H	8-bit immediate data
PUSHW imm16	PUSHW 1000H	16-bit immediate data
PUSHD imm32	PUSHD 20	32-bit immediate data
PUSHA	PUSHA	save 16-bit registers
PUSHAD	PUSHAD	save 32-bit registers
PUSHF	PUSHF	save 16-bit flag register
PUSHFD	PUSHFD	save 32-bit flag register

Note: The 80386/80486 is required to operate with 32-bit addresses and registers.

the PUSHA instruction. In the 80386/80486 a POPAD instruction reloads the 32-bit registers from the stack.

Suppose that a POP BX instruction executes. The first byte of data removed from the stack (the memory location addressed by SP in the stack segment) moves into register BL. The second byte is removed from stack segment memory location SP + 1, and placed into register BH. After both bytes are removed from the stack, the SP register increments by 2. Figure 4–13 shows how the POP BX instruction removes data from the stack and places them into register BX.

The opcodes used for the POP instruction, and all its variations, appear in Table 4–8. Note that a POP CS instruction is not a valid instruction in the 80286 instruction set. If we allow a POP CS instruction to execute, a portion of the address (CS) of the next instruction changes. This makes the POP CS instruction unpredictable and therefore not allowed.

Initializing the Stack

When the stack area is initialized, we load both the stack segment register (SS) and the stack pointer (SP) register. It is normal to designate an area of memory as the stack segment by loading SS with the bottom location of the stack segment.

For example, if the stack segment resides in memory locations 10000H–1FFFFH, we load SS with a 1000H. (Recall that we append to the rightmost end of the stack segment register a 0H for real mode addressing.) To start the stack at the top of this 64K byte stack segment, the stack pointer (SP) is loaded with a 0000H. Figure 4–14 shows how this value causes data to be pushed onto the top of the

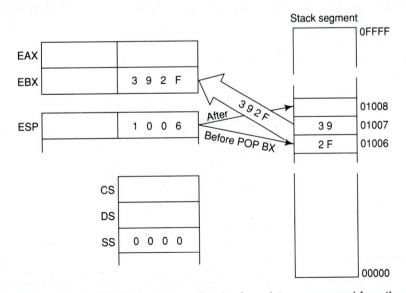

FIGURE 4–13 The POP BX instruction, showing how data are removed from the stack.

TABLE 4–8   The POP instructions

Symbolic	Example	Note
POP reg16	POP DI	16-bit register
POP reg32	POP EBX	32-bit register
POP mem16	POP WORD PTR[DI+2]	16-bit memory address
POP mem32	POP DATA3	32-bit memory address
POP seg	POP GS	any segment register
POPA	POPA	16-bit registers
POPAD	POPAD	32-bit registers
POPF	POPF	16-bit flag register
POPFD	POPFD	32-bit extended flag register

Note: The 80386/80486 is required to operate with 32-bit addresses and registers.

stack segment with a PUSH CX instruction. Remember that all segments are *cyclic* in nature—that is, the top location of a segment is *contiguous* with the bottom location of the segment.

In assembly language, a stack segment is set up as illustrated in Example 4–1. The first statement identifies the start of the stack segment, and the last statement identifies the end of the stack segment. The assembler and linker program places the correct stack segment address in SS and the length of the segment (top of the stack) into SP. There is no need to load these registers in your program unless you wish to change the initial values for some reason.

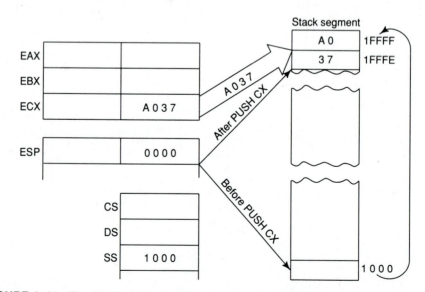

FIGURE 4–14   The PUSH CX instruction, showing the cyclic nature of the stack segment.

EXAMPLE 4–1

```
0000                    STACK_SEG      SEGMENT STACK

0000  0100[                    DW      100H DUP (?)
            ????
         ]
0200                    STACK_SEG      ENDS
```

An alternative method for defining the stack segment is used with one of the memory models for the MASM assembler only (refer to Appendix A). Other assemblers do not use models, or if they do, they are not the same as with MASM. Here (see Example 4–2) the .STACK statement, followed by the number of *bytes* allocated to the stack, defines the stack area. This is identical in effect to Example 4–1. The .STACK statement also initializes both SS and SP.

EXAMPLE 4–2

```
.MODEL SMALL
.STACK 200H                     ;set stack size
```

If the stack is not specified using either method, a warning will appear when the program is linked. The warning may be ignored if the stack size is 128 bytes or less. The system automatically assigns (through DOS) a 128-byte section of memory to the stack. This memory section is located in the program segment prefix (PSP), which is appended to the beginning of each program file. If you use more memory for the stack, you will erase information in the PSP that is critical to the operation of your program and the computer.

4–4   LOAD-EFFECTIVE ADDRESS

There are several load-effective address instructions in the 8086–80486 microprocessor instruction set. The LEA instruction loads any 16-bit register with the address determined by the addressing mode selected for the instruction. The LDS and LES variations load any 16-bit register with the offset address retrieved from a memory location and then load either DS or ES with a segment address retrieved from memory. In the 80386 and 80486 microprocessor LFS, LGS, and LSS are added to the instruction set and a 32-bit register can be selected to receive a 32-bit offset from memory. Table 4–9 lists the load-effective address instructions.

LEA

The LEA instruction loads a 16-bit register with the offset address of the data specified by the operand. As the first example in Table 4–9 shows, the operand address NUMB is loaded into register AX, not the contents of address NUMB.

TABLE 4–9 The load-effective address instructions

Symbolic	Function
LEA AX,NUMB	AX is loaded with the address of NUMB
LEA EAX,NUMB	EAX is loaded with the address of NUMB
LDS DI,LIST	DI and DS are loaded with the address stored at LIST
LDS EDI,LIST	EDI and DS are loaded with the address stored at LIST
LES BX,CAT	BX and ES are loaded with the address stored at CAT
LFS DI,DATA1	DI and FS are loaded with the address stored at DATA1
LGS SI,DATA5	SI and GS are loaded with the address stored at DATA5
LSS SP,MEM	SP and SS are loaded with the address stored at MEM

By comparing LEA with MOV, we observe the following difference: LEA BX,[DI] loads the offset address specified by [DI] (the contents of DI) into the BX register; MOV BX,[DI] loads the data stored at the memory location addressed by [DI] into register BX.

Earlier in the text, we presented several examples using the OFFSET pseudo-operation. The ***OFFSET directive*** performs the same function as an LEA instruction if the operand is a displacement. For example, the MOV BX,OFFSET LIST performs the same function as LEA BX,LIST. Both instructions load the offset address of memory location LIST into the BX register.

But why is the LEA instruction available if the OFFSET directive accomplishes the same task? First, OFFSET only functions with simple operands such as LIST. It may not be used for an operand such as [DI], LIST[SI], and so on. The OFFSET directive is more efficient than the LEA instruction for simple operands. It takes the microprocessor longer to execute the LEA BX,LIST than MOV BX,OFFSET LIST. The 80286, for example, requires 3 clocks to execute the LEA BX,LIST instruction and only 2 clocks to execute MOV BX,OFFSET LIST. The reason that the MOV BX,OFFSET LIST instruction executes faster is because the assembler calculates the offset address of LIST, while with the LEA instruction, the microprocessor does the calculation as it executes the instruction. The MOV BX,OFFSET LIST instruction is actually assembled as a move immediate instruction and is more efficient.

Suppose that the microprocessor executes an LEA BX,[DI] instruction and DI contains a 1000H. Because DI contains the offset address, the microprocessor transfers a copy of DI into BX. A MOV BX,DI instruction performs this task in less time and is often preferred to the LEA BX,[DI] instruction.

Another example is LEA SI,[BX+DI]. This instruction adds BX to DI and stores the sum in the SI register. The sum generated is a modulo-64K sum. (A modulo-64K sum drops any carry out of the 16-bit result.) If BX = 1000H and DI = 2000H, the offset address moved into SI is 3000H. If BX = 1000H and DI = FF00H, the offset address is 0F00H. Notice that the second result is a modulo-64K sum.

LDS, LES, LFS, LGS, and LSS

The LDS, LES, LFS, LGS, and LSS instructions load any 16-bit or 32-bit register with an offset address and the DS, ES, FS, GS, or SS segment register with a segment address. These instructions use any of the memory-addressing modes to access a 32-bit or 48-bit section of memory that contains both the segment and offset address. These instructions may not use the register addressing mode (MOD = 11). Note that the LFS, LGS, and LSS instructions are only available on 80386 and 80486 microprocessors, as are the 32-bit registers.

Figure 4–15 illustrates an example LDS BX,[DI] instruction. This instruction transfers the 32-bit number addressed by DI in the data segment into the BX and DS registers. The LDS, LES, LFS, LGS, and LSS instructions obtain a new far address from memory. The offset address appears first, followed by the segment address. This format is used for storing all 32-bit memory addresses.

A far address can be stored in memory by the assembler. For example, the ADDR DD FAR PTR FROG instruction stores the offset and segment address (*far address*) of FROG in 32 bits of memory at location ADDR. The DD directive tells the assembler to store a *double word* (32-bit number) in memory address ADDR.

In the 80386/80486, an LDS EBX,[DI] instruction loads EBX from the 4-byte section of memory addressed by DI in the data segment. Following this 4-byte offset is a word that is loaded to the DS register. Notice that instead of addressing a

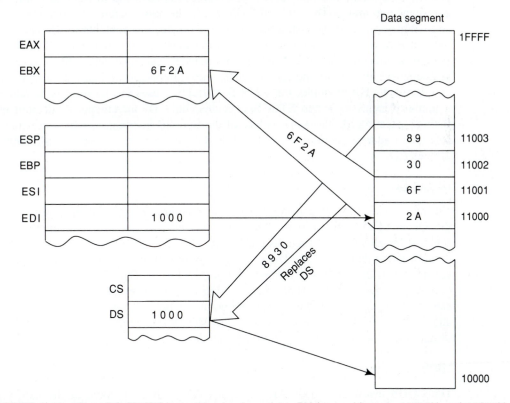

FIGURE 4–15 The LDS BX,[DI] instruction loads register BX from addresses 11000H and 11001H and register DS from locations 11002H and 11003H.

32-bit section of memory, the 80386/80486 addresses a 48-bit section of the memory whenever a 32-bit offset address is loaded to a 32-bit register.

4-5   STRING DATA TRANSFERS

There are five *string data transfer instructions*—LODS, STOS, MOVS, INS, and OUTS. Each string instruction allows data transfers that are either a single byte, word, or double word, or if repeated, a block of bytes, words, or double words. Before the string instructions are presented, the operation of the D flag-bit (*direction*), DI, and SI must be understood as they apply to the string instructions.

The Direction Flag

The direction flag (D) selects auto-increment (D = 0) or auto-decrement (D = 1) operation for the DI and SI registers during string operations. The direction flag is used only with the string instructions. The CLD instruction *clears* the D flag (D = 0) and the STD instruction *sets* it (D = 1). Therefore, the CLD instruction selects the auto-increment mode (D = 0) and STD selects the auto-decrement mode (D = 1).

Whenever a string instruction transfers a byte, the contents of DI and/or SI increment or decrement by 1. If a word is transferred, the contents of DI and/or SI increment or decrement by 2. Double-word transfers cause DI and/or SI to increment or decrement by 4. Only the actual registers used by the string instruction increment or decrement. For example, the STOSB instruction uses the DI register to address a memory location. When STOSB executes, only DI increments or decrements, without affecting SI. The same is true of the LODSB instruction, which uses the SI register to address memory data. LODSB only increments/decrements SI, without affecting DI.

DI and SI

During the execution of a string instruction, memory accesses occur through either the DI or SI register, or both. The DI offset address accesses data in the extra segment for all string instructions that use it. The SI offset address accesses data, by default, in the data segment. The segment assignment of SI may be changed with a segment override prefix as described later in this chapter. The DI segment assignment is *always* in the extra segment when a string instruction executes. This assignment cannot be changed. The reason that one pointer addresses data in the extra segment and the other in the data segment is so the MOVS instruction can move 64K bytes of data from one segment of memory to another.

LODS

The LODS instruction loads AL, AX, or EAX with data stored at the data segment offset address indexed by the SI register. (Note that only the 80386 and 80486 can use EAX.) After loading AL with a byte, AX with a word, or EAX with a double word, the contents of SI increment if D = 0, or decrement if D = 1. A 1 is added to or subtracted from SI for a byte-sized LODS, a 2 is added or subtracted for a word-sized LODS, and a 4 is added or subtracted for a double-word-sized LODS.

Table 4–10 lists the permissible forms of the LODS instruction. The LODSB (*loads a byte*) instruction causes a byte to be loaded into AL; the LODSW (*loads a word*) instruction causes a word to be loaded into AX; and the LODSD (*loads a double word*) instruction causes a double word to be loaded into EAX. Although rare, as an alternative to LODSB, LODSW, and LODSD the LODS instruction may be followed by a byte-, word-, or double-word-sized operand to select a byte, word, or double word transfer. Operands are often defined as bytes with DB, as words with DW, and as double words with DD. The DB pseudo-operation *defines byte(s),* the DW pseudo-operation *defines word(s),* and the DD pseudo-operation *defines double word(s).*

Figure 4–16 shows the effect of executing the LODSW instruction if the D flag = 0, SI = 1000H, and DS = 1000H. Here a 16-bit number, stored at memory locations 11000H and 11001H, moves into AX. Because D = 0, and this is a word transfer, the contents of SI increment by 2 *after* AX loads with memory data.

STOS

The STOS instruction stores AL, AX, or EAX at the extra segment memory location addressed by the DI register. (Note only the 80386/80486 can use EAX and double words.) Table 4–11 lists all forms of the STOS instruction. As with LODS, a STOS instruction may be appended with a B, W, or D for a byte, word, or double word transfers. The STOSB (*stores a byte*) instruction stores the byte in AL at the extra segment memory location addressed by DI. The STOSW (*stores a word*) instruction stores AX in the extra segment memory location addressed by DI. A double word is stored in the extra segment location addressed by DI with the STOSD (*stores a double word*) instruction. After the byte (AL), word (AX), or double word (EAX) is stored, the contents of DI increments or decrements.

TABLE 4–10 Forms of the LODS instruction

Symbolic	Function
LODSB	AL = [SI]; SI = SI ± 1
LODSW	AX = [SI]; SI = SI ± 2
LODSD	EAX = [SI]; SI = SI ± 4
LODS LIST	AL = [SI]; SI = SI ± 1 (if LIST is a byte)
LODS DATA1	AX = [SI]; SI = SI ± 2 (if DATA1 is a word)
LODS DATA4	EAX = [SI]; SI = SI ± 4 (if DATA4 is a double word)

Note: SI addresses data in the data segment by default for LODS, and only the 80386/80486 can use double words.

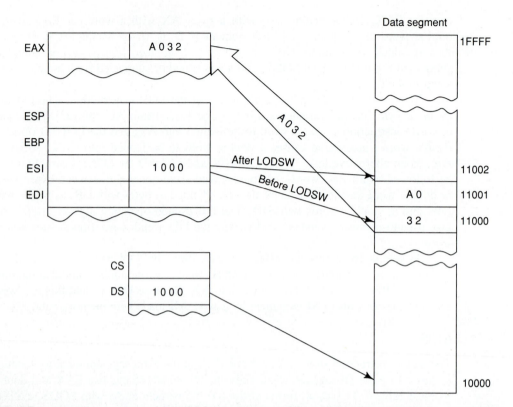

FIGURE 4–16 The operation of the LODSW instruction if DS = 1000H, SI = 1000H, D = 0, 11000H = 32, and 11001H = AO.

STOS with a REP. The **repeat prefix** (REP) may be added to any string data transfer instruction except the LODS instruction. The REP prefix causes CX to decrement by 1 each time the string instruction executes. After CX decrements, the string instruction repeats. If CX reaches a value of 0, the instruction terminates and the program continues with the next sequential instruction. Thus, if CX is loaded

TABLE 4–11 Forms of the STOS instruction

Symbolic	Function
STOSB	[DI] = AL; DI = DI ± 1
STOSW	[DI] = AX; DI = DI ± 2
STOSD	[DI] = EAX; DI = DI ± 4
STOS LIST	[DI] = AL; DI = DI ± 1 (if LIST is a byte)
STOS DATA1	[DI] = AX; DI = DI ± 2 (if DATA1 is a word)
STOS DATA4	[DI] = EAX; DI = DI ± 4 (if DATA4 is a double word)

Note: DI addresses data in the extra segment, and double words are only used by the 80386 or 80486 microprocessor.

with a 100, and a REP STOSB instruction executes, the microprocessor automatically repeats the STOSB instruction 100 times. Since the DI register is automatically incremented or decremented after each datum are stored, this instruction stores the contents of AL in a block of memory instead of a single byte of memory.

Suppose that 10 bytes of data, in an area of memory (BUFFER), must be cleared to 00H. This task is accomplished with a series of STOSB instructions (10 of them) or with one STOSB prefixed by an REP, if CX is a 10. The program listed in Example 4–3 clears memory area BUFFER.

EXAMPLE 4–3

```
                  ;using REP STOSB to clear a BUFFER
                  ;
0000 BF 0000 R            MOV   DI,OFFSET BUFFER   ;address BUFFER
0003 B9 000A              MOV   CX,10              ;load count
0006 FC                   CLD                      ;auto-increment
0007 B0 00                MOV   AL,0               ;clear AL
0009 F3/AA                REP STOSB                ;clear buffer
```

This short sequence of instructions addresses location BUFFER using the MOV DI,OFFSET BUFFER instruction. Though BUFFER is in the extra segment, the assembler still uses an offset address to address the memory. Notice how the REP prefix precedes the STOSB instruction in both assembly language and hexadecimal machine language. In machine language, the F3H is the REP prefix and AAH is the STOSB opcode.

A faster method of clearing this 10-byte buffer is to use the STOSW instruction with a count of five. Example 4–4 illustrates the same task as Example 4–3 except the count changes to a 5 and the STOSW instruction repeats instead of the STOSB instruction. Also, register AX is cleared instead of register AL.

EXAMPLE 4–4

```
                  ;using REP STOSW to clear a BUFFER
                  ;
0000 BF 0000 R            MOV   DI,OFFSET BUFFER   ;address BUFFER
0003 B9 0005              MOV   CX,5               ;load count
0006 FC                   CLD                      ;auto-increment
0007 B8 0000              MOV   AX,0               ;clear AX
000A F3/AB                REP STOSW                ;clear buffer
```

MOVS

The most useful string data transfer instruction is the MOVS instruction because it transfers data from one memory location to another. This is the only *memory-to-memory* transfer allowed in the 8086–80486 microprocessor. The MOVS instruction transfers a byte, word, or double word from the data segment location addressed by SI, to the extra segment location addressed by DI. As with the other string instructions, the pointers then increment or decrement as dictated by the direction flag. Table 4–12 lists all the permissible forms of the MOVS instruction. Note that only the source operand (SI), located in the data segment, may be overridden so another segment may be used. The destination operand (DI) must always be located in the extra segment.

Suppose that the contents of a 100-byte array must be transferred to another 100-byte array. The repeated MOVSB (*moves a byte*) instruction is ideal for this task, as illustrated in Example 4–5.

EXAMPLE 4–5

```
                        ;using REP MOVSB to transfer data from LIST1
                        ;into LIST2
                        ;
0000  BE 0000 R              MOV    SI,OFFSET LIST1    ;address LIST1
0003  BF 0064 R              MOV    DI,OFFSET LIST2    ;address LIST2
0006  B9 0064                MOV    CX,100            ;load count
0009FC                       CLD                      ;auto-increment
000A  F3/A4                  REP MOVSB                ;transfer data
```

INS

The INS (*input string*) instruction (not available on the 8086/8088 microprocessor) transfers a byte, word, or double word of data from an I/O device into the extra segment memory location addressed by the DI register. The *I/O address* is contained in the DX register. This instruction is useful for inputting a block of data from an external I/O device directly into the memory. One application transfers data from a disk drive to memory. Disk drives are often considered and interfaced as I/O devices in a computer system.

As with the prior string instructions, there are two basic forms of the INS. The INSB instruction inputs data from an 8-bit I/O device and stores it in the byte-sized memory location indexed by SI. The INSW instruction inputs 16-bit I/O data and stores it in a word-sized memory location. The INSD instruction inputs a double

TABLE 4–12 Forms of the MOVS instruction

Symbolic	Function
MOVSB	[DI] = [SI]; DI = DI ± 1; SI = SI ± 1 (byte transferred)
MOVSW	[DI] = [SI]; DI = DI ± 2; SI = SI ± 2 (word transferred)
MOVSD	[DI] = [SI]; DI = DI ± 4; SI = SI ± 4 (double word transferred)
MOVS BYTE1,BYTE2	[DI] = [SI]; DI = DI ± 1; SI = SI ± 1 (if BYTE1 and BYTE2 are bytes)
MOVS WORD1,WORD2	[DI] = [SI]; DI = DI ± 2; SI = SI ± 2 (if WORD1 and WORD2 are words)
MOVS DWORD1,DWORD2	[DI] = [SI]; DI = DI ± 4; SI = SI ± 4 (if DWORD1 and DWORD2 are double words)

word. These instructions can be repeated using the REP prefix. This allows an entire block of input data to be stored in the memory from an I/O device. Table 4–13 lists the various forms of the INS instruction.

Example 4–6 shows a short program that inputs 50 bytes from an I/O device whose address is 03ACH and stores the data in memory array LISTS. This software assumes that data are available from the I/O device at all times. Otherwise, the software must check to see if the I/O device is ready to transfer data, precluding the use of a REP prefix.

EXAMPLE 4–6

```
                    ;using REP INSB to input data to a memory array
                    ;
0000 BF 0000 R          MOV    DI,OFFSET LISTS      ;address array
0003 BA 03AC            MOV    DX,3ACH              ;address I/O
0006 FC                 CLD                         ;auto-increment
0007 B9 0032            MOV    CX,50                ;load count
000A F3/6C              REP INSB                    ;input data
```

OUTS

The OUTS (*output string*) instruction (not available on the 8086/8088 microprocessor) transfers a byte, word, or double word of data from the data segment memory location address by SI to an I/O device. The I/O device is addressed by the DX register as it was with the INS instruction. Table 4–14 shows the variations available for the OUTS instruction.

Example 4–7 shows a short program that transfers data from a memory array to an I/O device. This software assumes that the I/O device is always ready for data.

EXAMPLE 4–7

```
                    ;using REP OUTS to output data from a memory array
                    ;
0000 BE 0064 R          MOV    SI,OFFSET ARRAY      ;address array
0003 BA 03AC            MOV    DX,3ACH              ;address I/O
```

TABLE 4–13 Forms of the INS instruction

Symbolic	Function
INSB	[DI] = [DX]; DI = DI ± 1 (byte transferred)
INSW	[DI] = [DX]; DI = DI ± 2 (word transferred)
INSD	[DI] = [DX]; DI = DI ± 4 (double word transferred)
INS LIST	[DI] = [DX]; DI = DI ± 1 (if LIST is a byte)
INS DATA1	[DI] = [DX]; DI = DI ± 2 (if DATA1 is a word)
INS DATA4	[DI] = [DX]; DI = DI ± 4 (if DATA4 is a double word)

Note: [DX] indicates that DX contains the I/O device address. These instructions are not available on the 8086/8088 microprocessor, and only the 80386/80486 use double words.

TABLE 4–14 Forms of the OUTS instruction

Symbolic	Function
OUTSB	[DX] = [SI]; SI = SI ± 1 (byte transferred)
OUTSW	[DX] = [SI]; SI = SI ± 2 (word transferred)
OUTSD	[DX] = [SI]; SI = SI ± 4 (double word transferred)
OUTS LIST	[DX] = [SI]; SI = SI ± 1 (if LIST is a byte)
OUTS DATA1	[DX] = [SI]; SI = SI ± 2 (if DATA1 is a word)
OUTS DATA4	[DX] = [SI]; SI = SI ± 4 (if DATA4 is a double word)

Note: [DX] indicates that DX contains the I/O device address. These instructions are not available on the 8086/8088 microprocessor. Only the 80386/80486 use double words.

```
0006 FC              CLD                ;auto-increment
0007 B9 0064         MOV   CX,100       ;load count
000A F3/6E           REP OUTSB
```

4–6   MISCELLANEOUS DATA TRANSFER INSTRUCTIONS .

Don't be fooled by the word "miscellaneous"; these instructions are used in programs. The data transfer instructions detailed in this section are: XCHG, LAHF, SAHF, XLAT, IN, OUT, BSWAP, MOVSX, and MOVZX. Because the miscellaneous instructions will not be used as often as a MOV instruction, they have been grouped together and discussed in this section.

XCHG

The exchange instruction (XCHG) exchanges the contents of a register with the contents of any other register or memory location. The XCHG instruction cannot exchange segment registers or memory-to-memory data. Exchanges are byte, word, or double word sized (80386/80486 only) and use any addressing mode discussed in Chapter 3 except immediate addressing. Table 4–15 shows the forms available for the XCHG instruction.

The XCHG instruction using the 16-bit AX register with another 16-bit register is the most efficient exchange. This instruction occupies one byte of memory. Other XCHG instructions require 2 or more bytes of memory, depending on the addressing mode selected.

When using a memory addressing mode and the assembler, it doesn't matter which operand addresses memory. The XCHG AL,[DI] instruction is identical to the XCHG [DI],AL instruction as far as the assembler is concerned.

If the 80386/80486 microprocessor is available, the XCHG instruction can exchange double-word data. For example, the XCHG EAX,EBX instruction exchanges the contents of the EAX register with the EBX register.

TABLE 4–15 Forms of the XCHG instruction

Symbolic	Note
XCHG reg,reg	Exchanges byte-, word-, or double-word registers.
XCHG reg,mem	Exchanges byte-, word-, or double-word memory data with register data

Note: Only the 80386/80486 can use double word data.

LAHF and SAHF

The LAHF and SAHF instructions are seldom used because they were designed as *bridge* instructions. These instructions allowed 8085 (an early 8-bit microprocessor) software to be translated into 8086 software by a translation program. Because any software that required translation was probably completed many years ago, these instructions have little application today. The LAHF instruction transfers the rightmost 8 bits of the flag register into the AH register. The SAHF instruction transfers the AH register into the rightmost 8 bits of the flag register.

XLAT

The XLAT (*translate*) instruction converts the contents of the AL register into a number stored in a memory table. This instruction performs the direct table lookup technique often used to convert one code to another. An XLAT instruction first adds the contents of AL to BX to form a memory address within the data segment. It then copies the contents of this address into AL. This is the only instruction that adds an 8-bit number to a 16-bit number.

Suppose that a 7-segment LED display lookup table is stored in memory at address TABLE. The XLAT instruction then translates the BCD number in AL to a 7-segment code in AL. Example 4–8 provides a short program that converts a BCD code to 7-segment code. Figure 4–17 shows the operation of this example program if TABLE = 1000H, DS = 1000H, and the initial value of AL = 05H (a 5 BCD). After the translation, AL = 6DH.

EXAMPLE 4–8

```
                    ;using XLAT to convert from BCD to 7-segment code
                    ;
0000  BB 1000 R            MOV    BX,OFFSET TABLE    ;address TABLE
0003  D7                   XLAT                      ;convert
```

IN and OUT

Table 4–16 lists the forms of the IN and OUT instructions, which perform I/O operations. Notice that the contents of AL, AX, or EAX *only* are transferred between the I/O device and the microprocessor. An IN instruction transfers data from

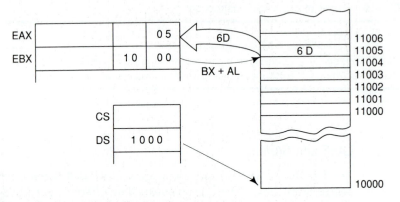

FIGURE 4–17 The operation of the XLAT instruction at the point just before the 6DH is loaded into AL.

an external I/O device to AL, AX, or EAX, and an OUT transfers data from AL, AX, or EAX to an external I/O device. (Note that only the 80386/80486 contain EAX.)

Two forms of I/O device (*port*) address exist for IN and OUT: fixed-port and variable-port. Fixed-port addressing allows data transfer between AL, AX, or EAX using an 8-bit I/O port address. It is called *fixed-port addressing* because the port number follows the instruction's opcode. Often instructions are stored in a ROM. A fixed-port instruction stored in a ROM has its port number permanently fixed because of the nature of read-only memory.

The port address appears on the address bus during an I/O operation. For the 8-bit fixed-port I/O instructions, the 8-bit port address is *zero extended* into a 16-bit

TABLE 4–16 IN and OUT instructions

Symbolic	Function
IN AL,p8	8-bit data are input to AL from port p8
IN AX,p8	16-bit data are input to AX from port p8
IN EAX,p8	32-bit data are input to EAX from port p8
IN AL,DX	8-bit data are input to AL from port DX
IN AX,DX	16-bit data are input to AX from port DX
IN EAX,DX	32-bit data are input to EAX from port DX
OUT p8,AL	8-bit data are sent to port p8 from AL
OUT p8,AX	16-bit data are sent to port p8 from AX
OUT p8,EAX	32-bit data are sent to port p8 from EAX
OUT DX,AL	8-bit data are sent to port DX from AL
OUT DX,AX	16-bit data are sent to port DX from AX
OUT DX,EAX	32-bit data are sent to port DX from EAX

Note: p8 = an 8-bit I/O port number, and DX = the 16-bit port address held in DX.

address. For example, if the IN AL,6AH instruction executes, data from I/O address 6AH is input to AL. The address appears as a 16-bit 006AH on pins A0–A15 of the address bus. Address bus bits A16–A19 (8086/8088), A16–A23 (80286/80386SX), A16–A24 (80386SL/80386SLC), or A16–A32 (80386/80486) are undefined for an IN or OUT instruction.

Variable-port addressing allows data transfers between AL, AX, or EAX and a 16-bit port address. It is called *variable-port addressing* because the I/O port number is stored in register DX, which can be changed (*varied*) during the execution of a program. The 16-bit I/O port address appears on the address bus pin connections A0–A15. The IBM PC uses a 16-bit port address to access its I/O space. The I/O space for a PC is located at I/O port 0000H–03FFH. Some plug-in adapter cards may use I/O addresses above 03FFH.

Figure 4–18 illustrates the execution of the OUT 19H,AX instruction, which transfers the contents of AX to I/O port 19H. Notice that the I/O port number appears as a 0019H on the 16-bit address bus and that the data from AX appear on the 16-bit data bus of the 8086, 80286, 80386SX, or 80386SL/80386SLC microprocessor. The system control signal, $\overline{IOWC}$ (I/O write control) is a logic zero to enable the I/O device.

MOVSX and MOVZX

The MOVSX (move and sign extend) and MOVZX (move and zero extend) instructions are found in the 80386 and 80486 instruction sets. These instructions move data and at the same time either sign extend or zero extend the data. Table 4–17 illustrates these instructions with several examples of each.

When a number is zero extended, the most significant part fills with zeros. For example, if an 8-bit 34H is zero extended into a 16-bit number, it becomes 0034H. Zero extension is often used to convert unsigned 8- or 16-bit numbers into unsigned 16- or 32-bit numbers using the MOVZX instruction.

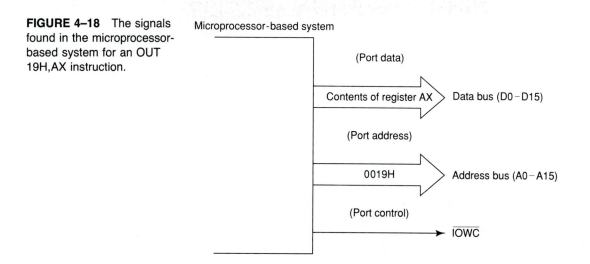

FIGURE 4–18 The signals found in the microprocessor-based system for an OUT 19H,AX instruction.

TABLE 4–17   The MOVSX and MOVZX instructions

Symbolic	Example	Note
MOVSX reg,reg	MOVSX CX,BL	converts the 8-bit contents of BL into a 16-bit number in CX by sign extension
	MOVSX ECX,AX	converts the 16-bit contents of AX into a 32-bit number in ECX by sign extension
MOVSX reg,mem	MOVSX BX,DATA	converts the 8-bit contents of DATA into a 16-bit number in BX by sign extension
	MOVSX EAX,[EDI]	converts the 16-bit contents of the data segment memory location addressed by EDI into a 32-bit number in EAX by sign extension
MOVZX reg,reg	MOVZX DX,AL	converts the 8-bit contents of AL into a 16-bit number in DX by zero extension
	MOVZX EBP,DI	converts the 16-bit contents of DI into a 32-bit number in EBP by zero extension
MOVZX reg,mem	MOVZX DX,DATA1	converts the 8-bit contents of DATA1 into a 16-bit number in DX by zero extension
MOVZX reg,mem	MOVZX EAX,DATA2	converts the 16-bit contents of DATA2 into a 32-bit number in EAX by zero extension

A number is sign extended when its sign-bit is copied into the most significant part. For example, if an 8-bit 84H is sign extended into a 16-bit number, it becomes FF84H. The sign-bit of an 84H is a one, which is copied into the most significant part of the sign extended result. Sign extension is most often used to convert 8- or 16-bit signed numbers into 16- or 32-bit signed numbers using the MOVSX instruction.

BSWAP

The byte swap instruction (BSWAP) is available only in the 80486 microprocessor. This instruction takes the contents of any 32-bit register and swaps the first byte with the fourth and the second with the third. For example, the BSWAP EAX instruction with EAX = 00112233H swaps bytes in EAX, so it results in EAX = 33221100H. Notice that the order of all 4 bytes is reversed by this instruction.

4–7   SEGMENT OVERRIDE PREFIX

The *segment override prefix,* which may be added to almost any 8086–80486 instruction in any memory addressing mode, allows the programmer to deviate from the default segment. The segment override prefix is an additional byte that appends to the front of an instruction to select an alternate segment register. About the only

TABLE 4–18 Instructions that include segment override prefixes

Symbolic	Segment Accessed	Normal Segment
MOV AX,DS:[BP]	Data segment	Stack segment
MOV AX,ES:[BP]	Extra segment	Stack segment
MOV AX,SS:[DI]	Stack segment	Data segment
MOV AX,CS:[SI]	Code segment	Data segment
MOV AX,ES:LIST	Extra segment	Data segment
LODS ES:DATA	Data segment	Extra segment
MOV EAX,FS:DATA2	FS segment	Data segment
MOV BL,GS:[ECX]	GS segment	Data segment

Note: Only the 80386 and 80486 allow the use of the FS and GS segments.

instructions that cannot be prefixed are the jump and call instructions that must use the code segment register for address generation. The segment override is also used to select the FS and GS segments in the 80386–80486 microprocessor.

To take an example, the MOV AX,[DI] instruction accesses data within the data segment by default. If required by a program, this can be changed by prefixing the instruction. Suppose that the data are in the extra segment instead of the data segment. This instruction addresses the extra segment if changed to MOV AX,ES:[DI].

Table 4–18 shows some altered instructions that address a different memory segment than normal. Each time we prefix an instruction with a segment override prefix, the instruction becomes one byte longer. Although this is no serious change to the length of the instruction, it does add to its execution time. Usually, we limit the use of the segment override prefix and remain in the default segments to write shorter and more efficient software.

4–8   ASSEMBLER DETAILS

The assembler* for the 8086–80486 microprocessor can be used in two ways. This section of the text presents both methods and explains how to organize a program's memory space using the assembler. It also explains the purpose and use of some of the more important directives used with this assembler. Appendix A provides additional detail about the assembler.

Directives

Before the format of an assembly language program is discussed, some details about the directives (pseudo-operations) that control the assembler must be presented. Some common assembly language directives appear in Table 4–19. **Directives** indicate how an operand or section of a program is to be processed by the assembler. Some di-

---

*The assembler used in this text is the Microsoft macro assembler, called MASM.

TABLE 4–19   Common assembler directives

Directive	Function
.286	Selects the 80286 instruction set
.286P	Selects the protected mode 80286 instruction set
.386	Selects the 80386 instruction set
.386P	Selects the 80386 protected mode instruction set
.486	Selects the 80486 instruction set
.486P	Selects the 80486 protected mode instruction set
.287	Selects the 80287 numeric coprocessor
.387	Selects the 80387 numeric coprocessor
ALIGN 2	Starts the data in a segment at word or double-word boundaries
ASSUME	Indicates the names of each segment to the assembler; it does not load the segment registers
AT	Indicates what physical segment address is used with the SEGMENT statement
BYTE	Indicates a byte-sized operand as in BYTE PTR or THIS BYTE
DB	Define byte(s) (8-bits)
DD	Define double word(s) (32-bits)
DQ	Define quad word(s) (64-bits)
DT	Define ten bytes(s) (80-bits)
DUP	Generates duplicates of characters or numbers
DW	Define word(s) (16-bits)
DWORD	Indicates a double word-sized operand as in THIS DWORD
END	Indicates the end of the program

rectives generate and store information in the memory, while others do not. The DB (define byte) directive stores bytes of data in the memory, while the BYTE PTR directive never stores data. The BYTE PTR directive indicates the size of the data referenced by a pointer or index register.

Note that the assembler by default accepts only 8086/8088 instructions unless the software is preceded by the .286 or .286P directive or one of the other microprocessor selection switches. The .286 directive tells the assembler to use the 80286 instruction set in the real mode, while the .286P directive tells the assembler to use the 80286 protected mode instruction set.

Storing Data in a Memory Segment.   The DB (define byte), DW (define word), and DD (define double word) directives are most often used with the 8086–80486 to define and store memory data. If a numeric coprocessor is present in the system, the DQ (define quad word) and DT (define ten bytes) directives are also common. These directives use the label to identify a memory location and the directive to indicate the size of the location.

Directive	Function
ENDM	Indicates the end of a macro sequence
ENDP	Indicates the end of a procedure
ENDS	Indicates the end of a segment
EQU	Equates data to a label
FAR	Specifies a far address as in JMP FAR PTR LISTS
MACRO	Defines the name, parameters, and start of a macro
NEAR	Specifies a near address as in JMP NEAR PTR HELP
OFFSET	Specifies an offset address
ORG	Sets the origin within a segment
PROC	Defines the beginning of a procedure
PTR	Indicates a memory pointer
SEGMENT	Defines the start of a memory segment
STACK	Indicates that a segment is a stack segment
STRUC	Defines the start of a data structure
THIS	Used with EQU to set a label to a byte, word, or double word
USES	An MASM version 6.0 directive that automatically saves registers used by a procedure
USE16	Directs the assembler to use the 16-bit instruction mode and data sizes for the 80386 and 80486 microprocessors
USE32	Directs the assembler to use the 32-bit instruction mode and data sizes for the 80386 and 80486 microprocessors
WORD	Acts as a word operand as in WORD PTR or THIS WORD

Note: Most of these directives function with most versions of the assembler.

Example 4–9 shows a memory segment that contains various forms of data definition directives. The first statement indicates the start of the segment and its symbolic name. The last statement of the segment contains the ENDS directive that indicates the end of the segment. The name of the segment (LIST_SEG), can be anything that the programmer desires to call it.

EXAMPLE 4–9

```
                        ;using DB, DW, and DD
                        ;
0000                    LIST_SEG      SEGMENT

0000 01 02 03           DATA_ONE      DB       1,2,3       ;define bytes
0003 45                               DB       45H         ;hexadecimal
0004 41                               DB       'A'         ;ASCII
0005 F0                               DB       11110000B   ;binary
```

```
0006 000C 000D    DATA_TWO    DW    12,13         ;define words
000A 0200                     DW    LIST1         ;symbolic
000C 2345                     DW    2345H         ;hexadecimal
000E 00000300    DATA_THREE   DD    300H          ;hexadecimal
0012 4007DF3B                 DD    2.123         ;real
0016 544269E1                 DD    3.34E+12      ;real
001A 00          LISTA        DB    ?             ;reserve 1 byte
001B 000A[       LISTB        DB    10 DUP (?)    ;reserve 10 bytes
          ??
              ]
0025 00                       ALIGN  2            ;set word boundary

0026 0100[       LISTC        DW    100H DUP (0)
        0000
              ]

0226 0016[       LIST_NINE    DD    22 DUP (?)
      ????????
                ]

027E 0064[       SIXES        DB    100 DUP (6)
          06
              ]

02E2            LIST_SEG     ENDS
                            END
```

This example shows various forms of data storage for bytes at DATA_ONE. More than one byte can be defined on a line in binary, hexadecimal, decimal, or ASCII code. The DATA_TWO label shows how to store various forms of word data. Double-words are stored at DATA_THREE and they include floating-point, single-precision real numbers.

Memory can be reserved for use in the future by using a ? as an operand for a DB, DW, or DD directive. When a ? is used in place of a numeric or ASCII value, the assembler sets aside a location and does not initialize it to any value. The DUP (duplicate) directive creates an array as shown in several ways in Example 4–9. A 10 DUP (?) reserves 10 locations of memory, but stores no specific value in any of the 10 locations. If a number appears within the () part of the DUP statement, the assembler initializes the reserved section of memory with data.

The ALIGN directive, used in this example, makes sure that the memory arrays are stored on word boundaries. An ALIGN 2 places data on word boundaries for an 80286 microprocessor using word data, and an ALIGN 4 places them on double-word boundaries for an 80386 or 80486 using double-word data. It is important that word-sized data is placed at word boundaries and double-word-sized data at double-word boundaries. If not, the 80286, 80386, or 80486 spends more time than necessary accessing these data types. A word stored at an odd-numbered memory location takes twice as long to access as a word stored on an even-numbered memory location.

EQU and THIS.   The equate directive (EQU) equates a numeric or ASCII value or a label to another label. Equates make a program clearer and simplifies debugging.

Example 4–10 shows several equate statements and a few instructions that show how they function in a program.

EXAMPLE 4–10

```
                        ;using equate
                        ;
= 000A                  TEN      EQU   10
= 0009                  NINE     EQU   9

0000  B0 0A                      MOV   AL,TEN
0002  04 09                      ADD   AL,NINE
```

The THIS directive always appears as THIS BYTE, THIS WORD, or THIS DWORD. In certain cases, data must be referred to as both a byte and a word. The assembler can only assign either a byte or a word address to a label. To assign a byte label to a word, we use the software listed in Example 4–11.

EXAMPLE 4–11

```
                        ;using THIS and ORG
                        ;
0000                    DATA_SEG      SEGMENT

0100                                  ORG   100H

= 0100                  DATA1    EQU   THIS BYTE
0100  0000              DATA2    DW    ?

0102                    DATA_SEG      ENDS

0000                    CODE_SEG      SEGMENT 'CODE'

                        ASSUME  CS:CODE_SEG,DS:DATA_SEG

0000  8A 1E 0100 R              MOV   BL,DATA1
0004  A1 0100 R                 MOV   AX,DATA2
0007  8A 3E 0101 R              MOV   BH,DATA1+1

000B                    CODE_SEG      ENDS
```

This example also illustrates how the ORG (origin) statement changes the starting address of the data in the data segment to location 100H. The ASSUME statement tells the assembler what names have been chosen for the code, data, extra, and stack segments. Without the assume statement, the assembler assumes nothing and uses a prefix on all instructions that address memory data.

PROC and ENDP.   The PROC and ENDP directives indicate the start and end of a procedure (*subroutine*). These directives force structure upon the program because the procedure is clearly defined. Both directives require a label to indicate the name of the subroutine. The PROC directive, which indicates the start of a procedure, must also be followed with either NEAR or FAR. A NEAR procedure is one that resides in the same code segment as the program. A FAR procedure may reside at any location in the memory system. We often call NEAR procedures *local,* and FAR procedures *global.* The term *global* denotes a procedure that can be used by any program, while *local* defines a procedure that is used only by the current program.

Example 4–12 shows a procedure that adds BX, CX, and DX and stores the sum in register AX. Although this procedure is short, and may not be that useful, it does illustrate how to use the PROC and ENDP directives to delineate the procedure.

EXAMPLE 4–12

```
                        ;procedure that adds BX, CX, and DX with the sum
                        ;stored in AX
                        ;
0000                    ADDEM   PROC  FAR

0000  03 D9                     ADD   BX,CX
0002  03 DA                     ADD   BX,DX
0004  8B C3                     MOV   AX,BX
0006  CB                        RET

0007                    ADDEM   ENDP
```

If version 6.0 of the Microsoft MASM assembler program is available, the PROC directive can specify and automatically save any registers used within the procedure. The USES statement indicates what registers are used by the procedure so the assembler can automatically save them before your procedure begins and restore them before the procedure ends with the RET instruction. For example, the ADDS PROC USES AX BX CX statement automatically pushes AX, BX, and CX on the stack before the procedure begins and pops them from the stack before the RET instruction executes at the end of the procedure. Example 4–13 illustrates a procedure written using MASM 6.0 that shows the USES statement. Note that the registers in the list are not separated by commas and the PUSH and POP instructions are not displayed in the procedure listing. The USES statement does not appear elsewhere in this text so compatibility with MASM version 5.10 can be maintained.

EXAMPLE 4–13

```
                        ;procedure that includes the USES directive to save
                        ;BX, CX, and DX on the stack.
                        ;
0000                    ADDS   PROC  NEAR USES BX CX DX

0003  03 D8                    ADD   BX,AX
0005  03 CB                    ADD   CX,BX
0007  03 D1                    ADD   DX,CX
0009  8B C2                    MOV   AX,DX
                               RET

000F                    ADDS   ENDP
```

Memory Organization

The assembler uses two basic formats for developing software. One method uses models and the other uses full-segment definitions. Memory models as presented in this section are unique to the MASM assembler program. The TASM assembler also uses memory models, but they differ from the MASM models. The full-segment definitions are common to most assemblers, including the *Intel assembler,* and are most often used for software development. The models are easier to use, but the full-segment definitions offer better control over the assembly language task and are recommended and used in other places in this text. We use full segments because they apply to all assemblers, whereas models do not.

Models. There are many models that can be used with the MASM assembler, from tiny to huge. Appendix A contains a table that lists all the models available for use with the assembler. To designate a model, we use the .MODEL statement followed by the size of the memory system. The tiny model requires that all software and data fit into one 64K byte memory segment and is useful for small programs. The small model requires that only one data segment be used with one code segment for a total of 128K bytes of memory. Other models are available up to the huge model.

Example 4–14 illustrates how the .MODEL statement defines the parameters of a short program that copies the contents of a 100-byte block of memory (LISTA) into a second 100-byte block of memory (LISTB).

EXAMPLE 4–14

```
                              .MODEL SMALL

                              .STACK 100H             ;define stack

                              .DATA                   ;define data

0000  0064[              LISTA   DB      100 DUP (?)
        ??
      ]

0064  0064[              LISTB   DB      100 DUP (?)
        ??
      ]

                              .CODE                   ;define code

0000  B8 —— R          HERE:   MOV     AX,@DATA        ;load ES, DS
0003  8E C0                    MOV     ES,AX
0005  8E D8                    MOV     DS,AX

0007  FC                       CLD                     ;move data
0008  BE 0000 R                MOV     SI,OFFSET LISTA
000B  BF 0064 R                MOV     DI,OFFSET LISTB
000E  B9 0064                  MOV     CX,100
0011  F3/A4                    REP MOVSB

0013                           .EXIT 0                 ;exit to DOS

                              END     HERE
```

Full-Segment Definitions. Example 4–15 illustrates the same program using full-segment definitions. This program appears longer but more structured than the model method of setting up a program. The first segment defined is the STACK_SEG that is clearly delineated with the SEGMENT and ENDS directives. Within these directives a DW 100 DUP (?) sets aside 100H words for the stack segment. Because the word STACK appears next to SEGMENT, the assembler and linker automatically load both the stack segment register (SS) and stack pointer (SP).

EXAMPLE 4–15

```
0000                    STACK_SEG      SEGMENT STACK

0000 0100[                             DW       100H DUP (?)
            ????
        ]

0200                    STACK_SEG      ENDS

0000                    DATA_SEG       SEGMENT 'DATA'

0000 0064[                     LISTA   DB       100 DUP (?)
          ??
        ]

0064 0064[                     LISTB   DB       100 DUP (?)
          ??
        ]

00C8                    DATA_SEG       ENDS

0000                    CODE_SEG       SEGMENT 'CODE'

                                ASSUME CS:CODE_SEG,DS:DATA_SEG,SS:STACK_SEG

0000                            MAIN   PROC     FAR

0000 B8 — R                            MOV      AX,DATA_SEG    ;load DS and ES
0003 8E C0                             MOV      ES,AX
0005 8E D8                             MOV      DS,AX

0007 FC                                CLD                     ;move data
0008 BE 0000 R                         MOV      SI,OFFSET LISTA
000B BF 0064 R                         MOV      DI,OFFSET LISTB
000E B9 0064                           MOV      CX,100
0011 F3/A4                             REP MOVSB

0013 B4 4C                             MOV      AH,4CH          ;exit to DOS
0015 CD 21                             INT      21H

0017                            MAIN   ENDP

0017                    CODE_SEG       ENDS

                                END    MAIN
```

Next the data are defined in the DATA_SEG. Here two arrays of data appear as LISTA and LISTB. Each array contains 100 bytes of space for the program. The names of the segments in this program can be changed to any name. We include the group name 'DATA' so the Microsoft program CODEVIEW can be effectively used to debug this software. If the group name is not placed in a program, CODEVIEW can still be used to debug a program, but the program will not be debugged in symbolic form. Other group names, such as 'STACK', 'CODE', and so forth are listed in Appendix A.

The CODE_SEG is organized as a far procedure because most software is procedure oriented. Before the program begins, the code segment contains the ASSUME statement. The ASSUME statement tells the assembler and linker the name used for the code segment (CS) is CODE_SEG; it also tells the assembler and linker that the data segment is DATA_SEG and the stack segment is STACK_SEG. Also notice we include the group name 'CODE' for the code segment. Other group names appear in Appendix A with the models.

After the program loads both the extra segment register and data segment register with the location of the data segment, it transfers 100 bytes from LISTA to LISTB. Following this is a sequence of two instructions that return control back to DOS (the disk operating system).

The last statement in the program is END MAIN. The END statement indicates the end of the program and the location of the first instruction executed. Here we want the machine to execute the main procedure, so a label follows the END directive. In a file that is linked to another file, there is no label on the END directive.

In the 80386/80486 microprocessor an additional directive is found attached to the code segment. The USE16 or USE32 directive tells the assembler to use 16- or 32-bit instruction modes for the microprocessor. Software developed for the DOS environment must use the USE16 directive for the 80386/80486 program to function correctly. In fact, any program designed to execute in the real mode must include the USE16 directive. Example 4–16 shows how the same software listed in Example 4–15 is formed for the 80386/80486 microprocessor.

EXAMPLE 4–16

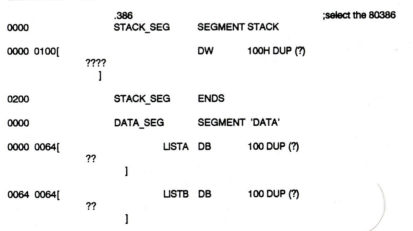

```
                        .386                                    ;select the 80386
0000                    STACK_SEG       SEGMENT STACK

0000 0100[                              DW      100H DUP (?)
        ????
     ]

0200                    STACK_SEG       ENDS

0000                    DATA_SEG        SEGMENT 'DATA'

0000 0064[                      LISTA   DB      100 DUP (?)
        ??
     ]

0064 0064[                      LISTB   DB      100 DUP (?)
        ??
     ]
```

```
00C8                    DATA_SEG    ENDS

0000                    CODE_SEG       SEGMENT USE16 'CODE'

                                ASSUME CS:CODE_SEG,DS:DATA_SEG,SS:STACK_SEG

0000                    MAIN   PROC    FAR

0000 B8 — R                    MOV      AX,DATA_SEG   ;load DS and ES
0003 8E C0                     MOV      ES,AX
0005 8E D8                     MOV      DS,AX

0007 FC                        CLD                      ;move data
0008 BE 0000 R                 MOV      SI,OFFSET LISTA
000B BF 0064 R                 MOV      DI,OFFSET LISTB
000E B9 0064                   MOV      CX,100
0011 F3/A4                     REP MOVSB

0013 B4 4C                     MOV      AH,4CH         ;exit to DOS
0015 CD 21                     INT      21H

0017                    MAIN   ENDP

0017                    CODE_SEG    ENDS

                               END      MAIN
```

A Sample Program

Example 4–17 provides a sample program that reads a character from the keyboard and displays it on the CRT screen. Although this program is trivial, it does illustrate a complete workable program that functions on any personal computer using DOS from the earliest 8088-based system to the latest 80486-based system. This program also illustrates the use of a few DOS function calls. Appendix A lists the DOS function calls with their parameters. The BIOS function calls allow the use of the keyboard, printer, disk drives, and everything else that is available in your computer system.

EXAMPLE 4–17

```
                        ;program that reads a key and displays key
                        ;an @ key ends the program
                        ;
0000                    CODE_SEG    SEGMENT 'CODE'

                               ASSUME  CS:CODE_SEG

0000                    MAIN   PROC    FAR

0000 B4 06                     MOV      AH,6          ;read key
0002 B2 FF                     MOV      DL,0FFH
0004 CD 21                     INT      21H
0006 74 F8                     JE       MAIN          ;if no key
```

```
0008 3C 40                      CMP     AL,'@'      ;test for @
000A 74 08                      JE      MAIN1       ;if @

000C B4 06                      MOV     AH,6        ;display key
000E 8A D0                      MOV     DL,AL
0010 CD 21                      INT     21H
0012 EB EC                      JMP     MAIN        ;repeat
0014              MAIN1:
0014 B4 4C                      MOV     AH,4CH          ;exit to DOS
0016 CD 21                      INT     21H

0018                    MAIN     ENDP

0018            CODE_SEG         ENDS

                                END     MAIN
```

This example program uses only a code segment because there is no data. A stack segment should appear, but has been left out because DOS automatically allocates a 128-byte stack for all programs. The only time that the stack is used in this example is for the INT 21H instructions that call a procedure in DOS. Note that when this program is linked, the linker signals that no stack segment is present. This warning may be ignored in this example because the stack is less than 128 bytes.

The program uses DOS functions 06H and 4CH. The function number is placed in AH before the INT 21H instruction executes. The 06H function reads the keyboard if DL = 0FFH or displays the ASCII contents of DL if it is not 0FFH. Upon close examination, the first section of the program moves a 06H into AH and a 0FFH into DL so a key is read from the keyboard. The INT 21H tests the keyboard and if no key is typed, it returns equal. The JE instruction tests the equal condition and jumps to MAIN if no key is typed.

When a key is typed, the program continues to the next step. This step compares the contents of AL with an @ symbol. Upon return from the INT 21H, the ASCII character of the typed key is found in AL. In this program if we type an @ symbol, the program ends. If we do not type an @ symbol the program continues by displaying the character typed on the keyboard with the next INT 21H instruction.

The second INT 21H instruction moves the ASCII character into DL so it can be displayed on the CRT screen. After displaying the character, a JMP executes. This causes the program to continue at MAIN where it repeats reading a key.

If the @ symbol is typed, the program continues at MAIN1, where it executes the DOS function code number 4CH. This causes the program to return to the DOS prompt (A>) so the computer can be used for other tasks.

More information about the assembler and its application appears in Appendix A and in the next several chapters. That appendix provides a complete overview of the assembler, linker, and DOS functions. It also provides a list of the BIOS (*basic I/O system*) functions. The information provided in the following chapters clarifies how to use the assembler for certain tasks at different levels of difficulty.

4–9   **SUMMARY**

1. Data movement instructions transfer data between registers, a register and memory, a register and the stack, memory and the stack, the accumulator and I/O, and the flags and the stack.
2. Data movement instructions include MOV, PUSH, POP, XCHG, XLAT, IN, OUT, LEA, LSD, LES, LAHF, SAHF, and the string instructions, LODS, STOS, MOVS, INS, and OUTS.
3. The first byte of an instruction contains the opcode, which specifies the operation performed by the microprocessor. The opcode may be preceded by an override prefix in some forms of instructions.
4. The D-bit, located in many instructions, selects the direction of data flow. If D = 0, the data flow from the REG field to the R/M field of the instruction. If D = 1, the data flow from the R/M field to the REG field.
5. The W-bit, found in most instructions, selects the size of the data transfer. If W = 0, the data are byte sized, and if W = 1, the data are word sized. In the 80386/80486 microprocessor W = 1 can also specify a 32-bit register.
6. MOD selects the addressing mode of operation for a machine language instruction's R/M field. If MOD = 00, there is no displacement; if it equals a 01, an 8-bit sign-extended displacement appears; if a 10, a 16-bit displacement occurs; and if a 11, a register is used instead of a memory location. In the 80386/80486 the MOD bits also specify a 32-bit displacement.
7. A 3-bit binary register code specifies the REG and R/M fields when the MOD = 11. The 8-bit registers are AH, AL, BH, BL, CH, CL, DH, and DL, and the 16-bit registers are AX, BX, CX, DX, SP, BP, DI, and SI. The 32-bit registers are EAX, EBX, ECX, EDX, ESP, EBP, EDI, and ESI.
8. When the R/M field depicts a memory mode, a 3-bit code selects one of the following modes: [BX+DI], [BX+SI], [BP+DI], [BP+SI], [BX], [BP], [DI], or [SI] for 16-bit instruction in the 8086–80486. In the 80386/80486 the R/M field specifies EAX, EBX, ECX, EDX, EBP, EDI, and ESI or one of the scaled-index modes of addressing memory data. If the scaled-index mode is selected (R/M = 100), an additional byte (scaled-index byte) is added to the instruction to specify the base register, index register, and the scaling factor.
9. All memory-addressing modes, by default, address data in the data segment unless BP addresses memory. The BP register addresses data in the stack segment.
10. The segment registers may be addressed only by the MOV, PUSH, or POP instruction. The MOV instruction may transfer a segment register to a 16-bit register or vice versa. We do not allow the MOV CS,reg or the POP CS instruction.
11. Data are transferred between a register or a memory location and the stack by the PUSH and POP instructions. Variations of these instructions allow immediate data to be pushed onto the stack, the flags to be transferred to and from the stack, and all the 16-bit registers to be transferred between the stack and the registers. When data are transferred to the stack, 2 bytes (8086–80286) always

move with the least significant byte placed at the SP location -1 byte and the most significant byte placed at the SP location -2 bytes. After placing the data on the stack, SP decrements by 2. In the 80386/80486, 4 bytes of data from a memory location or register may also be transferred to the stack.

12. Opcodes that transfer data between the stack and the flags are PUSHF and POPF. Opcodes that transfer all the 16-bit registers between the stack and the registers are PUSHA and POPA. In the 80386/80486 a PUSHFD and POPFD transfer the contents of the EFLAGS.

13. LEA, LDS, and LES instructions load a register or registers with an effective address. The LEA instruction loads any 16-bit register with an effective address, while LDS and LES load any 16-bit register and either DS or ES with the effective address. In the 80386/80486 additional instructions include LFS, LGS, and LSS, which load a 16-bit register and FS, GS, or SS.

14. String data transfer instructions use either DI or SI or both to address memory. The DI offset address is located in the extra segment and the SI offset address is located in the data segment.

15. The direction flag (D) chooses the auto-increment or auto-decrement mode of operation for DI and SI for string instructions. If we clear D with the CLD instruction, we select the auto-increment mode, and if we set D with STD, we select the auto-decrement mode. Either DI or SI or both increment/decrement by 1 for a byte operation and by 2 for a word operation.

16. LODS loads AL, AX, or EAX with data from the memory location addressed by SI; STOS stores AL, AX, or EAX in the memory location addressed by DI; and MOVS transfers a byte or a word from the memory location addressed by SI into the location addressed by DI.

17. INS inputs data from an I/O device addressed by DX and stores it in the memory location addressed by DI, and OUTS outputs the contents of the memory location addressed by SI and sends it to the I/O device addressed by DX.

18. The REP prefix may be attached to any string instruction to repeat it. The REP prefix repeats the string instruction the number of times found in register CX.

19. Translate (XLAT) converts the data in AL into a number stored at the memory location addressed by BX plus AL.

20. IN and OUT transfer data between AL, AX, or EAX and an external I/O device. The address of the I/O device is either stored with the instruction (fixed port) or in register DX (variable port).

21. The segment override prefix selects a different segment register for a memory location than the default segment. For example, the MOV AX,[BX] instruction uses the data segment, but the MOV AX,ES:[BX] instruction uses the extra segment because of the ES: prefix. The segment override prefix is the only way that the FS and GS segments are addressed in the 80386/80486 microprocessor.

22. The MOVZX (move and zero extend) and MOVSX (move and sign extend) instructions found in the 80386/80486 microprocessor increase the size of a byte to a word or the size of a word to a double word. The zero extend version increases the size of the number by inserting leading zeros. The sign extend

version increases the size of the number by copying the sign-bit into the more significant bits of the number.

23. Assembler directives DB (define byte), DW (define word), DD (define double word), and DUP (duplicate) store data in the memory system.
24. The EQU (equate) directive allows data or labels to be equated to labels.
25. The SEGMENT directive identifies the start of a memory segment and ENDS identifies the end of a segment.
26. The ASSUME directive tells the assembler what segment names you have assigned to CS, DS, ES, and SS. In the 80386/80486 it also indicates the segment name for FS and GS.
27. The PROC and ENDP directives indicate the start and end of a procedure.
28. Memory models can be used to shorten the program slightly, but they are more difficult to use for larger programs and programming in general. Memory models are not compatible with all assembler programs.

4–10   GLOSSARY

Assembler   A program that converts symbolic machine language instructions into hexadecimal machine language.

Directive   A special command for the assembler that directs the assembler operation.

Displacement   A distance represented as an 8-bit sign-extended value, a 16-bit offset, or a 32-bit offset used with the memory-addressing modes.

Fixed port   An 8-bit I/O address is called a fixed port because the port number is stored with the instruction in the memory as immediate data.

Machine language   The native binary code that the microprocessor understands and uses as the instructions that control its operation.

Microprocessor environment   The internal register set as saved on the stack by the PUSHA or PUSHAD instruction.

OFFSET directive   Specifies the offset address associated with an instruction.

Operand-size override prefix   A one-byte prefix added to the front of an 80386/80486 instruction that changes the size of the address to 16 bits when the microprocessor is operated in the 32-bit instruction mode or to 32 bits when it is operated in the 16-bit instruction mode.

Port   A term used to specify an I/O address. In the 8086–80486 microprocessor the port address can be either 8 or 16 bits in width. Ports are numbered from 0000H to FFFFH, with 8-bit ports numbered from 0000H to 00FFH.

Register-size override prefix   A one-byte prefix added to the front of an 80386/80486 instruction that changes the size of the register set to 16 bits when the microprocessor is operated in the 32-bit instruction mode or to 32 bits when it is operated in the 16-bit instruction mode.

Repeat prefix (REP)   A one-byte prefix added to the front of MOVS, STOS, INS, or OUTS to repeat the instruction by the number of times stored in the CX register.

Scaled-index byte   An additional byte added to the 80386/80486 instruction for the scaled-index addressing mode.

Segment override prefix   A byte added to the front of an instruction that changes the segment used in the address calculation from the default segment to any segment.

Sign extend   The act of increasing the size of a number by copying the sign-bit into more significant bits of the number.

String data transfer instruction   A LODS, MOVS, STOS, INS, or OUTS instruction that automatically adjusts the memory pointer or pointers after each instruction is executed.

Variable port   A 16-bit port address stored in register DX. It is called variable because the contents of DX and hence the port number can be varied.

Zero extend   The act of increasing the size of a number by adding leading zeros.

4–11   QUESTIONS

1. The first byte of an instruction is the _____ unless it contains an override prefix.
2. Describe the purpose of the D- and W-bits found in some machine language instructions.
3. The MOD field, in a machine language instruction, specifies what information?
4. If the register field (REG) of an instruction contains a 010 and W = 0, what register is selected, assuming that the instruction is a 16-bit mode instruction?
5. How are the 32-bit registers selected for the 80386/80486 microprocessor?
6. What memory-addressing mode is specified by R/M = 001 with MOD = 00 for a 16-bit instruction?
7. Identify the default segment register assigned to:
   a. SP
   b. BX
   c. DI
   d. BP
   e. SI
8. Convert an 8B07H from machine language to assembly language.
9. Convert an 8B1E004CH from machine language to assembly language.
10. If a MOV SI,[BX+2] instruction appears in a program, what is its machine language equivalent?
11. If a MOV ESI,[EAX] instruction appears in a program for the 80386/80486 microprocessor operated in the 16-bit instruction mode, what is its machine language equivalent?
12. What is wrong with a MOV CS,AX instruction?

13. PUSH and POP always transfer a _____-bit number between the stack and a register or memory location in the 8086–80286 microprocessor.
14. What segment register may not be popped from the stack?
15. What registers move onto the stack with the PUSHA instruction?
16. What registers move onto the stack for a PUSHAD instruction?
17. Describe the operation of each of the following instructions:
    a. PUSH AX
    b. POP ESI
    c. PUSH [BX]
    d. PUSHFD
    e. POP DS
    f. PUSH 4
18. Explain what happens when the PUSH BX instruction executes. Make sure to show where BH and BL are stored. (Assume that SP = 0100H and SS = 0200H.)
19. Repeat Question 18 for the PUSH EAX instruction.
20. The 16-bit POP instruction (except for POPA) increments SP by _____.
21. What values appear in SP and SS if the stack pointer addresses memory location 02200H?
22. Compare the operation of a MOV DI,NUMB instruction with an LEA DI,NUMB instruction.
23. What is the difference between a LEA SI,NUMB instruction and a MOV SI,OFFSET NUMB instruction?
24. Which is more efficient, a MOV with an OFFSET or an LEA instruction?
25. Describe how the LDS BX,NUMB instruction operates.
26. What is the difference between the LDS and LSS instructions?
27. Develop a sequence of instructions that move the contents of data segment memory locations NUMB and NUMB+1 into BX, DX, and SI.
28. What is the purpose of the direction flag?
29. Which instructions set and clear the direction flag?
30. The string instructions use DI and SI to address memory data in which memory segments?
31. Explain the operation of the LODSB instruction.
32. Explain the operation of the STOSW instruction.
33. Explain the operation of the OUTSB instruction.
34. What does the REP prefix accomplish and what type of instruction is it used with?
35. Develop a sequence of instructions that copies 12 bytes of data from an area of memory addressed by SOURCE into an area of memory addressed by DEST.
36. Where is the I/O address (port number) stored for an INSB instruction?
37. Select an assembly language instruction that exchanges the contents of the EBX register with the ESI register.
38. Would the LAHF and SAHF instructions normally appear in software?
39. Explain how the XLAT instruction transforms the contents of the AL register.

40. Write a short program that uses the XLAT instruction to convert the BCD numbers 0–9 into ASCII-coded numbers 30H–39H. Store the ASCII-coded data into a TABLE.
41. Explain what the IN AL,12H instruction accomplishes.
42. Explain how the OUT DX,AX instruction operates.
43. What is a segment override prefix?
44. Select an instruction that moves a byte of data from the memory location addressed by the BX register, in the extra segment, into the AH register.
45. Develop a sequence of instructions that exchanges the contents of AX with BX, ECX with EDX, and SI with DI.
46. What is an assembly language directive?
47. Describe the purpose of the following assembly language directives: DB, DW, and DD.
48. Select an assembly language directive that reserves 30 bytes of memory for array LIST1.
49. Describe the purpose of the EQU directive.
50. What is the purpose of the .386 directive?
51. What is the purpose of the .MODEL directive?
52. If the start of a segment is identified with .DATA, what type of memory organization is in effect?
53. If the SEGMENT directive identifies the start of a segment, what type of memory organization is in effect?
54. What does the INT 21H accomplish if AH contains a 4CH?
55. What directives indicate the start and end of a procedure?
56. Explain the purpose of the USES statement as it applies to a procedure with version 6.0 of MASM.
57. Develop a near procedure that stores AL into four consecutive memory locations, within the data segment, as addressed by the DI register.
58. Develop a far procedure that copies word-sized memory location CS:DATA1 into AX, BX, CX, DX, and SI.

CHAPTER 5

Arithmetic and Logic Instructions

INTRODUCTION

In this chapter, we examine the arithmetic and logic instructions found in the 8086–80486 instruction set. Arithmetic instructions include addition, subtraction, multiplication, division, comparison, negation, incrementation, and decrementation. Logic instructions include: AND, OR, Exclusive-OR, NOT, shifts, rotates, and the logical compare (TEST). Also presented are the 80386/80486 instructions: SHRD, SHLD, bit tests, and bit scans.

We also will introduce string comparison instructions, which are used for scanning tabular data and for comparing sections of memory. Both tasks perform efficiently with the string scan and string compare instructions.

If you are already familiar with an 8-bit microprocessor, you will recognize that the 8086–80486 instruction set is superior. Even if this is your first microprocessor, you will quickly learn that the microprocessor possesses a powerful and easy to use set of arithmetic and logic instructions.

5–1 CHAPTER OBJECTIVES

Upon completion of this chapter, you will be able to:

1. Use the arithmetic and logic instructions to accomplish simple binary, BCD, and ASCII arithmetic.
2. Use AND, OR, and Exclusive-OR to accomplish binary bit manipulation.
3. Use the shift and rotate instructions.
4. Explain the operation of the 80386/80486 double precision shift, bit test, and bit scan instructions.
5. Check the contents of a table for a match with the string instructions.

5-2 ADDITION, SUBTRACTION, AND COMPARISON

The bulk of the arithmetic instructions found in any microprocessor includes addition, subtraction, and comparison. The 8086–80486 microprocessor is no different. In this section we illustrate and define these instructions. We also show their use in manipulating register and memory data.

Addition

Addition takes many forms in the 8086–80486 microprocessor. This section details the use of the ADD instruction for 8-, 16-, and 32-bit binary addition. Another form of addition, called *add-with-carry* is introduced with the ADC instruction. Finally, the increment instruction (INC), a special type of addition instruction, is presented. In Section 5–3 we examine other forms of addition, such as BCD and ASCII.

Table 5–1 illustrates the addressing modes available to the ADD instruction. (The addressing modes include almost all those mentioned in Chapter 3.) Since there are over 1,000 variations of the ADD instruction in the instruction set, it is impossible to list them all in this table. The only types of addition not allowed are memory-to-memory and segment register. The segment registers can only be moved, pushed, or popped. Note that as with all other instructions, the 32-bit registers are only available with the 80386/80486 microprocessor.

Register Addition. Example 5–1 shows a simple program that uses register addition to add several registers. In this example, we add the contents of AX, BX, CX, and DX to form a 16-bit result that is stored in the AX register.

EXAMPLE 5-1

```
0000  03 C3          ADD    AX,BX
0002  03 C1          ADD    AX,CX
0004  03 C2          ADD    AX,DX
```

Whenever arithmetic and logic instructions execute, the contents of the flag register change. The flags denote the result of the arithmetic operation. Any ADD instruction modifies the contents of the sign, zero, carry, auxiliary carry, parity, and overflow flags. The flag bits never change for most of the data transfer instructions presented in Chapter 4.

Immediate Addition. Immediate addition is used whenever constant or known data are added. An 8-bit immediate addition appears in Example 5–2. In this example, we first load a 12H into DL using an immediate move instruction. Next we add a 33H to the 12H in DL using an immediate addition. After the addition, the sum (45H) moves into register DL and the flags change as follows:

$$Z = 0 \text{ (result not zero)}$$
$$C = 0 \text{ (no carry)}$$
$$A = 0 \text{ (no half-carry)}$$
$$S = 0 \text{ (result positive)}$$
$$P = 0 \text{ (odd parity)}$$
$$O = 0 \text{ (no overflow)}$$

EXAMPLE 5–2

```
0006 B2 12              MOV    DL,12H
0008 80 C2 33           ADD    DL,33H
```

TABLE 5–1 Addition instructions

Instruction	Comment
ADD AL,BL	AL = AL + BL
ADD CX,DI	CX = CX + DI
ADD EBP,EAX	EBP = EBP + EAX
ADD CL,44H	CL = CL + 44H
ADD BX,35AFH	BX = BX + 35AFH
ADD EDX,12345H	EDX = EDX + 00012345H
ADD [BX],AL	AL adds to contents of the data segment offset location addressed by BX, and the result is stored in the same memory location
ADD CL,[BP]	The contents of the stack segment offset location addressed by BP adds to CL, and the result is stored in CL
ADD AL,[EBX]	The contents of the data segment offset location addressed by EBX adds to AL, and the result is stored in AL
ADD BX,[SI+2]	The word-sized contents of the data segment location addressed by SI plus 2 adds to BX, and the result is stored in BX
ADD CL,TEMP	The contents of data segment location TEMP add to CL, with the result stored in CL
ADD BX,TEMP[DI]	The word-sized contents of the data segment location addressed by TEMP plus DI adds to BX, and the result is stored in BX
ADD [BX+DI],DL	The data segment memory byte addressed by BX + DI is the sum of that byte plus DL
ADD BYTE PTR [DI],3	Add a 3 to the contents of the byte-sized memory location addressed by DI within the data segment
ADD BX,[EAX+2*ECX]	The data segment memory word addressed by the sum of 2 times ECX and EAX add to BX

Memory-to-Register Addition. Suppose an application requires that memory data is added to the AL register. Example 5–3 shows an example that adds 2 consecutive bytes of data, stored at the data segment offset locations NUMB and NUMB + 1, to the AL register.

EXAMPLE 5–3

```
0000 BF 0000 R          MOV    DI,OFFSET NUMB     ;address NUMB
0003 B0 00              MOV    AL,0               ;clear sum
0005 02 05              ADD    AL,[DI]            ;add NUMB
0007 02 45 01           ADD    AL,[DI+1]          ;add NUMB+1
```


The sequence of instructions first loads the contents of the destination index register (DI) with offset address NUMB. The DI register, used in this example, addresses data in the data segment beginning at memory location **NUMB**. Next the ADD AL,[DI] instruction adds the contents of memory location NUMB to AL. This occurs because DI addresses memory location NUMB, and the instruction adds its contents to AL. Finally, the ADD AL,[DI + 1] instruction adds the contents of memory location NUMB + 1 to the AL register. After both ADD instructions execute, the result appears in the AL register as the sum of NUMB plus NUMB + 1.

Array Addition. Memory arrays are lists of data. Suppose that an array of data (ARRAY) contains 10 bytes numbered from element 0 through element 9.

Example 5–4 shows a program that adds the contents of array elements 3, 5, and 7. (The program and the array elements it adds are chosen to demonstrate the use of some of the addressing modes for the microprocessor.)

EXAMPLE 5–4

```
0000 B0 00              MOV    AL,0               ;clear sum
0002 BE 0003            MOV    SI,3               ;address element 3
0005 02 84 0002 R       ADD    AL,ARRAY[SI]       ;add element 3
0009 02 84 0004 R       ADD    AL,ARRAY[SI+2]     ;add element 5
000D 02 84 0006 R       ADD    AL,ARRAY[SI+4]     ;add element 7
```


This example clears AL to zero so it can be used to accumulate the sum. Next we load register SI with a 3 to initially address array element 3. The ADD AL,ARRAY[SI] instruction adds the contents of array element 3 to the sum in AL. The instructions that follow add array elements 5 and 7 to the sum in AL, using a 3 in SI plus a displacement of 2 to address element 5 and a displacement of 4 to address element 7.

Suppose that an array of data contains 16-bit numbers that add to form a 16-bit sum in register AX. Example 5–5 shows a short sequence of instructions written for the 80386/80486 microprocessor that uses scaled-index addressing to add elements

3, 5, and 7 of an area of memory called ARRAY. In this example, we load EBX with the address ARRAY, and ECX holds the array element number.

EXAMPLE 5–5

```
0000  66¦ BB 00000000 R      MOV   EBX,OFFSET ARRAY    ;address ARRAY
0006  66¦ B9 00000003        MOV   ECX,3               ;address element 3
000C  67& 8B 04 4B           MOV   AX,[EBX+2*ECX]      ;get element 3
0010  66¦ B9 00000005        MOV   ECX,5               ;address element 5
0016  67& 03 04 4B           ADD   AX,[EBX+2*ECX]      ;add element 5
001A  66¦ B9 00000007        MOV   ECX,7               ;address element 7
0020  67& 03 04 4B           ADD   AX,[EBX+2*ECX]      ;add element 7
```

Increment Addition.   *Increment addition* (INC) adds 1 to a register or a memory location. The INC instruction can add 1 to any register or memory location except a segment register. Table 5–2 illustrates some of the possible forms of the increment instruction available to the 8086–80486 microprocessor. As with other instructions presented thus far, it is impossible to show all variations of the INC instruction because of the large number available.

With indirect memory increments, the size of the data must be described using the BYTE PTR, WORD PTR, or DWORD PTR directives. The reason is that the assembler program cannot determine if, for example, the INC [DI] instruction is a byte-, word-, or double-word-sized increment. The INC BYTE PTR [DI] instruction clearly indicates byte-sized memory data, the INC WORD PTR [DI] instruction unquestionably indicates a word-sized memory data, and the INC DWORD PTR [DI] instruction indicates double-word-sized data.

Example 5–6 shows how the program of Example 5–3 is modified to use the increment instruction for addressing NUMB and NUMB+1. Here, an INC DI instruction changes the contents of register DI from offset address NUMB to offset

TABLE 5–2   Increment instructions

Instruction	Comment
INC BL	BL = BL + 1
INC SP	SP = SP + 1
INC EAX	EAX = EAX + 1
INC BYTE PTR [BX]	The byte contents of the memory location addressed by BX in the data segment increment
INC WORD PTR [SI]	The word contents of the memory location addressed by SI in the data segment increment
INC DWORD PTR [ECX]	The double-word contents of the data segment memory location addressed by ECX increment
INC DATA1	The contents of DATA1 increment

address NUMB+1. Both the programs of Examples 5–3 and 5–6 add the contents of NUMB and NUMB+1. The difference between these programs lies in the way that this data's address is formed through the contents of the DI register using the increment instruction.

EXAMPLE 5–6

```
0000 BF 0000 R          MOV     DI,OFFSET NUMB      ;address NUMB
0003 B0 00              MOV     AL,0                ;clear sum
0005 02 05              ADD     AL,[DI]             ;add NUMB
0007 47                 INC     DI                  ;address NUMB+1
0008 02 05              ADD     AL,[DI]             ;add NUMB+1
```

Increment instructions affect the flag bits, as do most other arithmetic and logic operations. The difference is that increment instructions do not affect the carry flag bit. Carry is not affected because we often use the increment instruction in programs that depend upon the contents of the carry flag.

Addition with Carry. An *addition-with-carry* instruction (ADC) adds the bit in the *carry* flag (C) to the operand data. This instruction mainly appears in software that adds numbers that are wider than 16 bits in the 8086–80286 or wider than 32-bits in the 80386/80486 microprocessor.

Table 5–3 lists several add-with-carry instructions with comments that explain their operation. Like the ADD instruction, ADC affects the flags after the addition.

Suppose that we write a program for the 8086–80286 which adds the 32-bit number in BX and AX to the 32-bit number in DX and CX. Figure 5–1 illustrates this addition so the placement and function of the carry flag can be understood.

TABLE 5–3 Add-with-carry instructions

Instruction	Comment
ADC AL,AH	AL = AL + AH + carry
ADC CX,BX	CX = CX + BX + carry
ADC EBX,EDX	EBX = EBX + EDX + carry
ADC [BX],DH	The byte contents of the memory location in the data segment addressed by BX is summed with DH and carry, and the result is stored in memory
ADC BX,[BP+2]	BX and the word contents of the memory location in the stack segment addressed by BP are summed with carry, and the result is stored in BX
ADC ECX,[EBX]	ECX and the double-word contents of the memory location in the data segment addressed by EBX are summed with carry, and the result is stored in ECX

Note: Only the 80386/80486 use 32-bit registers and addressing modes.

FIGURE 5–1

Addition-with-carry showing how the carry flag (C) links the two 16-bit additions into one 32-bit addition.

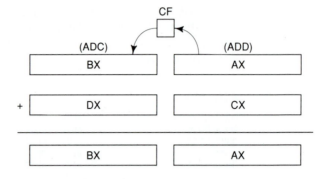

This addition cannot be performed without adding the carry flag bit because the 8086–80286 only adds 8- or 16-bit numbers. Example 5–7 shows how the addition occurs with a program. Here the contents of registers AX and CX add to form the least significant 16 bits of the sum. This addition may or may not generate a carry. A carry appears in the carry flag if the sum is greater than FFFFH. Because it is impossible to predict a carry, the most significant 16 bits of this addition are added with the carry flag using the ADC instruction. The ADC instruction adds the one or zero in the carry flag to the most significant 16 bits of the result. This program adds BX–AX to DX–CX, with the sum appearing in BX–AX.

EXAMPLE 5–7

```
0000  03 C1              ADD    AX,CX
0002  13 DA              ADC    BX,DX
```

Suppose the same program is rewritten for the 80386/80486 microprocessor, but modified to add two 64-bit numbers. The changes required are the use of the extended registers to hold the data and modifications of the instructions for the 80386/80486 microprocessor. These changes are shown in Example 5–8, which adds two 64-bit numbers.

EXAMPLE 5–8

```
0000  66¦ 03 C1          ADD    EAX,ECX
0003  66¦ 13 DA          ADC    EBX,EDX
```

Exchange and Add for the 80486 Microprocessor.   A new type of addition appears in the 80486 instruction set called exchange and add (XADD). The XADD instruction adds the source to the destination and stores the sum in the destination just as any addition. The difference is that after the addition takes place, the original value of the destination is copied into the source operand. This is one of the few instructions that change the source.

For example, if BL = 12H and DL = 02H, and the XADD BL,DL instruction executes, the BL register contains the sum of 14H and DL becomes 12H. The sum of 14H is generated and the original destination of 12H replaces the source. This instruction functions with any register size and any memory operand just as the normal ADD instruction does.

Subtraction

Many forms of subtraction (SUB) appear in the 8086–80486 instruction set. These forms use any addressing mode with 8-, 16-, or 32-bit data. A special form of subtraction (decrement) subtracts a 1 from any register or memory location. Section 5–5 shows how BCD and ASCII data subtract. As with addition, numbers that are wider than 16 bits or 32 bits must occasionally be subtracted. The subtract-with-borrow instruction (SBB) performs this type of subtraction.

Table 5–4 lists many addressing modes allowed with the subtract instruction (SUB). There are well over 1,000 possible subtraction instructions, far too many to list. About the only types of subtraction not allowed are memory-to-memory and segment register subtractions. Like other arithmetic instructions, the subtract instruction affects the flag bits.

Register Subtraction. Example 5–9 shows a program that performs register subtraction. This example subtracts the 16-bit contents of registers CX and DX from the contents of register BX. After each subtraction, the microprocessor modifies

TABLE 5–4 Subtraction instructions

Instruction	Comment
SUB CL,BL	CL = CL − BL
SUB AX,SP	AX = AX − SP
SUB ECX,EBP	ECX = ECX − EBP
SUB DH,6FH	DH = DH − 6FH
SUB AX,0CCCCH	AX = AX − 0CCCCH
SUB EAX,23456H	EAX = EAX − 00023456H
SUB [DI],CH	CH subtracts from the byte contents of the data segment memory location addressed by DI
SUB CH,[BP]	The byte contents of the stack segment memory location addressed by BP subtracts from CH
SUB AH,TEMP	The byte contents of the data segment memory location TEMP subtracts from AH
SUB DI,TEMP[BX]	The word contents of the data segment memory location addressed by TEMP plus BX subtracts from DI
SUB ECX,DATA1	The double-word contents of the data segment memory location addressed as DATA1 subtracts from ECX

Note: Only the 80386/80486 use 32-bit registers and addressing modes.

the contents of the flag register. The flags change for most arithmetic and logic operations.

EXAMPLE 5–9

```
0000  2B D9              SUB    BX,CX
0002  2B DA              SUB    BX,DX
```

Immediate Subtraction.   As with addition, the microprocessor also allows immediate operands for the subtraction of constant data. Example 5–10 presents a short program that subtracts a 44H from a 22H. Here, we first load the 22H into CH using an immediate move instruction. Next, the SUB instruction, using immediate data 44H, subtracts a 44H from the 22H. After the subtraction, the difference (DEH) moves into the CH register. The flags change as follows for this subtraction:

$$Z = 0 \text{ (result not zero)}$$
$$C = 1 \text{ (borrow)}$$
$$A = 1 \text{ (half-borrow)}$$
$$S = 1 \text{ (result negative)}$$
$$P = 1 \text{ (even parity)}$$
$$O = 0 \text{ (no overflow)}$$

EXAMPLE 5–10

```
0000  B5 22              MOV    CH,22H
0002  80 ED 44           SUB    CH,44H
```

Both carry flags (C and A) hold borrows after subtraction instead of carries, as after an addition. Notice in this example there is no *overflow*. This example subtracted a 44H (+68) from a 22H (+34) resulting in a DEH (−34). Because the correct 8-bit signed result is a −34, there is no overflow in this example. An 8-bit overflow only occurs if the signed result is greater than +127 or less than −128.

Decrement Subtraction.   *Decrement* subtraction (DEC) subtracts a 1 from a register or the contents of a memory location. Table 5–5 lists some decrement instructions that illustrate register and memory decrements.

The decrement indirect memory data instructions require BYTE PTR, WORD PTR, or DWORD PTR because the assembler cannot distinguish a byte from a word when an index register addresses memory. For example, DEC [SI] is vague, because the assembler cannot determine if the location addressed by SI is a byte or a word. Using DEC BYTE PTR [SI], DEC WORD PTR [DI], or DEC DWORD PTR [SI] reveals the size of the data.

Subtract with Borrow.   A subtraction-with-borrow (SBB) instruction functions as a regular subtraction, except the carry flag (C), which holds the *borrow*, also subtracts

TABLE 5–5 Decrement instructions

Instruction	Comment
DEC BH	BH = BH − 1
DEC SP	SP = SP − 1
DEC ECX	ECX = ECX − 1
DEC BYTE PTR [DI]	The byte contents of the data segment memory location addressed by DI decrements
DEC WORD PTR [BP]	The word contents of the stack segment memory location addressed by BP decrements
DEC DWORD PTR [EBX]	The double-word contents of the data segment memory location addressed by EBX decrements
DEC NUMB	Decrements the contents of data segment memory location NUMB. The way that NUMB is defined determines whether this is a byte or word decrement

Note: Only the 80386/80486 use 32-bit registers and addressing modes.

from the difference. The most common use for this instruction is for subtractions that are wider than 16 bits in the 8086–80286 microprocessor or wider than 32 bits in the 80386/80486. Wide subtractions require that borrows propagate through the subtraction, just as wide additions propagated the carry.

Table 5–6 lists many SBB instructions with comments that define their operation. Like the SUB instruction, SBB affects the flags. Notice that the subtract from

TABLE 5–6 Subtract-with-borrow instructions

Instruction	Comment
SBB AH,AL	AH = AH − AL − carry
SBB AX,BX	AX = AX − BX − carry
SBB EAX,EBX	EAX = EAX − EBX − carry
SBB CL,3	CL = CL − 3 − carry
SBB BYTE PTR [DI],3	3 and carry subtract from the byte contents of the data segment memory location addressed by DI
SBB [DI],AL	AL and carry subtract from the byte contents of the data segment memory location addressed by DI
SBB DI,[BP+2]	The word contents of the stack segment memory location addressed by BP plus 2 and the carry subtract from DI
SBB AL,[EBX+ECX]	The byte contents of the data segment memory location addressed by the sum of EBX and ECX and the carry subtract from AL

Note: Only the 80386/80486 use 32-bit registers and addressing modes.

21. What is the difference between the IMUL and MUL instructions?
22. Write a sequence of instructions that will cube the 8-bit number found in DL, assuming DL contains a 5 initially. Make sure your result is a 16-bit number.
23. Describe the operation of the IMUL BX,DX,100H instruction.
24. When 8-bit numbers divide, in which register is the dividend found?
25. When two 16-bit numbers divide, in which register is the quotient found?
26. What type of errors are detected during division?
27. Explain the difference between the IDIV and DIV instructions.
28. Where is the remainder found after an 8-bit division?
29. Write a short sequence of instructions that will divide the number in BL by the number in CL and then multiply the result by 2. The final answer must be a 16-bit number stored in the DX register.
30. What instructions are used with BCD arithmetic operations?
31. What instructions are used with ASCII arithmetic operations?
32. Explain how the AAM instruction converts from binary to BCD.
33. Develop a sequence of instructions that adds the 8-digit BCD number in AX and BX to the 8-digit BCD number in CX and DX. (AX and CX are the most significant registers. The result must be found in CX and DX after the addition.)
34. Select an AND instruction that will:
    a. AND BX with DX and save the result in BX
    b. AND 0EAH with DH
    c. AND DI with BP and save the result in DI
    d. AND 1122H with EAX
    e. AND the data addressed by BP with CX and save the result in memory
    f. AND the data stored in four words before the location addressed by SI with DX and save the result in DX
    g. AND AL with memory location WHAT and save the result at location WHAT
35. Develop a short sequence of instructions that will clear the three leftmost bits of DH without changing DH and store the result in BH.
36. Select an OR instruction that will:
    a. OR BL with AH and save the result in AH
    b. OR 88H with ECX
    c. OR DX with SI and save the result in SI
    d. OR 1122H with BP
    e. OR the data addressed by BX with CX and save the result in memory
    f. OR the data stored 40 bytes after the location addressed by BP with AL and save the result in AL
    g. OR AH with memory location WHEN and save the result in WHEN
37. Develop a short sequence of instructions that will set the rightmost five bits of DI without changing DI. Save the result in SI.
38. Select the XOR instruction that will:
    a. XOR BH with AH and save the result in AH
    b. XOR 99H with CL
    c. XOR DX with DI and save the result in DX

    d. XOR 1A23H with ESP

    e. XOR the data addressed by EBX with DX and save the result in memory

    f. XOR the data stored 30 words after the location addressed by BP with DI and save the result in DI

    g. XOR DI with memory location WELL and save the result in DI

39. Develop a sequence of instructions that will set the rightmost four bits of AX, clear the leftmost three bits of AX, and invert bits 7, 8, and 9 of AX.

40. Describe the difference between AND and TEST.

41. Select an instruction that tests bit position 2 of register CH.

42. What is the difference between a NOT and a NEG?

43. Select the correct instruction to perform each of the following tasks:

    a. shift DI right 3 places with zeros moved into the leftmost bit

    b. move all bits in AL left one place, making sure that a 0 moves into the rightmost bit position

    c. rotate all the bits of AL left 3 places

    d. rotate carry right one place through EDX

    e. move the DH register right one place making sure that the sign of the result is the same as the sign of the original number

44. What does the SCASB instruction accomplish?

45. For string instructions, DI always addresses data in the _____ segment.

46. What is the purpose of the D flag bit?

47. Explain what the REPE prefix does.

48. What condition or conditions will terminate the repeated string instruction REPNE SCASB?

49. Describe what the CMPSB instruction accomplishes.

50. Develop a sequence of instructions that will scan through a 300H byte section of memory called LIST searching for a 66H.

CHAPTER 6

Program Control Instructions

INTRODUCTION

The program control instructions direct the flow of a program and allow the flow of the program to change. A change in flow often occurs when decisions, made with the CMP or TEST instruction, are followed by a conditional jump instruction. This chapter explains the program control instructions including jumps, calls, returns, interrupts, and machine control instructions.

6–1   CHAPTER OBJECTIVES

Upon completion of this chapter, you will be able to:

1. Use both conditional and unconditional jump instructions to control the flow of a program.
2. Use the call and return instructions to include procedures in the program structure.
3. Explain the operation of the interrupts and interrupt control instructions.
4. Use machine control instructions to modify the flag bits.
5. Use ENTER and LEAVE to enter and leave programming structures.

6–2   THE JUMP GROUP

The main type of program control instruction, the *jump* (JMP), allows the programmer to skip sections of a program and branch to any part of the memory for the next instruction. A *conditional* jump instruction allows the programmer to make decisions based upon numerical tests. The results of these numerical tests are held in the flag bits, which are then tested by conditional jump instructions. Another instruction similar to the conditional jump, the conditional set, is explained with the conditional jump instructions in this section.

In this section of the text, we cover all jump instructions and illustrate their use with sample programs. We also revisit the LOOP and conditional LOOP instructions, first presented in Chapter 3, because they are also forms of the jump instruction.

Unconditional Jump (JMP)

Three types of unconditional jump instructions (refer to Figure 6–1) are available in the microprocessor's instruction set: short jump, near jump, and far jump. The *short jump* is a 2-byte instruction that allows jumps or branches to memory locations within +127 and −128 bytes from the memory location following the jump. The 3-byte *near jump* allows a branch or jump within ±32K bytes (anywhere) from the instruction in the current code segment. Finally, the 5-byte *far jump* allows a jump to any memory location within the entire memory system. The short and near jumps are often called *intrasegment jumps* and the far jumps are often called *intersegment jumps.*

In the 80386/80486 microprocessor, the near jump is within ±2G if the machine is operated in the protected mode and ±32K bytes if operated in the real mode. The protected mode 80386/80486 jump uses a 32-bit displacement that is not shown in Figure 6–1. The 80386/80486 far jump allows a jump to any location within the 4G byte address range of these microprocessors.

Short Jump.   Short jumps are called *relative jumps* because they can be moved anywhere in current code segment without a change. This is because a jump address is *not* stored with the opcode. Instead of a jump address, a distance, or *displacement*, follows the opcode. The short jump displacement is a *distance* represented by a 1-byte signed number whose value ranges between +127 and −128. The short jump instruction appears in Figure 6–2. When the microprocessor executes a short jump, the displacement sign extends and adds to the instruction pointer (IP/EIP) to

FIGURE 6–1   The three main forms of the JMP instruction. (Note that Disp is either an 8- or 16-bit signed displacement or distance.)

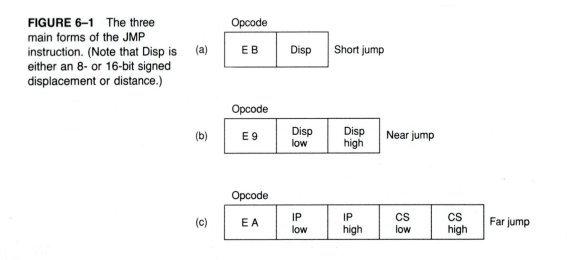

FIGURE 6-2 A short JMP to four memory locations beyond the address of the next instruction. (Note that this will skip the next 4 bytes of memory.)

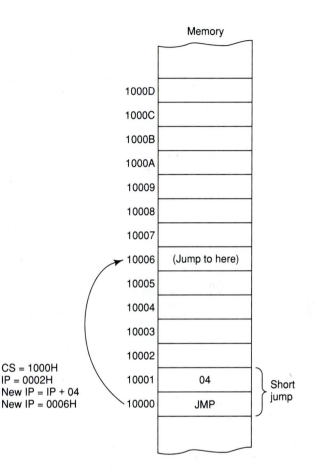

generate the jump address within the current code segment. The short jump instruction branches to this new address for the next instruction in the program.

Example 6-1 shows how short jump instructions pass control from one part of the program to another. It also illustrates the use of a label with the jump instruction. Notice how one jump (JMP SHORT NEXT) uses the SHORT directive to force a short jump, while the other does not. Most assembler programs choose the best form of the jump instruction, so the second (JMP START) jump instruction also assembles as a short jump. If we add the address of the next instruction (0009H) to the sign-extended displacement (0017H) of the first jump, the address of NEXT is location 0017H + 0009H, or 0020H.

EXAMPLE 6-1

```
0000 33 DB                    XOR    BX,BX

0002 B8 0001      START:      MOV    AX,1
0005 03 C3                    ADD    AX,BX
0007 EB 17                    JMP    SHORT NEXT
```

```
0020  8B D8          NEXT:     MOV   BX,AX
0022  EB DE                    JMP   START
```

Whenever a jump instruction references an address, a *label* identifies the address. The JMP NEXT is an example, which jumps to label NEXT for the next instruction. We never use an actual hexadecimal address with any jump instruction. The label NEXT must be followed by a colon (NEXT:) to allow an instruction to reference it for a jump. If a colon does not follow a label, you cannot jump to it. Note that the only time we ever use a colon after a label is when the label is used with a jump or call instruction.

Near Jump.   The near jump is similar to the short jump except the distance is farther. A near jump passes control to an instruction in the current code segment located within ±32K bytes from the near jump instruction or ±2G in the 80386/80486 operated in protected mode. The near jump is a 3-byte instruction that contains an opcode followed by a signed 16-bit displacement. (In the 80386/80486 the displacement is 32 bits and the near jump is 5 bytes in length.) The signed displacement adds to the instruction pointer (IP) to generate the jump address. Because the signed displacement is in the range of ±32K, a near jump can jump to *any* memory location within the current real mode code segment. The protected mode code segment in the 80386/80486 can be 4G bytes in length, so the 32-bit displacement allows a near jump to any location ±2G bytes. Figure 6–3 illustrates the operation of the real-mode near jump instruction.

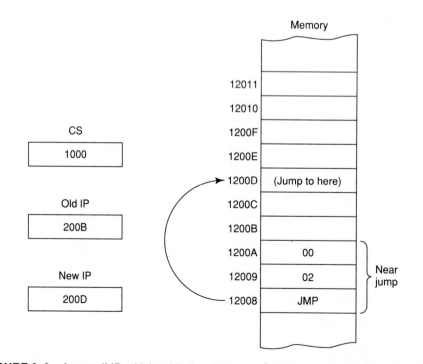

FIGURE 6–3   A near JMP, which adds the displacement to the contents of the IP register.

The near jump is also ***relocatable*** as was the short jump because it is also a relative jump. If the code segment moves to a new location in the memory, the distance between the jump instruction and the operand address remains the same. This allows a code segment to be relocated by moving it. This feature, with the relocatable data segments, makes the Intel family of microprocessors ideal for use in a general-purpose computer system. Software can be written and loaded anywhere in the memory, and it functions without modification because of the relative jumps and relocatable data segments. This is true of very few other microprocessors.

Example 6–2 shows the same basic program that appeared in Example 6–1, except the jump distance is greater. The first jump (JMP NEXT) passes control to the instruction at memory location 0200H within the code segment. Notice that the instruction assembles as an E9 0200 R. The letter R denotes the relocatable jump address of 0200H. The relocatable address of 0200H is for the assembler's internal use only. The actual machine language instruction assembles as an E9 F6 01, which *does not* appear in the assembler listing. The actual displacement is a 01F6H for this jump instruction. The assembler lists the actual jump address as 0200 R so the address is easier to interpret as we develop software. If we were to view the linked execution file in hexadecimal, we would see this jump has assembled as an E9 F6 01.

EXAMPLE 6–2

```
0000 33 DB                 XOR    BX,BX

0002 B8 0001    START:     MOV    AX,1
0005 03 C3                 ADD    AX,BX
0007 E9 0200 R             JMP    NEXT

0200 8B D8      NEXT:      MOV    BX,AX
0202 E9 0002 R             JMP    START
```

Far Jump. Far jumps (see Figure 6–4) obtain a new segment and offset address to accomplish the jump. Bytes 2 and 3 of this 5-byte instruction contain the new offset address, and byte 4 and 5 contain the new segment address. If the microprocessor (80286–80486) is operated in the protected mode, the segment address accesses a descriptor which contains the base address of the far jump segment and the offset address, which is either 16- or 32-bits.

Example 6–3 lists a short program that uses a far jump instruction. The far jump instruction sometimes appears with the FAR PTR directive as illustrated. Another way to obtain a far jump is to define a label as a far label. A label is far only if it is external to the current code segment. The JMP UP instruction in the example references a far label. The label UP is defined as a far label by the EXTRN UP:FAR directive. External labels appear in programs that contain more than one program file.

EXAMPLE 6–3

```
                          EXTRN    UP:FAR

0000 33 DB                XOR    BX,BX
```

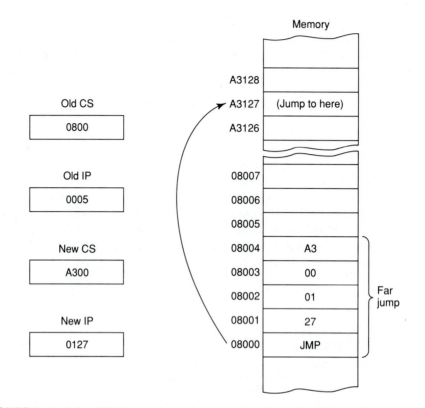

FIGURE 6–4 A far JMP that replaces the contents of both the CS and IP registers with the 4 bytes that follow the opcode.

```
0002  B8 0001      START:    MOV    AX,1
0005  03 C3                  ADD    AX,BX
0007  E9 0200 R             JMP    NEXT

0200  8B D8        NEXT:     MOV    BX,AX
0202  EA 0002 — R           JMP    FAR PTR START

0207  EA 0000 — E           JMP    UP
```

When the program files are joined, the linker inserts the address for the UP label into the JMP UP instruction. It also inserts the segment address in the JMP START instruction. The segment address in JMP FAR PTR START is listed as ____ R for relocatable, and the segment address in JMP UP is listed as ____ E for external. In both cases the ____ is filled in by the linker when it links or joins the program files.

Jumps with Register Operands. The jump instruction also can specify a 16- or 32-bit register as an operand. This automatically sets up the instruction as an indirect jump. The address of the jump is in the register specified by the jump instruction. Unlike the displacement associated with the near jump, the contents of the register are transferred directly into the instruction pointer. It does not add to the instruction pointer as with the short and near jumps. The JMP AX instruction, for example, copies the contents of the AX register into the IP when the jump occurs. This allows a jump to any location within the current code segment. In the 80386/80486 a JMP EAX instruction also jumps to any location within the current code segment; the difference is that in protected-mode the code segment can be 4G bytes in length, so a 32-bit offset address is needed.

Example 6–4 shows how the JMP AX instruction accesses a jump table. This sequence of instructions assumes that SI contains a 0, 1, or 2. Because the jump table contains 16-bit offset addresses, the contents of SI double to 0, 2, or 4, so a 16-bit entry in the table can be accessed. Next the offset address of the start of the jump table adds to SI to reference a jump address. The MOV AX,CS:[SI] instruction fetches an address from the code segment jump table; hence the JMP AX instruction jumps to the location in the jump table.

EXAMPLE 6–4

```
0000 03 F6              ADD    SI,SI              ;double SI
0002 81 C6 000B R       ADD    SI,OFFSET TABLE    ;add TABLE address
0006 2E: 8B 04          MOV    AX,CS:[SI]         ;get jump address
0009 FF E0              JMP    AX

000B 1000 R    TABLE:   DW     ZERO
000D 2000 R             DW     ONE
000F 3000 R             DW     TWO
```

This example fetches address ZERO if a 0 appears in SI. Once the address is fetched, the JMP AX instruction causes the program to branch to address ZERO. The same is true for an initial value of 1 or 2 in the SI register. These numbers access addresses ONE and TWO for their branch locations.

Indirect Jumps Using an Index. The jump instruction also may use the [] form of addressing to directly access the jump table. The jump table can contain offset addresses for near indirect jumps or segment and offset addresses for far indirect jumps. The assembler assumes that the jump is near unless the FAR PTR directive indicates a far jump instruction. Here we repeat Example 6–4 in Example 6–5 using the JMP CS:[SI] instead of JMP AX. This reduces the length of the program.

EXAMPLE 6–5

```
0000 03 F6              ADD    SI,SI              ;double SI
0002 81 C6 0009 R       ADD    SI,OFFSET TABLE    ;add TABLE address
0006 2E: FF 24          JMP    CS:[SI]

0009 1000 R    TABLE:   DW     ZERO
000B 2000 R             DW     ONE
000D 3000 R             DW     TWO
```

The mechanism used to access the jump table is identical with a normal memory reference. The JMP CS:[SI] instruction points to a jump address stored at the code segment memory location addressed by SI. It jumps to the address stored in the memory at this location. Both the register and indirect index jump instructions usually address a 16-bit offset. This means that both types of jumps are near jumps. If a JMP FAR PTR [SI] instruction executes, the microprocessor assumes that the jump table contains double-word 32-bit addresses (IP and CS).

Conditional Jumps and Conditional Sets

Conditional jumps are always *short jumps* in the 8086–80286 microprocessors. This limits the range of the jump to within $+127$ bytes and -128 bytes from the location following the conditional jump. In the 80386/80486, conditional jumps are either short or near jumps. This allows a conditional jump to any location within the current code segment in the 80386/80486 microprocessor. Table 6–1 lists all the conditional jump instructions with their test conditions.

The conditional jump instructions test the following flag bits: sign (S), zero (Z), carry (C), parity (P), and overflow (O). If the condition under test is true, a branch to the label associated with the jump instruction occurs. If the condition is false, the next sequential step in the program executes.

TABLE 6–1 Conditional jump instructions

Instruction	Condition Tested	Comment
JA	C = 0 and Z = 0	Jump if above
JAE	C = 0	Jump if above or equal to
JB	C = 1	Jump if below
JBE	C = 1 or Z = 1	Jump if below or equal to
JC	C = 1	Jump if carry set
JE or JZ	Z = 1	Jump if equal to or jump if zero
JG	Z = 0 and S = 0	Jump if greater than
JGE	S = 0	Jump if greater than or equal to
JL	S $\neq$ 0	Jump if less than
JLE	Z = 1 or S $\neq$ 0	Jump if less than or equal to
JNC	C = 0	Jump if carry cleared
JNE or JNZ	Z = 0	Jump if not equal to or jump if not zero
JNO	O = 0	Jump if no overflow
JNS	S = 0	Jump on no sign
JNP/JPO	P = 0	Jump if no parity/jump if parity odd
JO	O = 1	Jump on overflow
JP/JPE	P = 1	Jump on parity/jump on parity even
JS	S = 1	Jump on sign
JCXZ	CX = 0	Jump if CX = 0
JECXZ	ECX = 0	Jump if ECX = 0 (80386/80486 only)

The operation of most conditional jump instructions is straightforward because they often test just one flag bit, but some test more than one flag. Relative magnitude comparisons require more complicated conditional jump instructions that test more than one flag bit.

Because we use both signed and unsigned numbers, and the order of these numbers is different, there are two sets of magnitude comparison conditional jump instructions. Figure 6–5 shows the order of both signed and unsigned 8-bit numbers. The 16- and 32-bit numbers follow the same order as the 8-bit numbers except they are larger. Notice that an FFH is above the 00H in the set of unsigned numbers, but an FFH (−1) is less than 00H for signed numbers. Therefore, an unsigned FFH is *above* 00H, but a signed FFH is *less than* 00H.

When we compare signed numbers, we use JG, JE, JGE, JLE, JE, and JNE. The terms *greater than* and *less than* refer to signed numbers. When we compare unsigned numbers, we use JA, JB, JAE, JBE, JE, and JNE. The terms *above* and *below* refer to unsigned numbers.

The remaining conditional jumps test individual flag bits such as overflow and parity. Notice that JE has an alternative opcode JZ. All instructions have alternates, but many aren't used in programming because they don't make much sense. (The alternates appear in Appendix B with the instruction set listing.) For example, the JA instruction (jump above) has the alternative JNBE (jump not below or equal). A JA

FIGURE 6–5 Signed and unsigned numbers follow different orders.

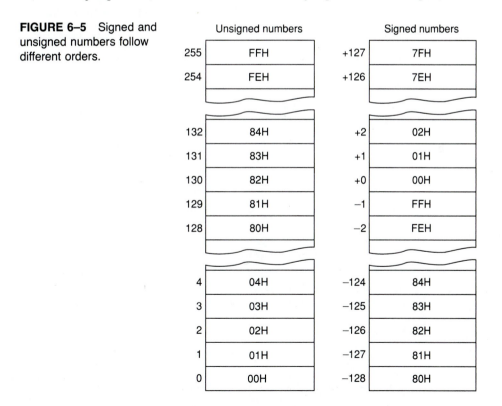

	Unsigned numbers		Signed numbers
255	FFH	+127	7FH
254	FEH	+126	7EH
132	84H	+2	02H
131	83H	+1	01H
130	82H	+0	00H
129	81H	−1	FFH
128	80H	−2	FEH
4	04H	−124	84H
3	03H	−125	83H
2	02H	−126	82H
1	01H	−127	81H
0	00H	−128	80H

functions exactly as a JNBE, but JNBE sounds and is awkward when compared to JA.

The most radical of the conditional jump instructions are JCXZ (jump if CX = 0) and JECXZ (jump if ECX = 0). These are the only conditional jump instructions that do not test the flag bits. Instead, JCXZ directly tests the contents of the CX register without affecting the flag bits and JECXZ tests the contents of the ECX register. If CX = 0, a jump occurs, and if CX ≠ 0, no jump occurs. Likewise for the JECXZ instruction, if ECX = 0, a jump occurs, and if CX ≠ 0, no jump occurs.

A program that uses JCXZ appears in Example 6–6. Here the SCASB instruction searches a table for a 0AH. Following the search, a JCXZ instruction tests CX to see if the count has become zero. If the count is zero, the 0AH is not found in the table. Another method used to test to see if the data are found is the JNE instruction. If JNE replaces JCXZ, it performs the same function. After the SCASB instruction executes, the flags indicate a not equal condition if the data was not found in the table.

EXAMPLE 6–6

```
                         ;procedure that searches a table of 100 bytes
                         ;for a 0AH
                         ;
0000                     SCAN    PROC  NEAR

0000 BE 0000 R                   MOV   SI,OFFSET TABLE    ;address TABLE
0003 B9 0064                     MOV   CX,100            ;load count
0006 B0 0A                       MOV   AL,0AH            ;load search data
0008 FC                          CLD
0009 F2/AE                       REPNE SCASB
000B E3 23                       JCXZ  NOT_FOUND
000D C3                          RET

000E                     SCAN    ENDP
```

The Conditional Set Instructions.   In addition to the conditional jump instructions, the 80386/80486 microprocessors also contain conditional set instructions. The conditions tested by conditional jumps are put to work with the conditional set instructions that set a byte to either a 1 or a 0 depending on the condition under test. Table 6–2 lists the available forms of the 80386/80486 conditional set instructions.

These instructions are useful where a condition must be tested at a point later in the program. For example, a byte can be set to indicate that the carry is cleared at a point in the program by using the SETNC MEM instruction. This instruction places a 01H into memory location MEM if carry is cleared and a 00H into MEM if carry is set. The contents of MEM can be tested at any point later in the program to determine if carry is cleared at the point where the SETNC MEM instruction was executed.

LOOP

The LOOP instruction is a combination of a decrement CX and a conditional jump. In the 8086–80286, LOOP decrements CX and if CX ≠ 0, it jumps to the address indicated by the label. If CX becomes a 0, the next sequential instruction executes. In the 80386/80486, LOOP decrements either CX or ECX depending upon the

TABLE 6–2 The conditional set instructions

Instruction	Condition Tested	Comment
SETB	C = 1	Set byte if below
SETAE	C = 0	Set byte if above or equal
SETBE	C = 1 or Z = 1	Set byte if below or equal
SETA	C = 0 and Z = 0	Set byte if above
SETE/SETZ	Z = 1	Set byte if equal/set byte if zero
SETNE/SETNZ	Z = 0	Set byte if not equal/set byte if not zero
SETL	S ≠ 0	Set byte if less than
SETLE	Z = 1 or S ≠ 0	Set byte if less than or equal
SETG	Z = 0 and S = 0	Set byte if greater than
SETGE	S = 0	Set byte if greater than or equal
SETS	S = 1	Set byte if sign (negative)
SETNS	S = 0	Set byte if no sign (positive)
SETC	C = 1	Set byte if carry
SETNC	C = 0	Set byte if no carry
SETO	O = 1	Set byte if overflow
SETNO	O = 0	Set byte if no overflow
SETP	P = 1	Set byte if parity (even)
SETNP	P = 0	Set byte if no parity (odd)

instruction mode. If the 80386/80486 is operated in the 16-bit instruction mode, LOOP uses CX, and if operated in the 32-bit instruction mode, LOOP uses ECX. This can be changed by the LOOPW (using CX) and LOOPD (using ECX) instructions in the 80386/80486 only.

Example 6–7 shows how data in one block of memory (BLOCK1) adds to data in a second block of memory (BLOCK2), using LOOP to control how many numbers add. The LODSW and STOSW instructions access the data in BLOCK1 and BLOCK2. The ADD AX,ES:[DI] instruction accesses the data in BLOCK2 located in the extra segment. The only reason that BLOCK2 is in the extra segment is that DI addresses extra segment data for the STOSW instruction.

EXAMPLE 6–7

```
                       ;procedure that adds word in BLOCK1 to BLOCK2
                       ;
0000                   ADDS    PROC  NEAR

0000 B9 0064                   MOV   CX,100            ;load count
0003 BE 0064 R                 MOV   SI,OFFSET BLOCK1  ;address BLOCK1
0006 BF 0000 R                 MOV   DI,OFFSET BLOCK2  ;address BLOCK2
0009                   AGAIN:
0009 AD                        LODSW                   ;get BLOCK1 data
000A 26: 03 05                 ADD   AX,ES:[DI]        ;add BLOCK2 data
000D AB                        STOSW                   ;store in BLOCK2
000E E2 F9                     LOOP  AGAIN             ;repeat 100 times
0010 C3                        RET

0011                   ADDS    ENDP
```

Conditional LOOPs.   As with REP, the LOOP instruction also has conditional forms: LOOPE and LOOPNE. The LOOPE (loop while equal) instruction jumps if CX $\neq$ 0 while an equal condition exists. It will exit the loop if the condition is not equal or if the CX register decrements to 0. The LOOPNE (loop while not equal) instruction jumps if CX $\neq$ 0 while a not equal condition exists. It will exit the loop if the condition is equal or if the CX register decrements to 0. In the 80386/80486 the conditional LOOP instruction can use either CX or ECX as the counter. The LOOPEW or LOOPED or LOOPNEW or LOOPNED override the instruction mode if needed.

As with the conditional repeat instructions, alternates exist for LOOPE and LOOPNE. The LOOPE instruction is the same as LOOPZ and the LOOPNE is the same as LOOPNZ. In most programs only the LOOPE and LOOPNE apply.

6–3   PROCEDURES

The procedure, or *subroutine*, is an important part of any computer system's architecture. A *procedure* is a group of instructions that usually performs one task; it is a reusable section of the software that is stored in memory once, but used as often as necessary. This saves memory space and makes it easier to develop software. The only disadvantage of a procedure is that it takes the computer a small amount of time to link to the procedure and return from it. The CALL instruction links to the procedure and the RET instruction returns from the procedure.

The stack stores the return address whenever a procedure is called during the execution of a program. The CALL instruction pushes the address of the instruction following it on the stack. The RET instruction removes an address from the stack so the program returns to the instruction following the CALL.

With the assembler, there are some finite rules for the storage of procedures. A procedure begins with the PROC directive and ends with the ENDP directive. Each directive appears with the name of the procedure. This structure makes it easy to locate the procedure in a program listing. The PROC directive is followed by the type of procedure: NEAR or FAR. Example 6–8 shows how the assembler requires the definition of both a near (intrasegment) and far (intersegment) procedure. In MASM version 6.0, the NEAR or FAR type can be followed by the USES statement. The USES statement allows any number of registers to be automatically pushed to the stack and popped from the stack within the procedure.

EXAMPLE 6–8

```
0000                    SUMS    PROC  NEAR

0000  03 C3                     ADD   AX,BX
0002  03 C1                     ADD   AX,CX
0004  03 C2                     ADD   AX,DX
0006  C3                        RET
```

```
0007                SUMS    ENDP

0007                SUMS1   PROC FAR

0007 03 C3                  ADD   AX,BX
0009 03 C1                  ADD   AX,CX
000B 03 C2                  ADD   AX,DX
000D CB                     RET

000E                SUMS1   ENDP
```

When we compare these two procedures, the only difference is the opcode of the return instruction. The near return instruction uses opcode C3H and the far return uses opcode CBH. A near return removes a 16-bit number from the stack and places it into the instruction pointer to return from the procedure in the current code segment. A far return removes 32 bits from the stack and places it into both IP and CS to return from the procedure to any memory location.

Most procedures that are to be used by all software (*global*) should be written as far procedures. Procedures that are used by a given task (*local*) are normally defined as near procedures.

CALL

The *CALL* instruction transfers the flow of the program to the procedure. The CALL instruction differs from the jump instruction because a CALL saves a return address on the stack. The return address returns control to the instruction that follows the CALL in a program when a RET instruction executes.

Near CALL. The near CALL instruction is 3 bytes long with the first byte containing the opcode and the second and third bytes containing the displacement or distance of ±32K in the 8086–80286. This is identical to the form of the near jump instruction. The 80386/80486 uses a 32-bit displacement when operated in the protected mode to allow a distance of ±2G bytes. When the near CALL executes, it first places the offset address of the next instruction on the stack. The offset address of the next instruction appears in the instruction pointer (IP or EIP). After saving this return address, it then adds the displacement from bytes 2 and 3 to the IP to transfer control to the procedure. There is no short CALL instruction.

Why save the IP or EIP on the stack? The instruction pointer always points to the next instruction in the program. For the CALL instruction, the contents of IP/EIP are pushed onto the stack so program control passes to the instruction following the CALL after a procedure ends. Figure 6–6 shows the return address (IP) stored on the stack, and the call to the procedure for the 8086–80286.

Far CALL. The far CALL instruction is like a far jump because it can CALL a procedure stored in any memory location in the system. The far CALL is a 5-byte instruction that contains an opcode followed by the next value for the IP and CS registers. Bytes 2 and 3 contain the new contents of the IP and bytes 4 and 5 contain the new contents for CS.

FIGURE 6–6 The effect of a near CALL instruction on the stack and the SP, SS, and IP registers. Notice how the old IP is stored on the stack.

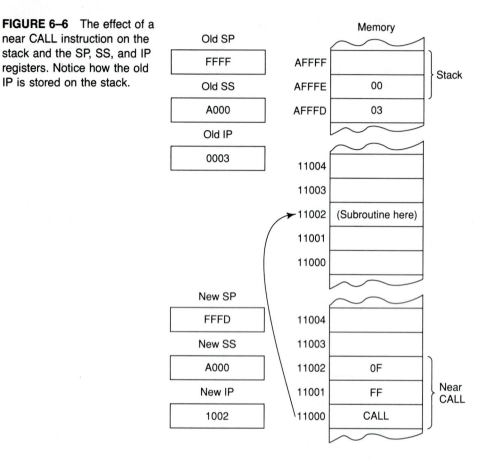

The far CALL instruction places the contents of both IP and CS on the stack before jumping to the address indicated by bytes 2–5 of the instruction. This allows the far CALL to call a procedure located anywhere in the memory and return from that procedure.

Figure 6–7 shows how the far CALL instruction calls a far procedure. Here the contents of IP and CS are pushed onto the stack. Next the program branches to the procedure.

CALLs with Register Operands. Like jumps, CALLs also may contain a register operand. An example is the CALL BX instruction. This instruction pushes the contents of IP onto the stack. It then jumps to the offset address, located in register BX, in the current code segment. This type of CALL always uses a 16-bit offset address stored in any 16-bit register except the segment registers.

Example 6–9 illustrates the use of the CALL register instruction to call a procedure that begins at offset address COMP. The OFFSET address COMP moves into the SI register and then the CALL SI instruction calls the procedure beginning at address COMP.

FIGURE 6–7 The far CALL instruction.

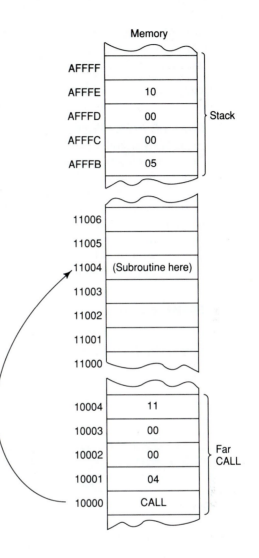

EXAMPLE 6–9

```
                    ;calling sequence
                    ;
0000 BE 0005 R              MOV   SI,OFFSET COMP
0003 FF D6                  CALL  SI
                    ;
                    ;procedure COMP
                    ;
0005                COMP    PROC  NEAR

0005 52                     PUSH  DX
0006 BA 03F8                MOV   DX,03F8H
0009 EC                     IN    AL,DX
```

```
000A 42                        INC    DX
000B EE                        OUT    DX,AL
000C 5A                        POP    DX
000D C3                        RET

000E                COMP       ENDP
```

CALLs with Indirect Memory Addresses.   The CALL with an indirect memory address is particularly useful whenever we need to choose different subroutines in a program. This selection process is often keyed with a number that addresses a CALL address in a lookup table.

Example 6–10 shows three separate subroutines referenced by the numbers 1, 2, and 3 in AL. The calling sequence adjusts the value of AL and extends it to a 16-bit number before adding it to the location of the lookup table. This references one of the three subroutines using the CALL CS:[BX+DI] instruction. The CS: prefix appears before the CALL instruction's operand because the TABLE is in the code segment in this example:

EXAMPLE 6–10

```
                   ;lookup table
                   ;
0000 0100 R        TABLE    DW    ONE
0002 0200 R                 DW    TWO
0004 0300 R                 DW    THREE
                   ;
                   ;calling sequence
                   ;
0006 4B                     DEC   BX              ;scale BX
0007 03 DB                  ADD   BX,BX           ;double BX
0009 BF 0000 R              MOV   DI,OFFSET TABLE ;address TABLE
000C 2E: FF 11              CALL  CS:[BX+DI]
                   ;
                   ;procedures
                   ;
0100               ONE      PROC  NEAR
                     .        .
                     .        .
0100               ONE      ENDP

0200               TWO      PROC  NEAR
                     .        .
                     .        .
0200               TWO      ENDP

0300               THREE    PROC  NEAR
                     .        .
                     .        .
0300               THREE    ENDP
```

The CALL instruction also can reference far pointers if the instruction appears as a CALL FAR PTR [SI]. This instruction retrieves a 32-bit address from the data segment memory location addressed by SI and uses it as the address of a far procedure.

RET

The *return* instruction (RET) removes either a 16-bit number (near return) from the stack and places it into IP or a 32-bit number (far return) and places it into IP and CS. The near and far return instructions are both defined in the procedure's PROC directive. This automatically selects the proper return instruction. In the 80386/80486 operated in the protected mode the far return removes 6 bytes from the stack; the first 4 contain the new value for EIP and the last 2 contain the new value for CS. The 80386/80486 protected mode near return removes 4 bytes from the stack and places them into EIP.

When IP/EIP or IP/EIP and CS are changed, the address of the next instruction is at a new memory location. This new location is the address of the instruction that immediately follows the most recent CALL to a procedure. Figure 6–8 shows how the CALL instruction links to a procedure and how the RET instruction returns in the 8086–80286 microprocessor.

There is one other form of the return instruction. This form adds a number to the contents of the stack pointer (SP) before the return. If the pushes must be deleted before a return, a 16-bit displacement adds to the SP before the return retrieves the return address from the stack. The effect of this is to delete stack data, or skip stack data.

Example 6–11 shows how this type of return erases the data placed on the stack by a few pushes. The RET 4 adds a 4 to SP before removing the return address from the stack. Since the PUSH AX and PUSH BX together place 4 bytes of data on the stack, this return effectively deletes AX and BX from the stack. This type of return appears rarely in programs.

EXAMPLE 6–11

```
0000              TESTS    PROC  NEAR

0000 50                    PUSH  AX
0001 53                    PUSH  BX
                             .      .
                             .      .
0030 C2 0004               RET    4

0033              TESTS    ENDP
```

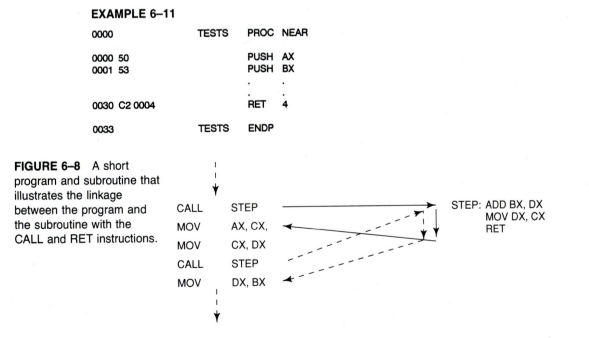

FIGURE 6–8 A short program and subroutine that illustrates the linkage between the program and the subroutine with the CALL and RET instructions.

```
CALL    STEP                                 STEP: ADD BX, DX
MOV     AX, CX,                                    MOV DX, CX
MOV     CX, DX                                     RET
CALL    STEP
MOV     DX, BX
```

6–4   INTRODUCTION TO INTERRUPTS

An *interrupt* is either a hardware-generated CALL (externally derived from a hardware signal) or a software-generated CALL (internally derived from an instruction). Either will interrupt the program by calling an *interrupt service procedure* or *interrupt handler*.

This section explains software interrupts, which are special types of CALL instructions in the 8086–80486 microprocessor. We cover the three types of software interrupt instructions (INT, INTO, and INT 3), provide a map of the interrupt vectors, and explain the purpose of the special interrupt return instruction (IRET).

Interrupt Vectors

An *interrupt vector* is a 4-byte number stored in the first 1,024 bytes of the memory (000000H–0003FFH) when the microprocessor operates in the real mode. In the protected mode, the vector table is replaced by an interrupt descriptor table that uses 8-byte descriptors to describe each of the interrupts. There are 256 different interrupt vectors. Each vector contains the address of an interrupt service procedure—the procedure called by an interrupt. Table 6–3 lists the interrupt vectors with a brief description and the memory location of each vector for the real mode.

TABLE 6–3   Interrupt vectors

Number	Address	Microprocessor	Function
0	0H–3H	8086–80486	Divide error
1	4H–7H	8086–80486	Single step
2	8H–BH	8086–80486	NMI (hardware interrupt)
3	CH–FH	8086–80486	Breakpoint
4	10H–13H	8086–80486	Interrupt on overflow
5	14H–17H	80286–80486	BOUND interrupt
6	18H–1BH	80286–80486	Invalid opcode
7	1CH–1FH	80286–80486	Coprocessor emulation interrupt
8	20H–23H	80386–80486	Double fault
9	24H–27H	80386	Coprocessor segment overrun
10	28H–2BH	80386–80486	Invalid task state segment
11	2CH–2FH	80386–80486	Segment not present
12	30H–33H	80386–80486	Stack fault
13	34H–37H	80386–80486	General protection fault
14	38H–3BH	80386–80486	Page fault
15	3CH–3FH		Reserved*
16	40H–43H	80286–80486	Floating-point error
17	44H–47H	80486SX	Alignment check interrupt
18–31	48H–7FH	8086–80486	Reserved*
32–255	80H–3FFH	8086–80486	User interrupts

*Some of these interrupts will appear on newer versions of the 8086–80486 when they become available.

Each vector contains a value for IP and CS that forms the address of the interrupt service procedure. The first 2 bytes contain the IP and the last 2 bytes contain the CS.

Intel reserves the first 32 interrupt vectors for the 8086–80486 and future products. The remaining interrupt vectors (32–255) are available for the user. Some reserved vectors are for errors that occur during the execution of software, such as the divide-error interrupt. Some vectors are reserved for the coprocessor. Still others occur for normal events in the system. In a personal computer, the reserved vectors are used for system functions as detailed later in this section. Vectors 1–6, 7, 9, 16, and 17 function in the real mode and protected mode; the remaining vectors only function in the protected mode.

Interrupt Instructions

The 8086–80486 has three different interrupt instructions available to the programmer: INT, INTO, and INT 3. In the real mode, each of these instructions fetches a vector from the vector table and then calls the procedure stored at the location addressed by the vector. In the protected mode, each of these instructions fetches an interrupt descriptor from the interrupt descriptor table. The descriptor specifies the address of the interrupt service procedure. The interrupt call is similar to a far CALL instruction because it places the return address (IP/EIP and CS) on the stack.

INTs. There are 256 different software interrupt instructions (INT) available to the programmer. Each INT instruction has a numeric operand whose range is 0 to 255 (00H–FFH). For example, the INT 100 uses interrupt vector 100, which appears at memory address 190H–193H. We calculate the address of the interrupt vector by multiplying the interrupt type number times 4. For example, the INT 10H instruction calls the interrupt service procedure whose address is stored beginning at memory location 40H (10H × 4) in the real mode. In protected mode, the interrupt descriptor is located by multiplying the type number by 8 instead of by 4 because each descriptor is 8 bytes long.

Each INT instruction is 2 bytes in length. The first byte contains the opcode and the second byte contains the vector type number. The only exception to this is INT 3, a 1-byte special software interrupt used for breakpoints.

Whenever a software interrupt instruction executes, it (1) pushes the flags onto the stack, (2) clears the T and I flag bits, (3) pushes CS onto the stack, (4) fetches the new value for CS from the vector, (5) pushes IP/EIP onto the stack, (6) fetches the new value for IP/EIP from the vector, and (7) jumps to the new location addressed by CS and IP/EIP. The INT instruction performs as a far CALL except that it not only pushes CS and IP onto the stack, but it also pushes the flags onto the stack. The INT instruction is a combination PUSHF and far CALL instruction.

Notice that when the INT instruction executes, it clears the interrupt flag (I), which controls the external hardware interrupt input pin INTR (interrupt request). When I = 0, the microprocessor disables the INTR pin and when I = 1, the microprocessor enables the INTR pin.

Software interrupts are most commonly used to call system procedures. The system procedures are common to all system and application software. The interrupts often control printers, video displays, and disk drives. The INT instruction replaces a far CALL because the INT instruction is 2 bytes in length whereas the far CALL is 5 bytes. Each time that the INT instruction replaces a far CALL it saves 3 bytes of memory in a program. This can amount to a sizable savings if the INT instruction appears often in a program.

IRET/IRETD. The interrupt return instruction (IRET) is used only with software or hardware interrupt service procedures. Unlike a simple return instruction (RET), the IRET instruction will (1) pop stack data back into the IP, (2) pop stack data back into CS, and (3) pop stack data back into the flag register. The IRET instruction accomplishes the same tasks as the POPF and RET instructions.

Whenever an IRET instruction executes it restores the contents of I and T from the stack. This is important because it preserves the state of these flag bits. If interrupts were enabled before an interrupt service procedure, they are *automatically* reenabled by the IRET instruction because it restores the flag register.

In the 80286–80486, IRETD instruction is used to return from an interrupt service procedure that is called in the protected mode. It differs from the IRET because it pops a 32-bit instruction pointer (EIP) from the stack. The IRET is used in the real mode and the IRETD is used in the protected mode.

INT 3. An INT 3 instruction is a special software interrupt designed to be used as a breakpoint. The difference between it and the other software interrupts is that INT 3 is a 1-byte instruction while the others are 2-byte instructions.

It is common to insert an INT 3 instruction in software to interrupt or break the flow of the software. This function is called a ***breakpoint***. A breakpoint occurs for any software interrupt, but because INT 3 is 1 byte long, it is easier to use for this function. Breakpoints help to debug faulty software.

INTO. Interrupt on overflow (INTO) is a conditional software interrupt that tests the overflow flag (O). If O = 0, the INTO instruction performs no operation, but if O = 1 and an INTO instruction executes, an interrupt occurs via vector type number 4.

The INTO instruction appears in software that adds or subtracts signed binary numbers. With these operations it is possible to have an overflow. Either the JO instruction or INTO instruction detects the overflow condition.

An Interrupt Service Procedure. Suppose that, in a particular system, we must add the contents of DI, SI, BP, and BX and save the sum in AX. Because this is a common task in this system, it is worthwhile to develop the task as a software interrupt. Example 6–12 shows this software interrupt. The main difference between this procedure and a normal far procedure is that it ends with the IRET instruction instead of the RET instruction.

EXAMPLE 6–12

```
0000                    INTS    PROC  FAR

0000  03 C3                     ADD   AX,BX
0002  03 C5                     ADD   AX,BP
```

```
0004  03 C7                      ADD    AX,DI
0006  03 C6                      ADD    AX,SI
0008  CF                         IRET

0009                INTS         ENDP
```

Interrupt Control

Although this section does not explain hardware interrupts, we introduce two instructions that control the INTR pin. The set interrupt flag instruction (STI) places a 1 into I, which enables the INTR pin. The clear interrupt flag instruction (CLI) places a 0 into I, which disables the INTR pin. The STI instruction enables INTR, and the CLI instruction disables INTR. In a software interrupt service procedure we usually enable the hardware interrupts as one of the first steps. This is accomplished by the STI instruction.

Interrupts in the Personal Computer

The interrupts found in the personal computer differ somewhat from the ones presented in Table 6–3. The reason that they differ is that the original personal computers are 8086/8088 based systems. This means that they only contained Intel specified interrupts 0–4. This design is carried forward so newer systems are compatible with the early personal computers.

Because the personal computer is operated in the real mode, the interrupt vector table is located at addresses 00000H–003FFH. The assignments used by computer system are listed in Table 6–4. Notice that these differ somewhat from the assignments in Table 6–3. Some of the interrupts shown in this table are used in example programs in later chapters. An example is the clock tick, which is extremely useful for timing events because it occurs a constant 18.2 times a second in all personal computers.

Interrupts 00H–1FH and 70H–77H are present in the computer no matter what operating system is installed. If DOS is installed, interrupts 20H–2FH are also present.

6–5   MACHINE CONTROL AND MISCELLANEOUS INSTRUCTIONS

The last category of real mode instructions found in the 8086–80486 microprocessor are the machine control and miscellaneous group. These instructions provide control of the carry bit, sample the $\overline{BUSY}/\overline{TEST}$ pin, and perform various other functions. Because many of these instructions are used in hardware control, they need only be explained briefly at this point. We cover most of these instructions in more detail in later chapters that deal with the hardware and programs that control the hardware.

Controlling the Carry Flag Bit

The carry flag (C) propagates the carry or borrow in multiple-word/double-word addition and subtraction. It also indicates errors in procedures. There are three

TABLE 6–4 Interrupt assignments for the personal computer

Number	Function
0	Divide error
1	Single step (debug)
2	Nonmaskable interrupt pin
3	Breakpoint
4	Arithmetic overflow
5	Print screen key and BOUND instruction
6	Illegal instruction error
7	Coprocessor not present interrupt
8	Timer tick (hardware) (Approximately 18.2 Hz)
9	Keyboard (hardware)
A	Hardware interrupt 2 (system bus) (cascade in AT)
B–F	Hardware interrupts 3–7 (system bus)
10	Video BIOS
11	Equipment environment
12	Conventional memory size
13	Direct disk service
14	Serial COM port service
15	Miscellaneous service
16	Keyboard service
17	Parallel port LPT service
18	ROM BASIC
19	Reboot
1A	Clock service
1B	Control-break handler
1C	User timer service
1D	Pointer for video parameter table
1E	Pointer for disk drive parameter table
1F	Pointer for graphics character pattern table
20	Terminate program
21	DOS services
22	Program termination handler
23	Control C handler
24	Critical error handler
25	Read disk
26	Write disk
27	Terminate and stay resident
28	DOS idle
2F	Multiplex handler
70–77	Hardware interrupts 8–15 (AT style computer)

instructions that control the contents of the carry flag: STC (*set carry*), CLC (*clear carry*), and CMC (*complement carry*).

Because the carry flag is seldom used, except with multiple-word addition and subtraction, it is available for other uses. The most common task for the carry flag is to indicate error upon return from a procedure. Suppose that a procedure reads data from a disk memory file. This operation can be successful or an error can occur such as file-not-found. Upon return from this procedure, if C = 1, an error has occurred and if C = 0, no error occurred. Most of the DOS and BIOS procedures use the carry flag to indicate error conditions.

WAIT

The WAIT instruction monitors the hardware $\overline{\text{BUSY}}$ pin on the 80286–80386 and the $\overline{\text{TEST}}$ pin on the 8086/8088. The name of this pin was changed in the 80286 microprocessor from $\overline{\text{TEST}}$ to $\overline{\text{BUSY}}$. If the WAIT instruction executes while the $\overline{\text{BUSY}}$ pin = 0, nothing happens and the next instruction executes. If the $\overline{\text{BUSY}}$ pin = 1 when the WAIT instruction executes, the microprocessor waits for the $\overline{\text{BUSY}}$ pin to return to a logic 0.

The $\overline{\text{BUSY}}$ or $\overline{\text{TEST}}$ pin of the microprocessor is usually connected to the $\overline{\text{BUSY}}$ or $\overline{\text{TEST}}$ pin of the 8087–80387 numeric coprocessor. This connection, with the WAIT instruction, allows the 8086–80386 to wait until the coprocessor finishes a task. Because the coprocessor is inside the 80486, the $\overline{\text{BUSY}}$ pin is not present.

HLT

The *halt* instruction (HLT) stops the execution of software. There are three ways to exit a halt: by an interrupt, by a hardware reset, or during a DMA operation. This instruction normally appears in a program to wait for an interrupt. It often synchronizes external hardware interrupts with the software system.

NOP

When the microprocessor encounters a no operation instruction (NOP), it takes a short time to execute. A NOP performs absolutely no operation and often pads software with space for future machine language instructions. If you are developing machine language programs, we recommend that you place NOPs into your program at 50-byte intervals. This is done in case you need to add instructions at some future point. A NOP also finds application in time delays that waste short periods of time.

LOCK Prefix

The LOCK prefix appends an instruction and causes the $\overline{\text{LOCK}}$ pin to become a logic 0. The $\overline{\text{LOCK}}$ pin often disables external bus masters or other system components. The LOCK prefix causes the lock pin to activate for the duration of a *locked* instruction. If we lock more than one sequential instruction, the $\overline{\text{LOCK}}$ pin remains a logic 0 for the duration of the sequence of locked instructions. The LOCK:MOV AL,[SI] instruction is an example of a locked instruction.

ESC

The escape (ESC) instruction passes information to the 8087–80387 numeric coprocessor. Whenever an ESC instruction executes, the microprocessor provides the memory address, if required, but otherwise performs a NOP. The 8087–80387 uses

6 bits of the ESC instruction to obtain its opcode and begin executing a coprocessor instruction.

The ESC opcode never appears in a program as ESC. In its place are a set of coprocessor instructions (FLD, FST, FMUL, etc.) that assemble as ESC instructions for the coprocessor. We provide more detail in the chapter that details the 8087–80387 coprocessor.

BOUND

The BOUND instruction is a compare instruction that can cause an interrupt (vector type number 5). This instruction compares the contents of any 16-bit or 32-bit register against the contents of two words or double words of memory: an upper and a lower boundary. If the value in the register compared with memory is *not* within the upper and lower boundary, a type 5 interrupt ensues. If it is within the boundary, the next instruction in the program executes.

For example, if the BOUND SI,DATA instruction executes, word-sized location DATA contains the lower boundary and word-sized location DATA + 2 bytes contains the upper boundary. If the number contained in SI is less than memory location DATA or greater than memory location DATA + 2 bytes, a type 5 interrupt occurs. Note that when this interrupt occurs the return address points to the BOUND instruction, not the instruction following BOUND. This differs from a normal interrupt where the return address points to the next instruction in the program.

ENTER and LEAVE

The ENTER and LEAVE instructions are used with stack frames. A stack frame is a mechanism used to pass parameters to a procedure through the stack memory. The stack frame also holds local memory variables for the procedure. Stack frames provide dynamic areas of memory for procedures in multiuser environments.

The ENTER instruction creates a stack frame by pushing BP onto the stack and then loading BP with the uppermost address of the stack frame. This allows stack frame variables to be accessed through the BP register. The ENTER instruction contains two operands; the first operand specifies the number of bytes to reserve for variables on the stack frame, and the second specifies the level of the procedure.

Suppose that an ENTER 8,0 instruction executes. This instruction reserves 8 bytes of memory for the stack frame and the zero specifies level 0. Figure 6–9 shows the stack frame set up by this instruction. Note that this instruction stores BP onto the top of the stack. It then subtracts 8 from the stack pointer, leaving 8 bytes of memory space for temporary data storage. The uppermost location of this 8-byte temporary storage area is addressed by BP. The LEAVE instruction reverses this process by reloading both SP and BP with their prior values.

Example 6–13 shows how the ENTER instruction creates a stack frame so two 16-bit parameters are passed to a system level procedure. Notice how the ENTER and LEAVE instructions appear in this program, and how the parameters pass through the stack frame to and from the procedure. This procedure uses two parameters that pass to it and returns two results through the stack frame.

FIGURE 6–9 The stack frame created by the ENTER 8,0 instruction. Notice that BP is stored beginning at the top of the stack frame. This is followed by an 8-byte area called a stack frame.

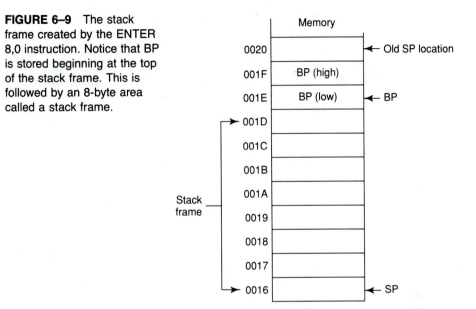

EXAMPLE 6–13

```
                        ;sequence used to call system software that
                        ;uses parameters stored in a stack frame
                        ;
0000  C8 0004 00                ENTER  4,0                  ;create 4 byte frame

0004  A1 00C8 R                 MOV    AX,DATA1
0007  89 46 FC                  MOV    [BP-4],AX            ;save para 1
000A  A1 00CA R                 MOV    AX,DATA2
000D  89 46 FE                  MOV    [BP-2],AX            ;save para 2

0010  E8 0100 R                 CALL   SYS                  ;call subroutine

0013  8B 46 FC                  MOV    AX,[BP-4]            ;get result 1
0016  A3 00C8 R                 MOV    DATA1,AX             ;save result 1
0019  8B 46 FE                  MOV    AX,[BP-2]            ;get result 2
001C  A3 00CA R                 MOV    DATA2,AX             ;save result 2

001F  C9                        LEAVE
                                   .      .
                                   .      .
                                (other software continues here)
                                   .      .
                                   .      .
                        ;system subroutine that uses the stack frame parameters
                        ;
0100                    SYS     PROC  NEAR

0100  60                        PUSHA

0101  8B 46 FC                  MOV    AX,[BP-4]            ;get para 1
0104  8B 5E FE                  MOV    BX,[BP-2]            ;get para 2
```

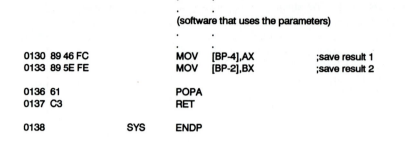

```
                                ·         ·
                                ·         ·
                              (software that uses the parameters)
                                ·         ·
                                ·         ·
0130  89 46 FC                MOV    [BP-4],AX          ;save result 1
0133  89 5E FE                MOV    [BP-2],BX          ;save result 2

0136  61                      POPA
0137  C3                      RET

0138              SYS         ENDP
```

6–6   SUMMARY

1. There are three types of unconditional jump instructions: short, near, and far. The short jump allows a branch to within +127 and −128 bytes. The near jump (using a displacement of ±32K) allows a jump to anywhere in the current code segment (intrasegment). The far jump allows a jump to any location in the memory (intersegment). The near jump in an 80386/80486 microprocessor is within ±2G bytes.

2. Whenever a label appears with a JMP instruction the label must be followed by a colon (LABEL:).

3. The displacement that follows a short or near jump is the distance from the next instruction to the jump location.

4. Indirect jumps are available in two forms: (a) jump to the location stored in a register and (b) jump to the location stored in a memory word (near indirect) or double word (far indirect).

5. Conditional jumps are all short jumps that test one or more of the flag bits: C, Z, O, P, or S. If the condition is true, a jump occurs, and if the condition is false, the next sequential instruction executes.

6. A special conditional jump instruction (LOOP) decrements CX and jumps to the label when CX is not 0. Other forms of looping include: LOOPE, LOOPNE, LOOPZ, and LOOPNZ. The LOOPE instruction jumps if CX is not 0, and if an equal condition exists. In the 80386/80486 microprocessor the LOOP instruction can also use the ECX register as a counter.

7. In the 80386/80486 microprocessor a group of conditional set instructions exists that either set a byte to 01H or clear it to 00H. If the condition under test is true, the operand byte is set to a 01H and if the condition under test is false, the operand byte is cleared to 00H.

8. Procedures are groups of instructions that perform one task and are used from any point in a program. The CALL instruction links to a procedure and the RET instruction returns from a procedure. In assembly language, the PROC directive defines the name and type of procedure. The ENDP directive declares the end of the procedure.

9. The CALL instruction is a combination of a PUSH and a JMP instruction. When CALL executes, it pushes the return address on the stack and then jumps to the procedure. A near CALL places the contents of IP on the stack and a far CALL places both IP and CS on the stack.

10. The RET instruction returns from a procedure by removing the return address from the stack and placing it into IP (near return) or IP and CS (far return).

11. Interrupts are either software instructions similar to CALL or hardware signals used to call procedures. This process interrupts the current program and calls a procedure. After the procedure, an IRET instruction returns control to the interrupted software.

12. Interrupt vectors are 4 bytes in length that contain the address (IP and CS) of the interrupt service procedure. The microprocessor contains 256 interrupt vectors in the first 1K byte of memory. The first 32 are defined by Intel, the remaining 224 are user interrupts.

13. Whenever an interrupt is accepted by the microprocessor, the flags, IP, and CS are pushed on the stack. Besides pushing the flags, the T and I flag bits are cleared to disable both the trace function and the INTR pin. The final event that occurs for the interrupt is that the interrupt vector is fetched from the vector table and a jump to the interrupt service procedure occurs.

14. Software interrupt instructions (INT) often replace system calls. Software interrupts save 3 bytes of memory each time they replace CALL instructions.

15. A special return instruction (IRET) must be used to return from an interrupt service procedure. The IRET instruction not only removes IP and CS from the stack, it also removes the flags from the stack.

16. Interrupt on an overflow (INTO) is a conditional interrupt that calls an interrupt service procedure if the overflow flag (O) = 1.

17. The interrupt enable flag (I) controls the INTR pin connection on the microprocessor. If the STI instruction executes, it sets I to enable the INTR pin. If the CLI instruction executes, it clears I to disable the INTR pin.

18. The carry flag bit (C) is cleared, set, and complemented by the CLC, STC, and CMC instructions.

19. The WAIT instruction tests the condition of the $\overline{\text{BUSY}}$ pin on the microprocessor. If $\overline{\text{BUSY}} = 0$, WAIT does not wait, but if $\overline{\text{BUSY}} = 1$, WAIT continues testing the $\overline{\text{BUSY}}$ pin until it becomes a logic 0.

20. The LOCK prefix causes the $\overline{\text{LOCK}}$ pin to become a logic 0 for the duration of the locked instruction. The ESC instruction passes information to the 80287–80387 numeric coprocessor.

21. The BOUND instruction compares the contents of any 16-bit register against the contents of two words of memory: an upper and a lower boundary. If the value in the register compared with memory is *not* within the upper and lower boundary, a type 5 interrupt ensues.

22. The ENTER and LEAVE instructions are used with stack frames. A stack frame is a mechanism used to pass parameters to a procedure through the stack memory. The stack frame also holds local memory variables for the procedure. The ENTER instruction creates the stack frame, and the LEAVE instruction

removes the stack frame from the stack. The BP register addresses stack frame data.

6–7   GLOSSARY

Breakpoint   An instruction or the point in a program where the program stops or is interrupted for debugging.

Call   An instruction that links to a procedure to be executed.

Conditional   A jump instruction that executes if the condition under test is true.

Debug   The process of removing an error from a program.

Displacement   A distance from the next location after a jump or call instruction.

Far   A type of jump or call that allows software to be accessed at any location in the memory system.

Global   A global procedure is a procedure that is accessed by all software.

Halt   Causes the microprocessor to stop executing instruction until either a reset or an interrupt occurs.

Handler   *See* Interrupt service procedure.

Interrupt   A hardware initiated call to a software procedure.

Interrupt service procedure   The procedure that is responsible for processing the interrupt. This is often called an interrupt handler.

Interrupt vector   A 4-byte section of memory that contains the address of the interrupt service procedure.

Intersegment   A term that applies to far jump and call instruction indicating that the jump or call can use a new segment.

Intrasegment   A term that implies that a jump or call remains within the same memory segment.

Jump   An instruction that branches to another instruction in the program.

Label   In assembly language, a label is a symbolic memory address that eases the task of writing a program.

Local   A procedure that is used by a unique application that may not apply to any other application.

Locked   Instructions can be locked to prevent external access to the microprocessor by using the lock prefix.

Near   A jump or call instruction that remains in the same memory segment.

Procedure   A reusable grouping of instructions to perform a single task that can be used many times in a program. Procedures are linked to with the CALL instruction and returned from with the RET instruction.

Relative   A relative jump is a short or near jump that can be moved to any point within the segment without change.

Relocatable   An instruction or data set that can be moved to any memory location without change.

Return   The act of coming back from a procedure.

Short   A type of jump instruction that can branch to within $+127$ and -128 bytes from the jump instruction.

Subroutine   *See* Procedure.

6–8   QUESTIONS

1. What is a short JMP?
2. What type of JMP is used when jumping anywhere in a segment?
3. Which JMP instruction allows the program to continue execution at any memory location in the system?
4. What JMP instruction is 5 bytes long?
5. What is the range of a near jump in the 80386/80486 microprocessor?
6. What can be said about a label that is followed by a colon?
7. The near jump modifies the program address by changing which register or registers?
8. The far jump modifies the program address by changing which register or registers?
9. Explain what the JMP AX instruction accomplishes. Also identify it as a near or a far jump instruction.
10. Contrast the operation of a JMP DI with a JMP [DI].
11. Contrast the operation of a JMP [DI] with a JMP FAR PTR [DI].
12. List the five flag bits tested by the conditional jump instructions.
13. Describe how the JA instruction operates.
14. When will the JO instruction jump?
15. What conditional jump instructions follow the comparison of signed numbers?
16. What conditional jump instructions follow the comparison of unsigned numbers?
17. Which conditional jump instructions test both the Z and C flag bits?
18. When does the JCXZ instruction jump?
19. Which SET instruction is used to set AL if the flag bits indicate a zero condition?
20. The 8086 LOOP instruction decrements register _____ and tests it for a 0 to decide if a jump occurs.
21. The 80486 LOOP instruction decrements register _____ and tests it for a 0 to decide if a jump occurs.
22. Explain how the LOOPE instruction operates.
23. Develop a short sequence of instructions that stores a 00H into 150H bytes of memory beginning at extra segment memory location DATA. You must use the LOOP instruction to help perform this task.

24. Develop a sequence of instructions that searches through a block of 100H bytes of memory. This program must count all the unsigned numbers that are above 42H and all that are below 42H. Byte-sized memory location UP must contain the count of numbers above 42H and location DOWN must contain the count of numbers below 42H.
25. What is a procedure?
26. Explain how the near and far CALL instructions function.
27. How does the near RET instruction function?
28. The last executable instruction in a procedure must be a _____.
29. What directive identifies the start of a procedure?
30. How is a procedure identified as near or far?
31. Explain what the RET 6 instruction accomplishes.
32. Write a near procedure that cubes the contents of the CX register. This procedure may not affect any register except CX.
33. Write a procedure that sums EAX, EBX, ECX, and EDX. If a carry occurs, place a logic 1 in EDI. If no carry occurs, place a 0 in EDI. The sum should be found in EAX after the execution of your procedure.
34. What is an interrupt?
35. What software instructions call an interrupt service procedure?
36. How many different interrupt types are available in the microprocessor?
37. What is the purpose of interrupt vector type number 0?
38. Illustrate the contents of an interrupt vector and explain the purpose of each part.
39. How does the IRET instruction differ from the RET instruction?
40. What is the IRETD instruction?
41. The INTO instruction only interrupts the program for what condition?
42. The interrupt vector for an INT 40H instruction is stored at what memory locations?
43. What instructions control the function of the INTR pin?
44. Which personal computer interrupt services the parallel LPT port?
45. Which personal computer interrupt services the keyboard?
46. What instruction tests the $\overline{\text{BUSY}}$ pin?
47. When will the BOUND instruction interrupt a program?
48. An ENTER 16,0 instruction creates a stack frame that contains _____ bytes.
49. What register moves to the stack when an ENTER instruction executes?
50. Which instruction passes opcodes to the numeric coprocessor?

CHAPTER 7

Keyboard/Display DOS and BIOS Functions

INTRODUCTION

This chapter develops programs and programming techniques using the MASM macro assembler, the DOS function calls, and the BIOS function calls. Some of the DOS and BIOS function calls are used in this chapter, but all are explained in complete detail in Appendix A. Please review the function calls as required as you read this chapter. The MASM assembler has already been explained and demonstrated in prior chapters, yet there are still many more features to learn at this point.

     Some programming techniques explained in this chapter include: macro sequences, keyboard and display manipulation, program modules, library files, and other important programming techniques. This chapter is meant as an introduction to programming, yet it provides valuable programming techniques that enable programs to be efficiently developed for the personal computer using either PCDOS* or MSDOS† as a springboard.

7–1   CHAPTER OBJECTIVES

Upon completion of this chapter, you will be able to:

1. Use the MASM assembler and linker program to create programs that contain more than one module.
2. Explain the use of EXTRN and PUBLIC as they apply to modular programming.

---

*PCDOS is a registered trademark of IBM Corporation.

†MSDOS is a registered trademark of Microsoft Corporation.

3. Set up a library file that contains commonly used subroutines.
4. Write and use MACRO and ENDM to develop macro sequences used with linear programming.
5. Develop programs using DOS function calls.
6. Differentiate a DOS function call from a BIOS function call.
7. Use modes 12H and 13H of the VGA display system.

7–2   MODULAR PROGRAMMING

Most programs are too large to be developed by one person. This means that programs are often developed by teams of programmers. The linker program is provided with MSDOS or PCDOS so programming modules can be linked together into a complete program. This section describes the linker, the linking task, library files, EXTRN, and PUBLIC as they apply to program modules and modular programming.

The Assembler and Linker

The *assembler program* converts a symbolic source module (file) into a hexadecimal object file. We have seen many examples of symbolic source files in prior chapters. Example 7–1 shows the assembler dialog that appears when a source module named FILE.ASM is assembled. Whenever you create a source file it must have an extension of ASM. Note that the extension is not typed into the assembler prompt when assembling a file. Source files are created using the editor that comes with the assembler or by almost any other editor or word processor that is capable of generating an ASCII file.

EXAMPLE 7–1

A>MASM

Microsoft (R) Macro Assembler Version 5.10
Copyright (C) Microsoft Corp 1981, 1989. All rights reserved.

Source filename [.ASM]: FILE
Object filename [FILE.OBJ]: FILE
Source listing [NUL.LST]: FILE
Cross reference [NUL.CRF]: FILE

The assembler program (MASM) asks for the source file name, the object file name, the list file name, and a cross-reference file name. In most cases the name for each of these will be the same as the source file. The object file (.OBJ) is not executable, but is designed as an input file to the linker. The source listing file (.LST) contains the assembled version of the source file and its hexadecimal machine language equivalent. The cross-reference file (.CRF) lists all labels and pertinent information required for cross referencing.

The **linker program** reads the object files, created by the assembler program, and links them together into a single execution file. An **execution file** is created with the filename extension EXE. Execution files are executed by typing the file name at the DOS prompt (A>). An example execution file is FROG.EXE that is executed by typing FROG at the DOS command prompt.

If a file is short enough, less than 64K bytes in length, it can be converted from an execution file to a **command file** (.COM). The command file is slightly different from an execution file in that the program must be originated at location 100H before it can execute. The program EXE2BIN is used for converting an execution file into a command file. The main advantage of a command file is that it loads off the disk into the computer much more quickly than an execution file. It also requires slightly less disk storage space than the execution file. If MASM version 6.0 is in use, the EXE2BIN program is not needed. The workbench editor program for version 6.0 creates the command file directly if the command file (.COM) option is selected. Workbench also builds the file by assembling it and linking it.

Example 7–2 shows the protocol involved to use the linker program to link the files FROG, WHAT, and DONUT. The linker also links the library files (LIBS) so the procedures located within LIBS can be used with the linked execution file. To invoke the linker, type LINK at the DOS command prompt as illustrated in Example 7–2. Note that before files can be linked, they must first be assembled and they must be *error free*.

EXAMPLE 7–2

A>LINK

Microsoft (R) Overlay Linker Version 3.64
Copyright (C) Microsoft Corp 1983-1988. All rights reserved.

Object Modules [.OBJ]: FROG+WHAT+DONUT
Run File [FROG.EXE]: FROG
List File [NUL.MAP]: FROG
Libraries [.LIB]: LIBS

In this example, after typing LINK, the linker program asks for the Object Modules, which are created by the assembler. In this example, we have three object modules: FROG, WHAT, and DONUT. If more than one object file exists, the main program file (FROG in this example) is typed first followed by any other supporting modules. (We use a plus sign to separate module names.)

After the program module names are typed, the linker suggests that the execution *(run-time)* file name is FROG.EXE. This may be selected by typing the same name or, if desired, by typing the enter key. It may also be changed to any other name at this point.

The list file is where a map of the program segments appears as created by the linking. If enter is typed, no list file is created, but if a name is typed, the list file appears on the disk.

Library files are entered in the last line. In this example we entered library file name LIBS. This library contains procedures used by the other program modules.

PUBLIC and EXTRN

The PUBLIC and EXTRN directives are very important to modular programming. We use PUBLIC to declare that labels of code, data, or entire segments are available to other program modules. We use EXTRN (external) to declare that labels are external to a module. Without these statements, we could not link modules together to create a program using modular programming techniques.

The PUBLIC directive is normally placed in the opcode field of an assembly language statement to define a label as public so it can be used by other modules. This label can be a jump address or a data address, or an entire segment can be made public. Example 7–3 shows the PUBLIC statement used to define some labels public to other modules. When segments are made public they are combined with other public segments that contain data with the same segment name.

EXAMPLE 7–3

```
                      DAT1     SEGMENT  PUBLIC          ;declare entire segment public

                               PUBLIC  DATA1           ;declare DATA1, DATA2 public
                               PUBLIC  DATA2

0000  0064[           DATA1    DB       100 DUP (?)     ;global
          ??
              ]
0064  0064[           DATA2    DB       100 DUP (?)     ;global
          ??
              ]

00C8                  DAT1     ENDS

0000                  CODES    SEGMENT 'CODE'

                               ASSUME  CS:CODES,DS:DAT1

                               PUBLIC  READ            ;declare READ public

0000                  READ     PROC FAR

0000  B4 06                    MOV     AH,6            ;read keyboard
0002  B2 FF                    MOV     DL,0FFH         ;no echo
0004  CD 21                    INT     21H
0006  74 F8                    JE      READ
0008  CB                       RET

0009                  READ     ENDP

0009                  CODES    ENDS

                               END
```

The EXTRN statement appears in both data and code segments to define labels as external to the segment. If data are defined as external, their size must be represented as BYTE, WORD, or DWORD. If a jump or call address is external, it must be represented as NEAR or FAR. Example 7–4 shows how the external statement is used to indicate that several labels are external to the program listed. Notice in this example that any external address or data is defined with the letter E in the hexadecimal assembled listing.

EXAMPLE 7–4

```
0000                DAT1    SEGMENT  PUBLIC              ;declare entire segment public

                            EXTRN  DATA1:BYTE
                            EXTRN  DATA2:BYTE
                            EXTRN  DATA3:WORD
                            EXTRN  DATA4:DWORD

0000                DAT1    ENDS

0000                CODES   SEGMENT 'CODE'

                            ASSUME  CS:CODES,ES:DAT1

                            EXTRN  READ:FAR

0000                MAIN    PROC  FAR

0000 B8 --- R               MOV   AX,DAT1
0003 8E C0                  MOV   ES,AX

0005 BF 0000 E              MOV   DI,OFFSET DATA1
0008 B9 000A                MOV   CX,10
000B                MAIN1:
000B 9A 0000 --- E          CALL  READ
0010 AA                     STOSB
0011 E2 F8                  LOOP  MAIN1
0013 CB                     RET

0014                MAIN    ENDP

0014                CODES   ENDS

                            END    MAIN
```

Libraries

Library files are collections of procedures that can be used by many different programs. These procedures are assembled and compiled into a library file by the LIB program that accompanies the MASM assembler program. Libraries allow common procedures to be collected into one place so they can be used by many different applications. The library file (FILENAME.LIB) is invoked when a program is linked with the linker program.

Why bother with library files? A library file is a good place to store a collection of related procedures. When the library file is linked with a program, only the procedures required by the program are removed from the library file and added to the program. If any amount of assembly language programming is to be accomplished efficiently, a good set of library files is essential.

Creating a Library File.   A library file is created with the LIB command typed at the DOS prompt. A library file is a collection of assembled .OBJ files that each perform one procedure. Example 7–5 shows two separate files (READ_KEY and ECHO) that will be used to structure a library file. Please notice that the name of the procedure must be declared PUBLIC in a library file and does not necessarily need to match the file name, although it does in this example.

EXAMPLE 7–5

```
                    ;The first library module is called READ_KEY.
                    ;This procedure reads a key from the keyboard
                    ;and returns with the ASCII character in AL.
                    ;
0000                LIB        SEGMENT 'CODE'

                               ASSUME  CS:LIB

                               PUBLIC  READ_KEY

0000                READ_KEY       PROC    FAR

0000 52                            PUSH    DX
0001                READ_KEY1:
0001 B4 06                         MOV     AH,6
0003 B2 0F                         MOV     DL,0FH
0005 CD 21                         INT     21H
0007 74 F8                         JE      READ_KEY1
0009 5A                            POP     DX
000A CB                            RET
000B                READ_KEY       ENDP

000B                LIB        ENDS

                               END
                    ;
                    ;This second library module is called ECHO
                    ;This procedure displays the ASCII character
                    ;in AL on the CRT screen.
                    ;
0000                LIB        SEGMENT 'CODE'

                               ASSUME  CS:LIB

                               PUBLIC  ECHO

0000                ECHO           PROC    FAR
0000 52                            PUSH    DX
0001 B4 06                         MOV     AH,6
0003 8A D0                         MOV     DL,AL
0005 CD 21                         INT     21H
0007 5A                            POP     DX
0008 CB                            RET
0009                ECHO           ENDP
```

0009   LIB   ENDS

                                    END

After each file is assembled, the LIB program is used to combine them into a library file. The LIB program prompts for information as illustrated in Example 7–6, where these files are combined to form the library IO.

EXAMPLE 7–6

A>LIB

Microsoft (R) Library Manager Version 3.10
Copyright (C) Microsoft Corp 1983-1988. All rights reserved.

Library name:IO
Library file does not exist. Create? Y
Operations:READ_KEY+ECHO
List file:IO

The LIB program begins with the copyright message from Microsoft, followed by the prompt *Library name:*. The library name chosen is IO for the IO.LIB file. Because this is a new file, the library program asks if we wish to create the library file. The *Operations:* prompt is where the library module names are typed. In this case we created a library using two procedure files (READ_KEY and ECHO). The list file shows the contents of the library; it is illustrated for this library in Example 7–7. The list file shows the size and names of the files used to create the library and also the public label (procedure name) that is used in the library file.

EXAMPLE 7–7

ECHO..............ECHO   READ_KEY..........READ_KEY

READ_KEY   Offset: 00000010H Code and data size: BH
  READ_KEY

ECHO   Offset: 00000070H Code and data size: 9H
  ECHO

If you must add additional library modules at a later time, type the name of the library file after invoking LIB. At the *Operations:* prompt, type the new module name preceded with a *plus sign* to add a new procedure. If you must delete a library module, use a *minus sign* before the operation file name.

Once the library file is linked to your program file only the library procedures actually used by your program are placed in the execution file. Don't forget to use the label EXTRN when specifying library calls from your program module.

Macros

A *macro* is a group of instructions that perform one task just as a procedure performs one task. The difference is that a procedure is accessed via a CALL instruction, while a macro is inserted in the program as a new opcode representing a sequence of instructions. A macro is a new opcode that you create. Macro sequences execute faster than procedures because there is no CALL and RET instruction to execute. The instructions of the macro sequence are placed in your program at the point where they are invoked. When software is developed using macro sequences it flows from top-down.

Macro sequences are ideal for systems that contain cache memory or systems that are required to execute software with maximum speed and efficiency. Programs that use macro sequences in place of procedures are often called *linear programs* because they flow from the top to the bottom without any calls. If a cache memory is used in the computer system, it is often desirable to develop programs that use macro sequences because the entire program can often be executed from the cache, significantly increasing its execution speed. If a procedure is called, the cache must often be reloaded, requiring additional time. If a macro is invoked, no additional time is needed because no jump to the procedure occurs. Cache memory looks ahead and often loads instructions before they are executed by the microprocessor. It does this by loading the cache with the next sequential section of four 32-bit double words. If a procedure is called, it is not in the next section of memory, which requires another section of the memory to be cached. This requires additional time on the part of the cache.

The MACRO and ENDM directives are used to delineate a macro sequence. The first statement of a macro is the MACRO statement that contains the name of the macro and any parameters associated with it. An example is MOVE MACRO A,B that defines the macro as MOVE. This new opcode uses two parameters, A and B. The last statement of a macro is the ENDM instruction on a line by itself without a label.

Example 7–8 shows how a macro is created and used in a program. This macro moves the word-sized contents of memory location B into word-sized memory location A. After the macro is defined in the example, it is used twice. The macro is *expanded* in this example so you can see how it assembles to generate the moves. A hexadecimal machine language statement followed by a 1 is a macro expansion statement. If a macro uses a second macro (nesting), the statement is followed by a 2, and so forth. The expansion statements were not typed in the source program. Notice that the comment in the macro is preceded with a ;; instead of ; as is customary.

EXAMPLE 7–8

```
MOVE     MACRO   A,B                    ;;moves word from B to A

         PUSH    AX
         MOV     AX,B
```

```
                                    MOV      A,AX
                                    POP      AX

                                    ENDM

                                    MOVE     VAR1,VAR2           ;use macro MOVE

0000 50            1                PUSH     AX
0001 A1 0002 R     1                MOV      AX,VAR2
0004 A3 0000 R     1                MOV      VAR1,AX
0007 58            1                POP      AX

                                    MOVE     VAR3,VAR4           ;use macro MOVE

0008 50            1                PUSH     AX
0009 A1 0006 R     1                MOV      AX,VAR4
000C A3 0004 R     1                MOV      VAR3,AX
000F 58            1                POP      AX
```

Local Variable in a Macro. Sometimes macros must contain local variables. A *local variable* is one that appears in the macro, but is not available outside the macro. To define a local variable, we use the LOCAL directive. Example 7–9 shows how a local variable, used as a jump address, appears in a macro definition. If this jump address is not defined as local, the assembler will flag it as a duplicate label error on the second and subsequent attempts to use the macro in a program.

EXAMPLE 7–9

```
                     READ    MACRO    A                  ;;reads keyboard
                             LOCAL    READ1              ;;define READ1 as local

                             PUSH     DX
                     READ1:
                             MOV      AH,6
                             MOV      DL,0FFH
                             INT      21H
                             JE       READ1
                             MOV      A,AL
                             POP      DX

                             ENDM

                             READ     VAR5               ;read key

0000 52          1           PUSH     DX
0001             1   ??0000:
0001 B4 06       1           MOV      AH,6
0003 B2 FF       1           MOV      DL,0FFH
0005 CD 21       1           INT      21H
0007 74 F8       1           JE       ??0000
0009 A2 0008 R   1           MOV      VAR5,AL
000C 5A          1           POP      DX

                             READ     VAR6               ;read key
```

```
000D 52          1              PUSH    DX
000E             1    ??0001:
000E B4 06       1              MOV     AH,6
0010 B2 FF       1              MOV     DL,0FFH
0012 CD 21       1              INT     21H
0014 74 F8       1              JE      ??0001
0016 A2 0009 R   1              MOV     VAR6,AL
0019 5A          1              POP     DX
```

This example reads a character from the keyboard and stores it into the byte-sized memory location indicated as a parameter with the macro. Notice how the local label READ1 is treated in the expanded macros. Local variables always appear as ??*nnnn,* where *nnnn* is a decimal number identifying it.

The LOCAL directive must always immediately follow the MACRO directive without any intervening blank lines or comments. If a comment or blank line appears between the MACRO and LOCAL statements, the assembler indicates an error and will not accept the variable as local.

Placing MACRO Definitions in Their Own Module.   Macro definitions can be placed in the program file as shown, or can be placed in their own macro module. A file can be created that contains only macros that are to be included with other program files. We use the INCLUDE directive to indicate that a program file will include a module that contains external macro definitions. Although this is not a library file, it for all practical purposes functions as a library of macro sequences. A macro include file is an ASCII file generated by an editor or word processor.

When macro sequences are placed in a file (often with the extension INC or MAC), they do not contain PUBLIC statements. If a file called MACRO.MAC contains macro sequences, the include statement is placed in the program file as INCLUDE C:\ASSM\MACRO.MAC. Notice that the macro file is on drive C, subdirectory ASSM in this example. The INCLUDE statement includes these macros just as if you had typed them into the file. No EXTRN statement is needed to access the macro statements that have been included.

The Modular Programming Approach

The modular programming approach often involves a team of people with each assigned to a different programming task. This allows the team manager to assign various portions of the program to different team members. Often the team manager develops the system flowchart or shell and then divides it into modules for team members.

A team member might be assigned the task of developing a macro definition file. This file might contain macro definitions that handle the I/O operations for the system. Another team member might be assigned the task of developing the procedures used for the system. In most cases the procedures are organized as a library file that is linked to the program modules. Finally several program files or modules might be used for the final system, each developed by different team members.

This approach requires clear communication between team members, as well as good documentation. Documentation is the key to making modules interface correctly. Also communication between members plays a key role in this approach.

7–3   USING THE KEYBOARD AND VIDEO DISPLAY

Today most programs make use of the keyboard and video display on a personal computer. This section of the text explains how to use the keyboard and video display connected to the IBM PC or compatible computer running under either MSDOS or PCDOS.

Reading the Keyboard with DOS Functions

The keyboard, in the personal computer, is often read via a DOS function call. A complete listing of the DOS function calls appears in Appendix A. This section uses INT 21H with various DOS function calls to read the keyboard. Data read from the keyboard is either in ASCII coded form or in extended ASCII coded form. The exact form depends on which keys are typed on the keyboard.

The ASCII coded data appear as outlined in Tables 1–3 and 1–4 in Chapter 1. Notice that these codes correspond to most of the keys on the keyboard. The extended ASCII codes listed in Table 1–4 apply to the printer or video screen. Also available through the keyboard are a different set of extended ASCII coded *keyboard* data. Table 7–1 lists most of the extended ASCII codes obtained with various keys and key combinations. These are different from the extended ASCII printer/video display codes in Chapter 1. Notice that most keys on the keyboard have alternative key codes. The function keys have four sets of codes selected by the function keys, the shift function keys, alternate function keys, and the control function keys.

There are three ways to read the keyboard. The first method reads a key and echoes (or displays) the key on the video screen. A second way just tests to see if a key is pressed, and if it is pressed it reads the key; otherwise it returns without any key. The third way allows an entire character line to be read from the keyboard.

Reading a Key with an Echo.   Example 7–10 shows how a key is read from the keyboard and *echoed* (sent) back out to the video display. Although this is the easiest way to read a key, it is also the most limited because it always echoes (displays) the character to the screen even if it is an unwanted character. The DOS function number 01H also responds to the control C key and exits the program to DOS if it is typed.

EXAMPLE 7–10

```
0000                    KEY     PROC    FAR

0000 B4 01                      MOV     AH,1            ;function 01H
0002 CD 21                      INT     21H             ;read key
0004 0A C0                      OR      AL,AL           ;test for 00H, clear carry
0006 75 03                      JNZ     KEY1
```

TABLE 7–1 Extended ASCII-coded keyboard data

Second	0	1	2	3	4	5	6	7	8
					First				
0	—	aQ	aD	aB	—	down arrow	cF3	aF9	a9
1	aESC	aW	aF	aN	—	page down	cF4	aF10	a0
2	—	aE	aG	aM	—	insert	cF5	—	a-
3	c2	aR	aH	a,	—	delete	cF6	—	a=
4	—	aT	aJ	a.	—	sF1	cF7	—	—
5	—	aY	aK	a/	—	sF2	cF8	—	F11
6	—	aU	aL	—	—	sF3	cF9	—	F12
7	—	aI	a;	a*	home	sF4	cF10	—	sF11
8	—	aO	a'	—	up arrow	sF5	aF1	a1	sF12
9	—	aP	a`	—	page up	sF6	aF2	a2	cF11
A	—	a[	—	a-	—	sF7	aF3	a3	cF12
B	—	a]	a\	—	left arrow	sF8	aF4	a4	aF11
C	—	aENT	aZ	—	—	sF9	aF5	a5	aF12
D	—	—	aX	—	right arrows	sF10	aF6	a6	—
E	aBS	aA	aC	—	a+	cF1	aF7	a7	—
F	sTAB	aS	—	—	end key	cF2	aF8	a8	—

Notes: a = alternate key, c = control key, and s = shift key.

216

```
0008 CD 21                     INT      21H              ;get extended
000A F9                        STC                       ;indicate extended
000B                 KEY1:
000B CB                        RET

000C                 KEY       ENDP
```

To read and echo a character, the AH register is loaded with DOS function number 01H. This is followed by the INT 21H instruction. (All DOS functions use AH to hold the function number before the INT 21H DOS function call.) Upon return from the INT 21H, the AL register contains the ASCII character typed and the video display also shows the typed character. The return from this DOS function call does not occur until a key is typed. If AL = 0 after the return, the INT 21H instruction must again be executed to obtain the extended ASCII coded character. This procedure in Example 7–10 returns with carry set (1) to indicate an extended ASCII character and carry cleared (0) to indicate a normal ASCII character.

Reading a Key with No Echo. The best single-character key-reading function is function number 06H. This function reads a key without an echo to the screen. It also returns with extended ASCII characters and *does not* respond to the control C key. This function uses AH for the function number (06H) and DL = 0FFH to indicate that the function call (INT 21H) will read the keyboard without an echo.

Example 7–11 shows a procedure that uses function number 06H to read the keyboard. This performs as Example 7–10 except that no character is echoed to the video display.

EXAMPLE 7–11

```
0000                 KEYS      PROC     FAR

0000 B4 06                     MOV      AH,6             ;function 06H
0002 B2 FF                     MOV      DL,0FFH
0004 CD 21                     INT      21H              ;read key
0006 74 F8                     JE       KEYS             ;if no key
0008 0A C0                     OR       AL,AL            ;test for 00H, clear carry
000A 75 03                     JNE      KEYS1
000C CD 21                     INT      21H              ;get extended
000E F9                        STC                       ;indicate extended
000F                 KEYS1:
000F CB                        RET

0010                 KEYS      ENDP
```

If you examine this procedure, there is one other difference from the procedure in Example 7–10. Function call number 06H returns from the INT 21H even if no key is typed, while function call 01H waits for a key to be typed. This is an important difference that should be noted. This feature allows software to perform other tasks between checking the keyboard for a character.

Read an Entire Line with Echo.   Sometimes it is advantageous to read an entire line of data with one function call. Function call number 0AH reads an entire line of information—up to 255 characters—from the keyboard. It continues to acquire keyboard data until the enter key (0DH) is typed. This function requires that AH = 0AH and DS:DX address the keyboard buffer (a memory area where the ASCII data are stored). The first byte of the buffer area contains the maximum number of keyboard characters read by this function. If the number typed exceeds this maximum number, the DOS function returns just as if the enter key were typed. The second byte of the buffer contains the count of the actual number of characters typed and the remaining locations in the buffer contain the ASCII keyboard data.

Example 7–12 shows how this function reads 2 lines of information into 2 memory buffers (BUF1 and BUF2). Before the call to the DOS function through procedure LINE, the first byte of the buffer is loaded with a 255, so that up to 255 characters can by typed. If you assemble and execute this program, the first line is accepted and so is the second. The only problem is that the second line appears on top of the first line. The next section of the text explains how to output characters to the video display to solve this problem and also display memory data on the video screen.

EXAMPLE 7–12

```
0000                    COD     SEGMENT 'CODE'

                                ASSUME  CS:COD,DS:COD

0000 0101[              BUF1    DB      257 DUP (?)
            ????
                 ]
0102 0101[              BUF2    DB      257 DUP (?)
            ????
                 ]

0204            MAIN    PROC    FAR

0204 8C C8              MOV     AX,CS
0206 8E D8              MOV     DS,AX

0208 BA 0000 R          MOV     DX,OFFSET BUF1      ;address buffer 1
020B C7 06 0000 R 00FF  MOV     BUF1,255            ;maximum count
0211 E8 0221 R          CALL    LINE                ;read first line

0214 BA 0102 R          MOV     DX,OFFSET BUF2      ;address buffer 2
0217 C7 06 0102 R 00FF  MOV     BUF2,255            ;maximum count
021D E8 0221 R          CALL    LINE                ;read second line

0220 CB                 RET

0221            MAIN    ENDP

0221            LINE    PROC    NEAR

0221 B4 0A              MOV     AH,0AH              ;function 0AH
0223 CD 21              INT     21H
0225 C3                 RET
```

0226		LINE	ENDP	
0226		COD	ENDS	
		END	MAIN	

Writing to the Video Display with DOS Functions

For almost any program written, data must be displayed on the video display. Video data are displayed in a number of different ways with DOS function calls. We use function 02H or 06H to display one character at a time or function 09H to display an entire string of characters. Because function 02H and 06H are identical, we tend to use function 06H because it is also used to read a key.

Displaying One ASCII Character. Both DOS functions 02H and 06H are explained together because they are identical for displaying ASCII data. Example 7–13 shows how this function displays a carriage return (0DH) and a line feed (0AH). Here a macro, called DISP (display), displays the carriage return and line feed. The combination of a carriage return and a line feed moves the cursor to the next line at the left margin of the video screen. This two-step process corrects the problem that occurred between the lines typed through the keyboard in Example 7–12.

EXAMPLE 7–13

```
                        DISP    MACRO   A               ;display A

                                MOV     AH,06H
                                MOV     DL,A
                                INT     21H

                                ENDM

                        DISP    0DH                     ;carriage return

0000 B4 06      1               MOV     AH,06H
0002 B2 0D      1               MOV     DL,0DH
0004 CD 21      1               INT     21H

                        DISP    0AH                     ;line feed

0006 B4 06      1               MOV     AH,06H
0008 B2 0A      1               MOV     DL,0AH
000A CD 21      1               INT     21H
```

Display a Character String. A character string is a series of ASCII coded characters that ends with a $ (24H) when used with DOS function call number 09H. Example 7–14 shows how a message is displayed at the current cursor position on the video display. Function call number 09H requires that DS:DX address the character string before executing the INT 21H.

EXAMPLE 7–14

```
0000                    COD     SEGMENT 'CODE'

                                ASSUME  CS:COD,DS:COD

0000 0D 0A 0A 54 68 69  MES     DB        0DH,0AH,0AH,'This is a test line','$'
     73 20 69 73 20 61
     20 74 65 73 74 20
     6C 69 6E 65 24

0017                    MAIN    PROC      FAR

0017 8C C8                      MOV       AX,CS               ;load DS
0019 8E D8                      MOV       DS,AX

001B B4 09                      MOV       AH,09H              ;function 09H
001D BA 0000 R                  MOV       DX,OFFSET MES
0020 CD 21                      INT       21H

0022 B4 4C                      MOV       AH,4CH              ;function 4CH
0024 CD 21                      INT       21H

0026                    MAIN    ENDP


0026                    COD     ENDS

                                END       MAIN
```

This example program can be entered into the assembler, linked, and executed to produce "This is a test line" on the video display. Notice that an additional DOS function call is appended to the end of this program. Function call 4CH returns the system to the DOS prompt at the end of the program. We often use function number 4CH to return to DOS. We can also use function 00H, but this function has no provision for errors and is considered obsolete. The AL register contains a 00H for no-error and other values that indicate errors when a program is terminated with DOS function number 4CH.

Consolidated Read Key and Echo Library Procedures.   Example 7–15 illustrates two procedures that could be assembled, linked, and added to a library file. The READ procedure reads a keyboard character and returns with either the ASCII or extended ASCII character in AL. If carry is set, AL contains the extended ASCII code and if carry is cleared, it contains the standard ASCII code. The ECHO procedure displays the ASCII coded character located in AL at the current cursor position.

EXAMPLE 7–15

```
0000                    LIB     SEGMENT 'CODE'

                                ASSUME  CS:LIB
```

```
                                      PUBLIC  READ
                                      PUBLIC  ECHO
                            ;
                            ;procedure that reads a key from the keyboard (no echo)
                            ;if CF = 0, AL = standard ASCII character
                            ;if CF = 1, AL = extended ASCII character
                            ;
0000                        READ    PROC    FAR

0000 52                             PUSH    DX              ;save DX
0001 B4 06                          MOV     AH,6            ;DOS function 06H
0003 B2 FF                          MOV     DL,0FFH
0005                        READ1:
0005 CD 21                          INT     21H
0007 74 FC                          JE      READ1           ;if no key
0009 0A C0                          OR      AL,AL           ;test for extended
000B 75 03                          JNZ     READ2           ;if standard ASCII

000D CD 21                          INT     21H             ;get extended
000F F9                             STC                     ;set carry
0010                        READ2:
0010 5A                             POP     DX
0011 CB                             RET

0012                        READ    ENDP
                            ;
                            ;procedure that displays the ASCII character in AL
                            ;
0012                        ECHO    PROC    FAR

0012 52                             PUSH    DX
0013 B4 06                          MOV     AH,6            ;DOS function 06H
0015 8A D0                          MOV     DL,AL           ;AL to DL
0017 CD 21                          INT     21H
0019 5A                             POP     DX
001A CB                             RET

001B                        ECHO    ENDP

001B                        LIB     ENDS

                                    END
```

Using the Video BIOS Functions Calls

In addition to the DOS function call INT 21H, we also have video BIOS *(basic I/O system)* function calls at INT 10H. The DOS function calls allow a key to be read and a character to be displayed with ease, but the cursor is difficult to position at the desired screen location. The video BIOS function calls allow more complete control over the video display than do the DOS function calls. The video BIOS function calls also require less time to execute than the DOS function calls.

Cursor Position.   Before any information is placed on the video screen, the position of the cursor should be known. This allows the screen to be cleared and the displayed information to be placed at any location on the video screen. The BIOS INT 10H function number 03H allows the cursor position to be read from the video interface.

TABLE 7–2   BIOS function
INT 10H

AH	Description	Parameters
02H	Sets cursor position	DH = Row
		DL = Column
		BH = Page number
03H	Reads cursor position	BH = Page number
		DH = Row
		DL = Column

The BIOS INT 10H function number 02H allows the cursor to be placed at any screen position. Table 7–2 shows the contents of various registers for both video BIOS INT 10H function calls 02H and 03H.

The page number, in register BH, should be 0 before setting the cursor position. Most software does not access the other pages (1–7) for the video display. If the display adapter is a VGA display, always use page 0. The page number is often ignored after a cursor read. The 0 page is available in the **CGA** *(color graphics adapter),* **EGA** *(enhanced graphics adapter),* and **VGA** *(variable graphics array)* text modes of operation. The other pages are available in some VGA and EGA modes and all CGA modes of operation.

The cursor position assumes that the left-hand page column is column 0 progressing across a line to column 79. The row number corresponds to the character line number on the screen. Row 0 is the uppermost line while row 24 is the last line on the screen. This assumes that the text mode selected for the video adapter is 80 characters per line by 25 lines (80 × 25). Other text modes are available, such as 40 × 25 and 96 × 43.

Example 7–16 shows how the INT 10H BIOS function call is used to clear the video screen. This is just one method of clearing the screen. Notice that the first function call positions the cursor to row 0 and column 0, which is called the **home** position. Next we use the DOS function call to write 2000 (80 characters per line × 25 character lines) blank spaces (20H) on the video display, then we again home the cursor.

EXAMPLE 7–16

```
0000                    CODE    SEGMENT 'CODE'

                        ASSUME  CS:CODE

0000            MAIN    PROC    FAR

                HOME    MACRO

                        MOV     AH,2        ;;set cursor position
                        MOV     BH,0        ;;page 0
                        MOV     DX,0        ;;row 0, column 0
                        INT     10H

                        ENDM
```

```
                              HOME
0000 B4 02        1           MOV     AH,2
0002 B7 00        1           MOV     BH,0
0004 BA 0000      1           MOV     DX,0
0007 CD 10        1           INT     10H

0009 B9 07D0                  MOV     CX,2000
000C B4 06                    MOV     AH,6
000E B2 20                    MOV     DL,' '          ;space
0010              MAIN1:
0010 CD 21                    INT     21H             ;display space
0012 E2 FC                    LOOP    MAIN1           ;repeat 1920 times

                              HOME
0014 B4 02        1           MOV     AH,2
0016 B7 00        1           MOV     BH,0
0018 BA 0000      1           MOV     DX,0
001B CD 10        1           INT     10H

001D B4 4C                    MOV     AH,4CH          ;exit to DOS
001F CD 21                    INT     21H

0021              MAIN        ENDP

0021              CODE        ENDS

                              END     MAIN
```

If this example is assembled, linked, and executed, a problem surfaces. This program is far too slow to be useful in most cases. To correct this situation, another video BIOS function call is used. We can use the scroll function (06H) to clear the screen at a much higher speed.

Function 06H is used with a 00H in AL to blank the entire screen. This allows Example 7–16 to be rewritten so that the screen clears at a much higher speed. See Example 7–17 for a better clear and home cursor program. Here function call number 08H reads the character attributes for blanking the screen. Next, they are positioned in the correct registers and DX is loaded with the screen size, 4FH (79) and 19H (25). If this program is assembled, linked, executed, and compared with Example 7–16 there is a big difference in the speed at which the screen is cleared. Please refer to Appendix A for other video BIOS INT 10H function calls that may prove useful in your applications. Also listed in Appendix A are a complete listing of all the INT functions available in most computers.

EXAMPLE 7–17

```
0000              CODE        SEGMENT 'CODE'

                              ASSUME  CS:CODE

0000              MAIN        PROC    FAR

                  HOME        MACRO
```

```
                              MOV    AH,2           ;;set cursor position
                              MOV    BH,0           ;;page 0
                              MOV    DX,0           ;;row 0, column 0
                              INT    10H

                              ENDM

0000 32 FF                    XOR    BH,BH          ;page 0
0002 B4 08                    MOV    AH,8           ;read attributes
0004 CD 10                    INT    10H

0006 8A DF                    MOV    BL,BH
0008 8A FC                    MOV    BH,AH
000A 2B C9                    SUB    CX,CX
000C BA 194F                  MOV    DX,194FH
000F B8 0600                  MOV    AX,0600H       ;clear page 0
0012 CD 10                    INT    10H

                              HOME
0014 B4 02         1          MOV    AH,2
0016 B7 00         1          MOV    BH,0
0018 BA 0000       1          MOV    DX,0
001B CD 10         1          INT    10H

001D B4 4C                    MOV    AH,4CH         ;exit to DOS
001F CD 21                    INT    21H

0021              MAIN   ENDP

0021              CODE   ENDS

                         END    MAIN
```

Direct Memory Access for the Video Text Screen

Although probably the least preferred, the quickest method of accessing the video text display is a direct access to the video text memory. The video memory for text modes 0–3 (the most commonly used modes) begins at memory location B800:0000. Each video display text character is stored in two memory locations. The first location contains the ASCII coded character, and the second contains the video display attribute of the character. The problem with direct access is that the cursor does not move, but it is useful if the screen data must be saved and reloaded without affecting the cursor.

The attribute byte is illustrated in Appendix A along with the video BIOS INT 10H function calls. Also included in the appendix is a list of the foreground and background colors that apply to the attribute byte. To illustrate the way that the direct access to video text memory functions, Example 7–18 is included here. This example sets the video mode to 3, using an INT 10H mode set function. It then proceeds to blank the display by directly accessing the video memory. After the screen is blanked it displays a message across the first line of the screen using a bright-green, blinking text on a red background to illustrate the use of the background color and blink option found in the attribute byte.

EXAMPLE 7–18

```
0000                    DATA      SEGMENT

0000 54 68 69 73 20 69 MES1  DB    'This is a test message.',0
     73 20 61 20 74 65
     73 74 20 6D 65 73
     73 61 67 65 2E 00

0018                    DATA      ENDS

0000                    CODE      SEGMENT 'CODE'
                                  ASSUME  CS:CODE,DS:DATA

0000                    MAIN      PROC      FAR

0000 B8 -- R                     MOV       AX,DATA        ;address data segment
0003 8E D8                       MOV       DS,AX
0005 FC                          CLD                      ;select autoincrement

0006 B8 0003                     MOV       AX,0003H       ;select video mode 3
0009 CD 10                       INT       10H            ;this blanks the screen

000B B4 CA                       MOV       AH,11001010B   ;green text, red background
000D BE 0000 R                   MOV       SI,OFFSET MES1

0010 BB B800                     MOV       BX,0B800H      ;address video memory
0013 8E C3                       MOV       ES,BX
0015 33 FF                       XOR       DI,DI
0017            MAIN1:                                    ;display MES1
0017 AC                          LODSB                    ;get ASCII
0018 0A C0                       OR        AL,AL
001A 74 03                       JZ        MAIN2          ;if finished
001C AB                          STOSW                    ;store ASCII and attribute
001D EB F8                       JMP       MAIN1
001F            MAIN2:
001F B4 06                       MOV       AH,6           ;wait for any key
0021 B2 FF                       MOV       DL,0FFH
0023 CD 21                       INT       21H
0025 74 F8                       JZ        MAIN2

0027 B8 4C00                     MOV       AX,4C00H       ;exit to DOS
002A CD 21                       INT       21H

002C           MAIN      ENDP

002C           CODE      ENDS

                         END      MAIN
```

7–4 DATA CONVERSIONS

In computer systems, data is seldom in the correct form. One main task of the system is to convert data from one form to another. This section of the chapter describes conversions between binary and ASCII. Binary data are removed from a register or memory and converted to ASCII for the video display. In many cases, ASCII data are converted to binary as they are typed on the keyboard. We also explain converting between ASCII and hexadecimal data.

Converting from Binary to ASCII

Conversion from binary to ASCII is accomplished in two ways: (a) by the AAM instruction if the number is less than 100, or (b) by a series of decimal divisions (divide by 10). Both techniques are presented in this section.

The AAM instruction converts the value in AX into a two-digit unpacked BCD number in AX. If the number in AX is 0062H (98 decimal) before AAM executes, AX contains a 0908H after AAM executes. This is not ASCII code, but it is converted to ASCII code by adding a 3030H to AX. Example 7–19 illustrates a procedure that processes the binary value in AL (0–99) and displays it on the video screen as decimal. This procedure blanks a leading zero, which occurs for the numbers 0–9, with an ASCII space code.

EXAMPLE 7–19

```
0000                    DISP    PROC    FAR

0000 52                         PUSH    DX          ;save DX
0001 32 E4                      XOR     AH,AH       ;blank AH
0003 D4 0A                      AAM                 ;convert to BCD
0005 80 C4 20                   ADD     AH,20H      ;add 20H
0008 80 FC 20                   CMP     AH,20H      ;test for leading zero
000B 74 03                      JE      DISP1       ;if leading zero
000D 80 C4 10                   ADD     AH,10H      ;convert to ASCII
0010                    DISP1:
0010 50                         PUSH    AX
0011 8A D4                      MOV     DL,AH       ;display first digit
0013 B4 06                      MOV     AH,6
0015 CD 21                      INT     21H
0017 58                         POP     AX
0018 04 30                      ADD     AL,30H      ;convert to ASCII
001A 8A D0                      MOV     DL,AL
001C B4 06                      MOV     AH,6        ;display second digit
001E CD 21                      INT     21H
0020 5A                         POP     DX          ;restore DX
0021 CB                         RET

0022                    DISP    ENDP
```

The reason that AAM converts any number between 0 and 99 to a two-digit unpacked BCD number is because it divides AX by 10. The result is left in AX; AH contains the quotient and AL the remainder. This same scheme of dividing by 10 can be expanded to convert any whole number from binary to an ASCII coded character string that can be displayed on the video screen. The algorithm for converting from binary to ASCII is:

1. Divide by 10 and save the remainder on the stack as a significant BCD digit.
2. Repeat step 1 until the quotient is a 0.

3. Retrieve each remainder and add a 30H to convert to ASCII before displaying or printing.

Example 7–20 shows how the unsigned 16-bit contents of AX is converted to ASCII and displayed on the video screen as an unsigned integer. Here we divide AX by 10 and save the remainder on the stack after each division for later conversion to ASCII. The reason that data are stored on the stack is because the least significant digit is returned first by the division. The stack is used to reverse the order of the data so it can be displayed correctly from the most significant digit to the least. After all the digits have been converted by division, the result is displayed on the video screen by removing the remainders from the stack and converting them to ASCII code. This procedure also blanks any leading zeros that occur.

EXAMPLE 7–20

```
0000                    DISPX   PROC    FAR

0000 52                         PUSH    DX          ;save BX, CX, and DX
0001 51                         PUSH    CX
0002 53                         PUSH    BX

0003 33 C9                      XOR     CX,CX       ;clear CX
0005 BB 000A                    MOV     BX,10       ;load 10
0008            DISPX1:
0008 33 D2                      XOR     DX,DX       ;clear DX
000A F7 F3                      DIV     BX
000C 52                         PUSH    DX          ;save remainder
000D 41                         INC     CX          ;count remainder
000E 0B C0                      OR      AX,AX       ;test quotient
0010 75 F6                      JNZ     DISPX1      ;if not zero
0012            DISPX2:
0012 5A                         POP     DX          ;display number
0013 B4 06                      MOV     AH,6
0015 80 C2 30                   ADD     DL,30H      ;convert to ASCII
0018 CD 21                      INT     21H
001A E2 F6                      LOOP    DISPX2      ;repeat

001C 5B                         POP     BX          ;restore BX, CX, and DX
001D 59                         POP     CX
001E 5A                         POP     DX
001F CB                         RET

0020                    DISPX   ENDP
```

Converting from ASCII to Binary

Conversions from ASCII to binary usually start with keyboard data entry. If a single key is typed, the conversion is accomplished by subtracting a 30H from the number. If more than one key is typed, conversion from ASCII to binary still requires that 30H be subtracted, but there is one additional step. After subtracting 30H, the number is added to the result after the prior result is first multiplied by a 10. The algorithm used to convert ASCII to binary is:

1. Begin with a binary result of 0.
2. Subtract 30H from the character typed on the keyboard to convert it to BCD.

3. Multiply the binary result by 10 and add the new BCD digit.
4. Repeat steps 2 and 3 until the character typed is not an ASCII coded number of 30H–39H.

Example 7–21 illustrates a procedure that implements the ASCII-to-binary conversion algorithm. Here the binary number returns in the AX register as a 16-bit result. If a larger result is required, the procedure must be reworked for 32-bit arithmetic. Each time this procedure is called, it reads a number from the keyboard until any key other than 0 through 9 is typed. It then returns with the binary equivalent in the AX register.

EXAMPLE 7–21

```
0000                    READN   PROC    FAR

0000 53                         PUSH    BX              ;save BX and CX
0001 51                         PUSH    CX
0002 B9 000A                    MOV     CX,10           ;load 10
0005 33 DB                      XOR     BX,BX           ;clear result
0007            READN1:
0007 B4 06                      MOV     AH,6            ;read key
0009 B2 FF                      MOV     DL,0FFH
000B CD 21                      INT     21H
000D 74 F8                      JE      READN1          ;wait for key

000F 3C 30                      CMP     AL,'0'          ;test against 0
0011 72 18                      JB      READN2          ;if below 0
0013 3C 39                      CMP     AL,'9'          ;test against 9
0015 77 14                      JA      READN2          ;if above 9

0017 8A D0                      MOV     DL,AL           ;echo key
0019 CD 21                      INT     21H

001B 2C 30                      SUB     AL,'0'          ;convert to BCD

001D 50                         PUSH    AX
001E 8B C3                      MOV     AX,BX           ;multiply by 10
0020 F7 E1                      MUL     CX
0022 8B D8                      MOV     BX,AX           ;save product
0024 58                         POP     AX
0025 32 E4                      XOR     AH,AH
0027 03 D8                      ADD     BX,AX           ;add BCD to product
0029 EB DC                      JMP     READN1          ;repeat
002B           READN2:
002B 8B C3                      MOV     AX,BX           ;move binary to AX
002D 59                         POP     CX              ;restore BX and CX
002E 5B                         POP     BX
002F CB                         RET

0030                    READN   ENDP
```

Displaying and Reading Hexadecimal Data

Hexadecimal data is easier to read from the keyboard and display than decimal data. Hexadecimal data is not used at the applications level, but at the system level.

System level data is often hexadecimal and must be either displayed in hexadecimal form or read from the keyboard as hexadecimal data.

Reading Hexadecimal Data. Hexadecimal data appears as 0 to 9 and A to F. The ASCII codes obtained from the keyboard for hexadecimal data are 30H to 39H for the numbers 0 through 9 and 41H to 46H (A–F) or 61H to 66H (a–f) for the letters. To be useful, a procedure that reads hexadecimal data must be able to accept both lowercase (a–f) and uppercase (A–F) letters.

Example 7–22 shows two procedures: one converts the contents of the data in AL from an ASCII-coded character to a single hexadecimal digit, while the other reads a 4-digit hexadecimal number from the keyboard and returns with it in register AX. The second procedure can be modified to read any size hexadecimal number from the keyboard. Notice how the lowercase letters are converted to uppercase (subtract 20H) in the first procedure.

EXAMPLE 7–22

```
0000                    CONV    PROC    NEAR

0000 3C 39                      CMP     AL,'9'
0002 76 08                      JBE     CONV2           ;if a number
0004 3C 61                      CMP     AL,'a'
0006 72 02                      JB      CONV1           ;if uppercase
0008 2C 20                      SUB     AL,20H          ;convert to uppercase
000A                    CONV1:
000A 2C 07                      SUB     AL,7
000C                    CONV2:
000C 2C 30                      SUB     AL,'0'
000E C3                         RET

000F                    CONV    ENDP

000F                    READH   PROC    FAR

000F 51                         PUSH    CX              ;save BX and CX
0010 53                         PUSH    BX
0011 B9 0004                    MOV     CX,4            ;load shift count
0014 8B F1                      MOV     SI,CX           ;load count
0016 33 DB                      XOR     BX,BX           ;clear result
0018                    READH1:
0018 B4 06                      MOV     AH,6            ;read key
001A B2 FF                      MOV     DL,0FFH
001C CD 21                      INT     21H
001E 74 F8                      JE      READH1          ;wait for key
0020 8A D0                      MOV     DL,AL           ;echo
0022 CD 21                      INT     21H
0024 E8 0000 R                  CALL    CONV            ;convert to hexadecimal
0027 D3 E3                      SHL     BX,CL           ;shift result
0029 02 D8                      ADD     BL,AL           ;add AL to result
002B 4E                         DEC     SI
002C 75 EA                      JNZ     READH1          ;repeat 4 times
002E 5B                         POP     BX              ;restore BX and CX
002F 59                         POP     CX
0030 CB                         RET

0031                    READH   ENDP
```

Displaying Hexadecimal Data.   To display hexadecimal data, a number is separated into 4-bit segments that are converted into hexadecimal digits. Conversion is accomplished by adding a 30H to the numbers 0 to 9 and a 37H to the letters A to F.

A procedure that displays the contents of the AX register on the video display appears in Example 7–23. Here the number is rotated left so the leftmost digit is displayed first. Because AX contains a 4-digit hexadecimal number, the procedure displays 4 hexadecimal digits. This procedure can be modified to display wider hexadecimal numbers.

EXAMPLE 7–23

```
0000                    DISPH   PROC    FAR

0000 51                         PUSH    CX              ;save CX and DX
0001 52                         PUSH    DX
0002 B1 04                      MOV     CL,4            ;load rotate count
0004 B5 04                      MOV     CH,4            ;load digit count
0006            DISPH1:
0006 D3 C0                      ROL     AX,CL           ;position number
0008 50                         PUSH    AX              ;save it
0009 24 0F                      AND     AL,0FH          ;get hex digit
000B 04 30                      ADD     AL,30H          ;adjust it
000D 3C 39                      CMP     AL,'9'          ;test against 9
000F 76 02                      JBE     DISPH2          ;if 0 – 9
0011 04 07                      ADD     AL,7            ;adjust it
0013            DISPH2:
0013 B4 06                      MOV     AH,6
0015 8A D0                      MOV     DL,AL
0017 CD 21                      INT     21H             ;display digit
0019 58                         POP     AX              ;restore AX
001A FE CD                      DEC     CH
001C 75 E8                      JNZ     DISPH1          ;repeat

001E 5A                         POP     DX              ;restore CX and DX
001F 59                         POP     CX
0020 CB                         RET

0021                    DISPH   ENDP
```

Using Lookup Tables for Data Conversions

Lookup tables are often used to convert from one data form to another. A lookup table is formed in the memory as a list of data that is referenced by a procedure to perform conversions. In the case of many lookup tables, the XLAT instruction is used to look up data in a table. This is provided so that the table contains 8-bit-wide data and its length is less than or equal to 256 bytes.

Converting from BCD to 7-Segment Code.   One simple application that uses a lookup table is BCD to 7-segment code conversion. Example 7–24 illustrates a lookup table that contains the 7-segment codes for the numbers 0 to 9. These codes are used with the 7-segment display pictured in Figure 7–1. This 7-segment display uses active high (logic 1) inputs to light a segment. The code is arranged so that the *a* segment is in bit position 0 and the *g* segment is in bit position 6. Bit position 7 is zero in this example, but it can be used for displaying a decimal point.

FIGURE 7–1 The 7-segment display.

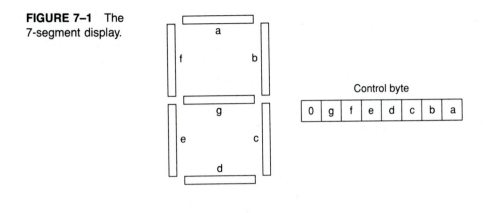

EXAMPLE 7–24

```
0000                    SEG7    PROC    FAR

0000 53                         PUSH    BX
0001 BB 0008 R                  MOV     BX,OFFSET TABLE
0004 2E: D7                     XLAT    CS:TABLE                ;see text
0006 5B                         POP     BX
0007 CB                         RET

0008 3F                 TABLE   DB      3FH                     ;0
0009 06                         DB      6                       ;1
000A 5B                         DB      5BH                     ;2
000B 4F                         DB      4FH                     ;3
000C 66                         DB      66H                     ;4
000D 6D                         DB      6DH                     ;5
000E 7D                         DB      7DH                     ;6
000F 07                         DB      7                       ;7
0010 7F                         DB      7FH                     ;8
0011 6F                         DB      6FH                     ;9

0012                    SEG7    ENDP
```

The procedure that performs the conversion contains only two instructions and assumes that AL contains the BCD digit to be converted to 7-segment code. One of the instructions addresses the lookup table by loading its address into BX, and the other performs the conversion and returns the 7-segment code in AL.

Because the lookup table is located in the code segment, and the XLAT instruction accesses the data segment by default, the XLAT instruction includes a segment override. Notice that a dummy operand (TABLE) is added to the XLAT instruction so the (CS:) code segment override prefix can be added to the instruction. Normally XLAT does not contain an operand unless its default segment must be overridden. The LODS and MOVS instructions are also overridden in the same manner as XLAT by using a dummy operand.

Using a Lookup Table to Access ASCII Data.   Some programming techniques require that numeric codes be converted to ASCII character strings. For example, suppose that you need to display the days of the week for a calendar program. Because the number of ASCII characters in each day is different, some type of lookup table must be used to reference the ASCII coded days of the week.

Example 7–25 shows a table that references ASCII coded character strings located in the code segment. Each character string contains an ASCII coded day of the week. The table references each day of the week. The procedure that accesses the day of the week uses the AL register and the numbers 0 to 6 to refer to Sunday through Saturday. If AL contains a 2 when this procedure is called, the word "Tuesday" is displayed on the video screen.

EXAMPLE 7–25

```
0000                    DAYS    PROC    FAR

0000 52                         PUSH    DX                          ;save DX and SI
0001 56                         PUSH    SI
0002 BE 001B R                  MOV     SI,OFFSET DTAB              ;address DTAB
0005 32 E4                      XOR     AH,AH                       ;clear AH
0007 03 C0                      ADD     AX,AX                       ;double AX
0009 03 F0                      ADD     SI,AX                       ;modify table address
000B 2E: 8B 14                  MOV     DX,CS:[SI]                  ;get string address
000E 8C C8                      MOV     AX,CS                       ;change data segment
0010 1E                         PUSH    DS
0011 8E D8                      MOV     DS,AX
0013 B4 09                      MOV     AH,9
0015 CD 21                      INT     21H                         ;display string
0017 1F                         POP     DS
0018 5E                         POP     SI                          ;restore DX and SI
0019 5A                         POP     DX
001A CB                         RET

001B 0029 R 0031 R     DTAB    DW      SUN,MON,TUE,WED,THU,FRI,SAT
     0039 R 0042 R
     004D R 0057 R
     005F R

0029 53 75 6E 64 61     SUN     DB      'Sunday $'
     79 20 24

0031 4D 6F 6E 64 61     MON     DB      'Monday $'
     79 20 24

0039 54 75 65 73 64     TUE     DB      'Tuesday $'
     61 79 20 24

0042 57 65 64 6E 65     WED     DB      'Wednesday $'
     73 64 61 79 20 24

004D 54 68 75 72 73     THU     DB      'Thursday $'
     64 61 79 20 24

0057 46 72 69 64 61     FRI     DB      'Friday $'
     79 20 24

005F 53 61 74 75 72     SAT     DB      'Saturday $'
     64 61 79 20 24

0069                    DAYS    ENDP
```

This procedure first accesses the table by loading the table address into the SI register. Next the number in AL is converted into a 16-bit number and doubled because each table entry is 2 bytes in length. This index is then added to SI to address the correct entry in the lookup table. The address of the ASCII character string is now loaded into DX by the MOV DX,CS:[SI] instruction.

Before the INT 21H DOS function is called, the DS register is placed on the stack and loaded with the segment address of CS. This allows DOS function number 09H (display a string) to be used to display the day of the week. This procedure converts the numbers 0 to 6 to the days of the week.

An Example Program Using Data Conversions

A program example will serve to combine some of the data conversion DOS functions discussed thus far. Suppose that you must display the time and date on the video screen. An example program, Example 7–26, displays the time as 10:45 P.M. and the date as Tuesday, May 14,1991. The program is relatively short because it calls a procedure that displays the time and a second that displays the date.

EXAMPLE 7–26

```
0000                    STAC    SEGMENT STACK

0000 0100[              DW      256 DUP (?)              ;set up stack
         ????
              ]

0200                    STAC    ENDS

0000                    DAT     SEGMENT

0000 0026 R 002F R 0038 R    DAY    DW      SUN,MON,TUE,WED,THU,FRI,SAT
     0042 R 004E R 0059 R
     0062 R
000E 006D R 0076 R 0080 R    MONT   DW      JAN,FEB,MAR,APR,MAY,JUN,JUL,AUG,SEP,OCT,NOV,DC
     0087 R 008E R 0093 R
     0099 R 009F R 00A7 R
     00B2 R 00BB R 00C5 R

0026 53 75 6E 64 61 79        SUN    DB      'Sunday, $'
     2C 20 24
002F 4D 6F 6E 64 61 79        MON    DB      'Monday, $'
     2C 20 24
0038 54 75 65 73 64 61        TUE    DB      'Tuesday, $'
     79 2C 20 24
0042 57 65 64 6E 65 73        WE     DB      'Wednesday, $'
     64 61 79 2C 20 24
004E 54 68 75 72 73 64        THU    DB      'Thursday, $'
     61 79 2C 20 24
0059 46 72 69 64 61 79        FRI    DB      'Friday, $'
     2C 20 24
0062 53 61 74 75 72 64        SAT    DB      'Saturday, $'
     61 79 2C 20 24

006D 4A 61 6E 75 61 72        JAN    DB      'January $'
     79 20 24
```

```
0076  46 65 62 72 75 61        FEB     DB      'February $'
      72 79 20 24
0080  4D 61 72 63 68 20        MAR     DB      'March $'
      24
0087  41 70 72 69 6C 20        APR     DB      'April $'
      24
008E  4D 61 79 20 24           MAY     DB      'May $'
0093  4A 75 6E 65 20 24        JUN     DB      'June $'
0099  4A 75 6C 79 20 24        JUL     DB      'July $'
009F  41 75 67 75 73 74        AUG     DB      'August $'
      20 24
00A7  53 65 70 74 65 6D        SEP     DB      'September $'
      62 65 72 20 24
00B2  4F 63 74 6F 62 65        OCT     DB      'October $'
      72 20 24
00BB  4E 6F 76 65 6D 62        NOV     DB      'November $'
      65 72 20 24
00C5  44 65 63 65 6D 62        DC      DB      'December $'
      65 72 20 24

00CF  0D 0A 24                 CRLF    DB      13,10,'$'
00D2  2E 4D 2E 20 20 24        MES1    DB      '.M. $'
00D8  2C 20 31 39 24           MES2    DB      ', 19$'
00DD  2C 20 32 30 24           MES3    DB      ', 20$'

00E2                  DAT      ENDS

0000                  COD      SEGMENT 'CODE'

                               ASSUME  CS:COD,DS:DAT,SS:STAC

0000                  MAIN     PROC    FAR                 ;main program

0000  B8 —— R                  MOV     AX,DAT              ;load DS
0003  8E D8                    MOV     DS,AX

0005  BA 00CF R                MOV     DX,OFFSET CRLF      ;get new line
0008  B4 09                    MOV     AH,9
000A  CD 21                    INT     21H

000C  E8 001D R                CALL    TIMES               ;display time
000F  E8 007E R                CALL    DATES               ;display date

0012  BA 00CF R                MOV     DX,OFFSET CRLF      ;get new line
0015  B4 09                    MOV     AH,9
0017  CD 21                    INT     21H

0019  B4 4C                    MOV     AH,4CH              ;exit to DOS
001B  CD 21                    INT     21H

001D                  MAIN     ENDP

001D                  TIMES    PROC    NEAR                ;display time XX:XX A.M.

001D  B4 2C                    MOV     AH,2CH              ;get time
001F  CD 21                    INT     21H

0021  B7 41                    MOV     BH,'A'              ;set AM

0023  80 FD 0C                 CMP     CH,12               ;test against 12
0026  72 05                    JB      TIMES1              ;if AM
```

```
0028  B7 50                      MOV     BH,'P'              ;set PM
002A  80 ED 0C                   SUB     CH,12               ;adjust time


002D                  TIMES1:
002D  0A ED                      OR      CH,CH               ;test for 0 hours
002F  75 02                      JNE     TIMES2              ;if not 0 hours
0031  B5 0C                      MOV     CH,12               ;replace with 12 hours
0033                  TIMES2:
0033  8A C5                      MOV     AL,CH               ;get hours
0035  32 E4                      XOR     AH,AH               ;clear AH
0037  D4 0A                      AAM                         ;convert to BCD
0039  0A E4                      OR      AH,AH               ;test tens of hours
003B  74 09                      JZ      TIMES3              ;if no tens of hours
003D  50                         PUSH    AX
003E  8A C4                      MOV     AL,AH
0040  04 30                      ADD     AL,'0'              ;convert to ASCII
0042  E8 0075 R                  CALL    DISP                ;display tens of hours
0045  58                         POP     AX
0046                  TIMES3:
0046  04 30                      ADD     AL,'0'              ;convert to ASCII
0048  E8 0075 R                  CALL    DISP                ;display units of hours
004B  B0 3A                      MOV     AL,':'              ;display colon
004D  E8 0075 R                  CALL    DISP

0050  8A C1                      MOV     AL,CL               ;get minutes
0052  32 E4                      XOR     AH,AH               ;clear AH
0054  D4 0A                      AAM                         ;convert to BCD

0056  05 3030                    ADD     AX,3030H            ;convert to ASCII
0059  50                         PUSH    AX
005A  8A C4                      MOV     AL,AH
005C  E8 0075 R                  CALL    DISP                ;display tens of minutes
005F  58                         POP     AX
0060  E8 0075 R                  CALL    DISP                ;display units of minutes

0063  B0 20                      MOV     AL,' '              ;display space
0065  E8 0075 R                  CALL    DISP

0068  8A C7                      MOV     AL,BH               ;display A or P
006A  E8 0075 R                  CALL    DISP

006D  BA 00D2 R                  MOV     DX,OFFSET MES1      ;display .M.
0070  B4 09                      MOV     AH,9
0072  CD 21                      INT     21H
0074  C3                         RET

0075                  TIMES     ENDP


0075                  DISP      PROC    NEAR                ;display ASCII

0075  50                         PUSH    AX
0076  B4 06                      MOV     AH,6
0078  8A D0                      MOV     DL,AL
007A  CD 21                      INT     21H
007C  58                         POP     AX
007D  C3                         RET

007E                  DISP      ENDP

007E                  DATES     PROC    NEAR                ;display date
```

```
007E  B4 2A              MOV    AH,2AH              ;get date
0080  CD 21              INT    21H
0082  52                 PUSH   DX                  ;save month and day

0083  32 E4              XOR    AH,AH               ;clear AH
0085  03 C0              ADD    AX,AX               ;double AX
0087  BE 0000 R          MOV    SI,OFFSET DAY       ;address day table
008A  03 F0              ADD    SI,AX
008C  8B 14              MOV    DX,[SI]             ;get string address
008E  B4 09              MOV    AH,9
0090  CD 21              INT    21H                 ;display day

0092  5A                 POP    DX
0093  52                 PUSH   DX
0094  8A C6              MOV    AL,DH               ;get month
0096  FE C8              DEC    AL
0098  32 E4              XOR    AH,AH
009A  03 C0              ADD    AX,AX
009C  BE 000E R          MOV    SI,OFFSET MONT      ;address month table
009F  03 F0              ADD    SI,AX
00A1  8B 14              MOV    DX,[SI]             ;get string address
00A3  B4 09              MOV    AH,9
00A5  CD 21              INT    21H                 ;display month

00A7  5A                 POP    DX                  ;get day
00A8  8A C2              MOV    AL,DL
00AA  32 E4              XOR    AH,AH               ;clear AH
00AC  D4 0A              AAM                        ;convert to BCD
00AE  0A E4              OR     AH,AH               ;test tens of day
00B0  74 09              JZ     DATES1              ;if zero
00B2  50                 PUSH   AX
00B3  8A C4              MOV    AL,AH
00B5  04 30              ADD    AL,'0'              ;convert to ASCII
00B7  E8 0075 R          CALL   DISP                ;display tens of day
00BA  58                 POP    AX


00BB          DATES1:
00BB  04 30              ADD    AL,'0'              ;convert to ASCII
00BD  E8 0075 R          CALL   DISP                ;display units of day
00C0  BA 00D8 R          MOV    DX,OFFSET MES2
00C3  81 F9 07D0         CMP    CX,2000             ;test for year 2000
00C7  72 06              JB     DATES2              ;if year 19XX
00C9  83 E9 64           SUB    CX,100
00CC  BA 00DD R          MOV    DX,OFFSET MES3
00CF          DATES2:
00CF  B4 09              MOV    AH,9                ;display 19 or 20
00D1  CD 21              INT    21H
00D3  81 E9 076C         SUB    CX,1900             ;adjust year
00D7  8B C1              MOV    AX,CX
00D9  D4 0A              AAM                        ;convert to BCD
00DB  05 3030            ADD    AX,3030H            ;convert to ASCII
00DE  50                 PUSH   AX
00DF  8A C4              MOV    AL,AH
00E1  E8 0075 R          CALL   DISP                ;display tens of year
00E4  58                 POP    AX
00E5  E8 0075 R          CALL   DISP                ;display units of year
00E8  C3                 RET

00E9          DATES    ENDP
```

```
00E9              COD     ENDS

                  END     MAIN
```

The time is available from DOS using INT 21H function call number 2CH. This function returns with the hours in CH and minutes in CL. Also available are seconds in DH and hundredths of seconds in DL. The date is available using INT 21H function call number 2AH. This leaves the day of the week in AL, the year in CX, the day of the month in DH, and the month in DL.

The example procedure uses two ASCII lookup tables that convert the day and month to ASCII character strings. It also uses the AAM instruction to convert from binary to BCD for the time and date. Data display is handled in two ways: by character string (function 09H) and by single character (function 06H).

The memory consists of three segments: stack (STAC), data (DAT), and code (COD). The data segment contains the character strings used with the procedures that display time and date. The code segment contains MAIN, TIMES, DATES, and DISP procedures. The MAIN procedure is a FAR procedure because it is the program, or main module. The other procedures are NEAR or local procedures used by MAIN.

7–5   GRAPHIC DISPLAYS (VGA)

Because most modern systems contain some form of a VGA graphic display, a section on this topic is included as an introduction to graphic display systems. This topic alone could demand an entire textbook, so only the basics are explained and applied here as a launching point for those interested in graphic displays.

The Basic VGA Display System

The basic VGA display system operates in several modes, but in this text we shall concentrate on the most common: the 16 color, 640×480 display and the 256 color, 320×200 display. Other resolutions are available, but not on all display adapters. These basic modes are used for many applications, not counting the ones that require the display of extremely high-resolution video images. To display a high-resolution video image we often use the 256 color, 800×600 or 1024×768 display modes, which are not available to all VGA displays. The most common resolutions available on more advanced VGA display adapters are: 256 color, 640×400; 16 color, 800×600; 256 color, 800×600; 16 color, 1024×768; and 256 color, 1024×768. In all cases these displays are bit-mapped displays instead of character-mode displays. This means that the display data are sent to the video adapter as a series of bits, or bit combinations, instead of ASCII characters as with the DOS function calls described earlier in this chapter.

The 16 color, 640×480 VGA graphics mode is mode 12H when specified using the video BIOS. The memory organization for this mode is illustrated in Figure 7–2.

FIGURE 7–2 The bit planes in a 16 color, 640 × 480 VGA display.

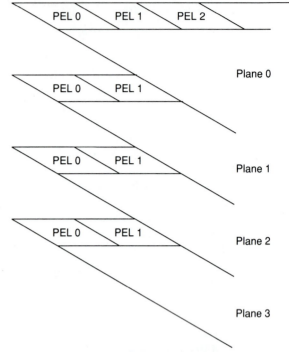

Notice that the memory is organized in four **bit planes.** Each byte in a bit plane represents eight *picture elements (PELS)* on the video screen. This means that the 640 × 480 display uses 38,400 bytes of memory in each bit plane to address the 307,200 PELS found on this display. The first scanning line of 640 bits is stored in the first 80 bytes of a video memory plane. The second scanning line is stored in the next 80 bytes, and so forth. The combination of the four bit planes is used to specify one of 16 colors for each of the PELS. To change one PEL on the video display, 4 bits are changed, one in each bit plane, to represent the new color for the PEL. Table 7–3 lists the color codes used for a standard VGA display using four bit planes. If all four bit planes are cleared, a black is displayed for the PEL.

Video memory exists at memory locations A0000H through AFFFFH for the VGA graphics modes. (The text modes use video memory beginning at location B8000H or B0000H, depending upon the mode selected.) The VGA graphics memory is 64K bytes in size and requires bit planes to address the 153,600 bytes of memory required for a 16 color, 640 × 480 display. Many video cards contain 256K bytes of memory that is addressed in sections (pages) through the 64K byte memory window at locations A0000H–AFFFFH. Each time a new bit plane is selected, the adapter internally addresses a separate 64K byte section of memory that appears at location A0000H–AFFFFH.

TABLE 7–3 Default color codes for the 16 color, 640 × 480 VGA graphics display

Code	Color
0000	Black
0001	Blue
0010	Green
0011	Cyan
0100	Red
0101	Magenta
0110	Brown
0111	White
1000	Dark gray
1001	Light blue
1010	Light green
1011	Light cyan
1100	Light red
1101	Light magenta
1110	Yellow
1111	Bright white

When the 256 color, 320 × 200 mode is selected, a byte in the video memory selects a single color (1 of 256) for a single PEL. This graphics display mode (13H) uses 64,000 bytes of memory, slightly less than 64K bytes, which are directly addressed at locations A0000H–AEA5FH. Although this display mode has a lower resolution, it displays 256 colors instead of 16. Note that this area of memory is organized differently than for the 16 color, 640 × 480 VGA display mode. The first byte (A000:0000) holds the upper left PEL instead of 8 PELs.

Programming in the 256 Color, 320 × 200 Mode

Software for this **bit-mapped** 256 color, 320 × 200 display mode is fairly easy to write because each byte represents a single PEL on the video display. Before any information is displayed, the display adapter (assumes VGA is present) must be switched to mode 13H, the 256 color, 320 × 200 mode. This is accomplished with video BIOS function 00H. To select a new video mode, place a 00H into AH and the new mode number in AL. This is followed by the video BIOS function call INT 10H to select the new mode.

Once in the new mode, a program can begin to display graphics information. Example 7–27 shows a short program that switches the display to the 256 color, 320 × 200 VGA mode and then displays a vertical color bar pattern on the screen to show the 256 colors programmed by default into the VGA adapter. Here 2 PELs are used for each color across the entire width of the display. These colors displayed on the screen are the default colors that are programmed into the VGA display adapter.

EXAMPLE 7–27

```
0000                    CODE    SEGMENT 'CODE'
                                ASSUME CS:CODE

0000                    COLOR   PROC    FAR

0000 B8 A000                    MOV     AX,0A000H           ;address video memory
0003 8E C0                      MOV     ES,AX
0005 BF 0000                    MOV     DI,0
0008 FC                         CLD                         ;select auto-increment

0009 B4 00                      MOV     AH,0
000B B0 13                      MOV     AL,13H              ;select mode 13H
000D CD 10                      INT     10H                 ;this also clears the screen
000F BB 00C8                    MOV     BX,200              ;scan line count
0012                    COLOR1:
0012 B9 00A0                    MOV     CX,320/2            ;column count/2
0015 B0 00                      MOV     AL,0                ;get color number
0017                    COLOR2:
0017 AA                         STOSB                       ;store 2 PELS
0018 AA                         STOSB
0019 FE C0                      INC     AL                  ;change to new color
001B E2 FA                      LOOP    COLOR2              ;repeat for one scan line

001D 4B                         DEC     BX                  ;repeat 200 times
001E 75 F2                      JNZ     COLOR1
0020                    COLOR3:
0020 B4 06                      MOV     AH,6                ;wait for any key
0022 B2 FF                      MOV     DL,0FFH
0024 CD 21                      INT     21H
0026 74 F8                      JZ      COLOR3

0028 B4 4C                      MOV     AH,4CH              ;exit to DOS after key
002A CD 21                      INT     21H

002C                    COLOR   ENDP

002C                    CODE    ENDS

                                END     COLOR
```

What if you want to change the default colors? This is accomplished by reprogramming a series of *palette registers* located on the video display card. There are 256 different locations in the palette that each represent a video display color. Each picture element is composed of the three basic video colors of red, green, and blue. These are the primary colors of light.

To display a color we use a combination of these three primary colors. The palette memory contains three 6-bit numbers (one for each primary color) that change the brightness of each primary color. The program in Example 7–27 used the default palette, which has colors set so that the first 16 are compatible with other display modes, the second 16 colors are shades of gray, and the remaining are various colors selected by the manufacturer of the video display adapter.

Video BIOS INT 10H function AH = 10H, AL = 10H, and BX = color number (00H–FFH) allows a palette location (color number) to be selected and changed to CH = green, CL = blue, and DH = red. The color values in CH, CL, and DH are 6-bit numbers ranging from 00H–3FH. A value of 00H is off and 3FH is maximum brightness. For example, a very bright cyan (blue-green) is CH = 3FH, CL = 3FH, and DH = 00H. Varying these color amplitudes and various combinations vary the brightness and hue of each color number.

To illustrate how the palette is changed, Example 7–28 contains a program that alters color numbers 80H–BFH to every possible brightness of red. It then displays a vertical color bar pattern, containing 64 bars, to illustrate each intensity of red. This program can be changed to display other colors by modifying the values loaded into the palette registers.

EXAMPLE 7–28

```
0000                    CODE    SEGMENT 'CODE'
                                ASSUME CS:CODE

0000                    COLOR   PROC    FAR

0000  B8 A000                   MOV     AX,0A000H           ;address video memory
0003  8E C0                     MOV     ES,AX
0005  BF 0000                   MOV     DI,0
0008  FC                        CLD                         ;select auto-increment

0009  B4 00                     MOV     AH,0
000B  B0 13                     MOV     AL,13H              ;select mode 13H
000D  CD 10                     INT     10H                 ;this also clears the screen

000F  BD 0040                   MOV     BP,64               ;count
0012  B6 00                     MOV     DH,0                ;set red to zero
0014  BB 0080                   MOV     BX,80H              ;intialize palette number
0017            COLORA:
0017  B8 1010                   MOV     AX,1010H            ;change palette
001A  B9 0000                   MOV     CX,0                ;no green or blue
001D  CD 10                     INT     10H
001F  43                        INC     BX                  ;next palette
0020  FE C6                     INC     DH                  ;next red
0022  4D                        DEC     BP
0023  75 F2                     JNZ     COLORA              ;repeat 64 times

0025  BB 00C8                   MOV     BX,200              ;scan line count
0028            COLOR1:
0028  B9 0040                   MOV     CX,320/5            ;column count/5
002B  B0 80                     MOV     AL,80H              ;get color number
002D            COLOR2:
002D  AA                        STOSB                       ;store 5 PELS
002E  AA                        STOSB
002F  AA                        STOSB
0030  AA                        STOSB
0031  AA                        STOSB
0032  FE C0                     INC     AL                  ;change to new color
0034  E2 F7                     LOOP    COLOR2              ;repeat for one scan line
```

```
0036 4B                        DEC      BX                      ;repeat 200 times
0037 75 EF                     JNZ      COLOR1
0039                COLOR3:

0039 B4 06                     MOV      AH,6                    ;wait for any key
003B B2 FF                     MOV      DL,0FFH
003D CD 21                     INT      21H
003F 74 F8                     JZ       COLOR3

0041 B4 4C                     MOV      AH,4CH                  ;exit to DOS after key
0043 CD 21                     INT      21H

0045                COLOR      ENDP

0045                CODE       ENDS

                               END      COLOR
```

To utilize this graphics display, procedures are needed to draw shapes and forms on the screen. This text cannot show all variations of these programs, but will show one at this point. Example 7–29 shows a procedure that draws a box on the 256 color, 320×200 VGA display. The box is any size and can be placed at any location on the screen. Parameters used to transfer size, color, and location are passed through registers to this procedure. To illustrate the operation of the BOX procedure, a program is also included in the example that draws one box on the video display screen.

EXAMPLE 7–29

```
0000                CODE       SEGMENT 'CODE'
                               ASSUME CS:CODE

0000                MAIN       PROC     FAR

0000 FC                        CLD                              ;select auto-increment

0001 B4 00                     MOV      AH,0
0003 B0 13                     MOV      AL,13H                  ;select mode 13H
0005 CD 10                     INT      10H                     ;this also clears the screen

0007 B0 02                     MOV      AL,2                    ;use color 02H
0009 B9 0064                   MOV      CX,100                  ;starting column number
000C BE 000A                   MOV      SI,10                   ;starting row number
000F BD 004B                   MOV      BP,75                   ;size
0012 E8 000D                   CALL     BOX                     ;display box
0015                MAIN1:
0015 B4 06                     MOV      AH,6                    ;wait for any key
0017 B2 FF                     MOV      DL,0FFH
0019 CD 21                     INT      21H
001B 74 F8                     JZ       MAIN1

001D B8 4C00                   MOV      AX,4C00H                ;exit to DOS
0020 CD 21                     INT      21H
```

```
0022                    MAIN    ENDP

0022                    BOX     PROC    NEAR

0022 BB A000                    MOV     BX,0A000H        ;address video RAM
0025 8E C3                      MOV     ES,BX
0027 50                         PUSH    AX
0028 B8 0140                    MOV     AX,320           ;find starting PEL
002B F7 E6                      MUL     SI
002D 8B F8                      MOV     DI,AX            ;address start of BOX
002F 03 F9                      ADD     DI,CX
0031 58                         POP     AX
0032 57                         PUSH    DI               ;save start
0033 8B CD                      MOV     CX,BP
0035                    BOX1:
0035 F3/ AA                     REP STOSB                ;draw top line
0037 8B CD                      MOV     CX,BP
0039 83 E9 02                   SUB     CX,2             ;adjust CX

003C                    BOX2:
003C 5F                         POP     DI
003D 81 C7 0140                 ADD     DI,320           ;address next row
0041 57                         PUSH    DI
0042 AA                         STOSB                    ;draw PEL
0043 03 FD                      ADD     DI,BP
0045 83 EF 02                   SUB     DI,2
0048 AA                         STOSB                    ;draw PEL
0049 E2 F1                      LOOP    BOX2

004B 5F                         POP     DI
004C 81 C7 0140                 ADD     DI,320           ;address last row
0050 8B CD                      MOV     CX,BP
0052 F3/ AA                     REP STOSB
0054 C3                         RET

0055                    BOX     ENDP

0055                    CODE    ENDS

                                END     MAIN
```

Programming in the 16 Color, 640 × 480 Mode

Programming the 256 color, 320 × 200 mode was fairly easy. The 16 color mode requires more effort because of the way the memory is organized in bit planes. To plot a single dot we need to:

1. Select and address a single bit (PEL) through the graphics address register (GAR) and bit mask register (BMR)
2. Address and set the map mask register (MMR) to 0FH and write 0 (black) to the address containing the PEL to clear the old color from the PEL

3. Set the desired PEL color through the MMR
4. Write the PEL

The Bit Mask Register.   The bit mask register (BMR) selects the bit or bits that are modified when a byte is written to the display adapter memory. Each bit position of the display adapter memory represents a PEL in the 16 color, 640 × 480 mode. Memory from locations A0000H through A95FFH store 38,400 bytes that represent the 307,200 PELs. Memory location A0000H holds the first 8 PELs, location A0001H holds the next 8 PELs, and so forth. If only the leftmost PEL is changed, the bit mask register is programmed with an 80H before any information is written to memory location A0000H. Likewise other bits or multiple bits can be modified by changing the bit mask register contents.

The BMR is accessed by programming the graphics address register (GAR) with an index of 8 to select the BMR. The bit mask is then programmed. The I/O address of the GAR is 03CEH, and the BMR uses I/O address 03CFH. Example 7–30 shows how the BMR is programmed so all 8 bits (8 PELs) change together.

EXAMPLE 7–30

```
0000 BA 03CE          MOV     DX,3CEH          ;graphics address register
0003 B0 08            MOV     AL,8             ;select bit mask register
0005 EE               OUT     DX,AL

0006 BA 03CF          MOV     DX,03CFH         ;bit mask register
0009 B0 FF            MOV     AL,0FFH          ;select all 8 bits
000B EE               OUT     DX,AL
```

The Map Mask Register.   The map mask register (MMR) selects the bit planes that are enabled for a write operation. If all bit planes are enabled we write color F (bright white by default). The only color we cannot write in this manner is black (color 0). To write a new color to a PEL or up to 8 PELs, if the bit mask is FFH, we set all four bit planes (0FH) and write a 00H to memory. This clears it to the color black. Next we select the color to be written by placing the binary color number (0–F) into the MMR and write the new PELs to memory.

Access to the MMR is provided by using index 2 for sequence address register (port 3C4H). Once the MMR is accessed, we select bit planes by writing to the MMR at I/O port 3C5H. Example 7–31 shows how to write the first (leftmost) PEL on the video screen and set it to color 2 (green). The video memory location must be read before it is changed. If the location is not read, the result is unpredictable. The read operation loads the video byte into an internal latch so it can be changed before it is rewritten to the video memory by the display adapter.

EXAMPLE 7–31

```
0000 B4 00            MOV     AH,0             ;set mode to 12H
0002 B0 12            MOV     AL,12H
0004 CD 10            INT     10H
```

```
0006 BA 03CE                    MOV     DX,3CEH                ;graphics address register
0009 B0 08                      MOV     AL,8                   ;select bit mask register
000B EE                         OUT     DX,AL

000C BA 03CF                    MOV     DX,3CFH                ;bit mask register
000F B0 80                      MOV     AL,080H                ;select leftmost bit
0011 EE                         OUT     DX,AL

0012 BA 03C4                    MOV     DX,3C4H                ;sequence address register
0015 B0 02                      MOV     AL,2                   ;select map mask register
0017 EE                         OUT     DX,AL

0018 BA 03C5                    MOV     DX,3C5H                ;map mask register
001B B0 0F                      MOV     AL,0FH                 ;enable all planes
001D EE                         OUT     DX,AL

001E B8 A000                    MOV     AX,0A000H              ;address video memory
0021 8E D8                      MOV     DS,AX
0023 BF 0000                    MOV     DI,0
0026 8A 05                      MOV     AL,[DI]                ;must read first
0028 C6 05 00                   MOV     BYTE PTR [DI],0        ;clear old color

002B B0 02                      MOV     AL,02H                 ;select color 2
002D EE                         OUT     DX,AL

002E C6 05 FF                   MOV     BYTE PTR [DI],0FFH     ;write memory
```

Suppose a procedure is required that plots any PEL on the video display. Example 7–32 shows such a procedure and a program that displays some dots at various points and in various colors on the video screen. Look very closely for the single red dot below and to the right of the cyan line when you execute this program.

EXAMPLE 7–32

```
0000                   CODE     SEGMENT 'CODE'
                                ASSUME CS:CODE

0000                   MAIN     PROC    FAR

0000 B8 A000                    MOV     AX,0A000H              ;address video RAM
0003 8E D8                      MOV     DS,AX
0005 FC                         CLD                            ;select auto-increment

0006 B4 00                      MOV     AH,0                   ;set mode to 12H
0008 B0 12                      MOV     AL,12H
000A CD 10                      INT     10H

000C B9 000A                    MOV     CX,10
000F BB 000A                    MOV     BX,10                  ;row address
0012 BE 0064                    MOV     SI,100                 ;column address
0015 B2 03                      MOV     DL,3                   ;color 3 (cyan)
0017                   MAIN1:                                  ;plot 10 dots
0017 E8 001B                    CALL    DOT
001A 46                         INC     SI
001B E2 FA                      LOOP    MAIN1

001D BB 0028                    MOV     BX,40                  ;row address
0020 BE 00C8                    MOV     SI,200                 ;column address
```

```
0023 B2 04                      MOV     DL,4            ;color 4 (red)
0025 E8 000D                    CALL    DOT
0028            MAIN2:
0028 B4 06                      MOV     AH,6            ;wait for key
002A B2 FF                      MOV     DL,0FFH
002C CD 21                      INT     21H
002E 74 F8                      JZ      MAIN2

0030 B8 4C00                    MOV     AX,4C00H        ;exit to DOS
0033 CD 21                      INT     21H

0035            MAIN    ENDP

0035            DOT     PROC    NEAR

0035 51                 PUSH    CX
0036 52                 PUSH    DX              ;save color

0037 8B C3              MOV     AX,BX
0039 BD 0280            MOV     BP,640          ;find video address
003C F7 E5              MUL     BP
003E 03 C6              ADD     AX,SI

0040 83 D2 00           ADC     DX,0
0043 BD 0008            MOV     BP,8
0046 F7 F5              DIV     BP              ;get byte address
0048 8B F8              MOV     DI,AX
004A B0 80              MOV     AL,80H          ;form bit mask
004C 83 FA 00           CMP     DX,0
004F 74 04              JE      DOT1
0051 8A CA              MOV     CL,DL
0053 D2 E8              SHR     AL,CL           ;adjust bit mask
0055           DOT1:
0055 50                 PUSH    AX              ;save bit mask
0056 BA 03CE            MOV     DX,3CEH         ;graphics address register
0059 B0 08              MOV     AL,8            ;select bit mask register
005B EE                 OUT     DX,AL

005C BA 03CF            MOV     DX,3CFH         ;bit mask register
005F 58                 POP     AX              ;get bit mask
0060 EE                 OUT     DX,AL

0061 BA 03C4            MOV     DX,3C4H         ;sequence address register
0064 B0 02              MOV     AL,2            ;select map mask register
0066 EE                 OUT     DX,AL

0067 BA 03C5            MOV     DX,3C5H         ;map mask register
006A B0 0F              MOV     AL,0FH          ;enable all planes
006C EE                 OUT     DX,AL

006D 8A 05              MOV     AL,[DI]         ;must read first
006F C6 05 00           MOV     BYTE PTR [DI],0 ;clear old color

0072 58                 POP     AX              ;select color
0073 50                 PUSH    AX
0074 EE                 OUT     DX,AL

0075 C6 05 FF           MOV     BYTE PTR [DI],0FFH      ;write memory

0078 5A                 POP     DX
0079 59                 POP     CX
007A C3                 RET
```

007B		DOT	ENDP
007B		CODE	ENDS
		END	MAIN

Text with Graphics. Displaying text in a graphics display mode such as mode 12H (16 color, 640 × 480) is also difficult. We can sometimes use the DOS function call 21H to place text on the screen, but only if we can allow video information under the text to change to both the text color and its background color selected by the video attribute byte. A better way to display text in a graphics mode is to use a lookup table that contains the graphic display characters so they appear without a different background color.

The video BIOS ROM contains 8 × 8, 8 × 14, 9 × 14, 8 × 16, and 9 × 16 character sets. These character sets, or your own, may be used as a basis for displaying text in a graphic display mode. Figure 7–3 shows the format of the 8 × 8 character and a few example characters. We most often use 8 × 8 characters for the 640 × 480 resolution display. The character set is obtained from the video BIOS ROM via the INT 10H function, which returns the address of the character set in ES:BP.

Example 7–33 shows how the 8 × 8 standard character set is copied from the video BIOS ROM into the data segment at memory location CHAR. Because each ASCII character requires 8 bytes of memory and because the ASCII code contains 128 characters, the size of the character table is 1,024 bytes in length. Note that BH = 03H to obtain the 8 × 8 standard character set. The extended character set (code 128–255) is obtained in the same fashion except BH = 04H.

EXAMPLE 7–33

0000		DATA	SEGMENT		
0000 0400 [			CHAR	DB	1024 DUP (?)
	00				
	]				
0400		DATA	ENDS		

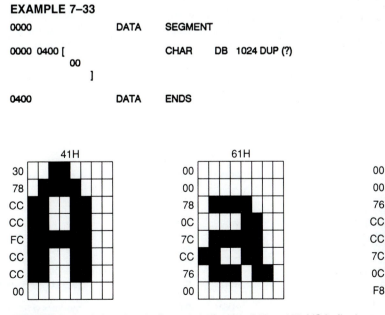

FIGURE 7–3 A few 8 × 8 characters for the 640 × 480 VGA display.

```
0000                    CODE    SEGMENT 'CODE'
                                ASSUME CS:CODE,DS:DATA

0000                    GETC    PROC    FAR

0000 FC                         CLD                             ;select increment

0001 B4 11                      MOV     AH,11H                  ;get ROM character set
0003 B0 30                      MOV     AL,30H
0005 B7 03                      MOV     BH,03H                  ;select 8 x 8
0007 CD 10                      INT     10H

0009 8C C0                      MOV     AX,ES                   ;address memory
000B 8E D8                      MOV     DS,AX
000D B8 — R                     MOV     AX,DATA
0010 8E C0                      MOV     ES,AX
0012 BF 0000 R                  MOV     DI,OFFSET CHAR
0015 8B F5                      MOV     SI,BP

0017 B9 0400                    MOV     CX,1024                 ;load count
001A F3/ A4                     REP MOVSB                       ;copy character set

001C CB                         RET

001D                    GETC    ENDP

001D                    CODE    ENDS

                                END     GETC
```

Once the character set is fetched from the video BIOS, it is used to display text on the graphics display. The procedure listed in Example 7–34 uses the character table stored in the data segment in Example 7–33 to display ASCII text on a graphics display. This procedure assumes the display uses 9 scanning lines for each character line and 80 characters appear across the screen on a character line. This allows 53 text lines with each containing 80 characters to be displayed on the graphics mode screen. Before the procedure is called, AL = ASCII to be displayed, DH = character line (0–52), DL = character column (0–79), and BL = text color (0–F). This procedure does not display a background color for the text. Instead, it superimposes the text on top of whatever graphics presentation appears on the video display. This same procedure can display any character set as long as it is an 8 × 8.

EXAMPLE 7–34

```
0021                    PUTC    PROC    NEAR                    ;display character

0021 52                         PUSH    DX
0022 56                         PUSH    SI
0023 BE 0000 R                  MOV     SI,OFFSET CHAR
0026 32 E4                      XOR     AH,AH                   ;adjust ASCII code
0028 D1 E0                      SHL     AX,1                    ;multiply by 8
002A D1 E0                      SHL     AX,1
002C D1 E0                      SHL     AX,1
002E 03 F0                      ADD     SI,AX                   ;adjust CHAR address

0030 52                         PUSH    DX                      ;address video byte
0031 8A C6                      MOV     AL,DH
```

```
0033 32 E4                    XOR      AH,AH
0035 B9 02D0                  MOV      CX,720
0038 F7 E1                    MUL      CX
003A 8B F8                    MOV      DI,AX
003C 5A                       POP      DX
003D 8A C2                    MOV      AL,DL
003F 32 E4                    XOR      AH,AH
0041 03 F8                    ADD      DI,AX

0043 B9 0008                  MOV      CX,8              ;set count for 8 bytes
0046              PUTC1:
0046 BA 03CE                  MOV      DX,3CEH           ;address bit mask register
0049 B0 08                    MOV      AL,8
004B EE                       OUT      DX,AL
004C 42                       INC      DX                ;select bit mask register
004D AC                       LODSB                      ;get mask pattern
004E EE                       OUT      DX,AL

004F BA 03C4                  MOV      DX,3C4H           ;address map mask register
0052 B0 02                    MOV      AL,2
0054 EE                       OUT      DX,AL
0055 42                       INC      DX
0056 B0 0F                    MOV      AL,0FH            ;select all bit planes
0058 EE                       OUT      DX,AL
0059 26: 8A 05                MOV      AL,ES:[DI]        ;read byte
005C 32 C0                    XOR      AL,AL             ;write black
005E AA                       STOSB
005F 4F                       DEC      DI
0060 8A C3                    MOV      AL,BL             ;get color
0062 EE                       OUT      DX,AL             ;set bit planes to color
0063 B0 FF                    MOV      AL,0FFH
0065 AA                       STOSB
0066 4F                       DEC      DI
0067 83 C7 50                 ADD      DI,80
006A E2 DA                    LOOP     PUTC1             ;repeat 8 times

006C 5E                       POP      SI
006D 5A                       POP      DX
006E C3                       RET

006F              PUTC         ENDP
```

A program such as Example 7–35 illustrates how to display a graphics screen and place text on top of the graphics. This program selects VGA mode 12H and then displays a cyan screen. Next it places two lines of text in two different colors on top of the cyan screen. The first line is bright white, and the second line is bright blue. Experiment with this program by changing the background screen to another color and also try changing the text color.

EXAMPLE 7–35

```
0000                 STACK       SEGMENT STACK

0000 0400 [                      DW        1024 DUP (?)
            0000
          ]

0800                 STACK       ENDS

0000                 DATA        SEGMENT
```

```
0000 0400 [            CHAR      DB        1024 DUP (?)
         00
            ]

0400 54 68 69 73 20 69 MES1     DB        'This is a test, only a test.',0
     73 20 61 20 74 65
     73 74 2C 20 6F 6E
     6C 79 20 61 20 74
     65 73 74 2E 00
041D 54 68 69 73 20 6C MES2     DB        'This line is blue, above is white.',0
     69 6E 65 20 69 73
     20 62 6C 75 65 2C
     20 61 62 6F 76 65
     20 69 73 20 77 68
     69 74 65 2E 00

0440              DATA     ENDS

0000              CODE     SEGMENT 'CODE'
                           ASSUME CS:CODE,DS:DATA,SS:STACK

0000              GETC     PROC NEAR

0000 FC                    CLD                          ;select increment

0001 1E                    PUSH      DS
0002 06                    PUSH      ES

0003 B4 11                 MOV       AH,11H             ;get ROM character set
0005 B0 30                 MOV       AL,30H
0007 B7 03                 MOV       BH,03H             ;select 8 x 8
0009 CD 10                 INT       10H

000B 8C C0                 MOV       AX,ES              ;address memory
000D 8E D8                 MOV       DS,AX
000F B8 —— R               MOV       AX,DATA
0012 8E C0                 MOV       ES,AX
0014 BF 0000 R             MOV       DI,OFFSET CHAR
0017 8B F5                 MOV       SI,BP

0019 B9 0400               MOV       CX,1024            ;load count
001C F3/ A4                REP MOVSB                    ;copy character set

001E 07                    POP       ES
001F 1F                    POP       DS

0020 C3                    RET

0021              GETC     ENDP

0021              PUTC     PROC      NEAR               ;display character

0021 52                    PUSH      DX
0022 56                    PUSH      SI

0023 BE 0000 R             MOV       SI,OFFSET CHAR

0026 32 E4                 XOR       AH,AH              ;adjust ASCII code
0028 D1 E0                 SHL       AX,1               ;multiply by 8
002A D1 E0                 SHL       AX,1
002C D1 E0                 SHL       AX,1
002E 03 F0                 ADD       SI,AX              ;adjust CHAR address
```

```
0030 52                    PUSH    DX
0031 8A C6                 MOV     AL,DH
0033 32 E4                 XOR     AH,AH
0035 B9 02D0               MOV     CX,720
0038 F7 E1                 MUL     CX
003A 8B F8                 MOV     DI,AX
003C 5A                    POP     DX
003D 8A C2                 MOV     AL,DL
003F 32 E4                 XOR     AH,AH
0041 03 F8                 ADD     DI,AX

0043 B9 0008               MOV     CX,8            ;set count for 8 bytes
0046              PUTC1:
0046 BA 03CE               MOV     DX,3CEH         ;address bit mask register
0049 B0 08                 MOV     AL,8
004B EE                    OUT     DX,AL
004C 42                    INC     DX              ;select bit mask register
004D AC                    LODSB                   ;get mask pattern
004E EE                    OUT     DX,AL

004F BA 03C4               MOV     DX,3C4H         ;address map mask register
0052 B0 02                 MOV     AL,2
0054 EE                    OUT     DX,AL
0055 42                    INC     DX
0056 B0 0F                 MOV     AL,0FH          ;select all bit planes
0058 EE                    OUT     DX,AL
0059 26: 8A 05             MOV     AL,ES:[DI]      ;read byte

005C 32 C0                 XOR     AL,AL           ;write black
005E AA                    STOSB
005F 4F                    DEC     DI
0060 8A C3                 MOV     AL,BL           ;get color
0062 EE                    OUT     DX,AL           ;set bit planes to color
0063 B0 FF                 MOV     AL,0FFH
0065 AA                    STOSB
0066 4F                    DEC     DI
0067 83 C7 50              ADD     DI,80
006A E2 DA                 LOOP    PUTC1           ;repeat 8 times

006C 5E                    POP     SI
006D 5A                    POP     DX
006E C3                    RET

006F              PUTC    ENDP

006F              DISP_S  PROC    NEAR            ;display string

006F AC                    LODSB                   ;get character
0070 0A C0                 OR      AL,AL           ;test for end of string
0072 74 07                 JE      DISP_S1
0074 E8 FFAA               CALL    PUTC            ;display character
0077 FE C2                 INC     DL              ;address next column
0079 EB F4                 JMP     DISP_S
007B              DISP_S1:
007B C3                    RET

007C              DISP_S  ENDP

007C              DISP_G  PROC    NEAR            ;set display to cyan

007C B9 9600               MOV     CX,38400        ;set count
007F BF 0000               MOV     DI,0            ;address video
```

```
0082 BA 03CE              MOV      DX,3CEH          ;address bit mask register
0085 B0 08                MOV      AL,8
0087 EE                   OUT      DX,AL
0088 42                   INC      DX               ;select bit mask
0089 B0 FF                MOV      AL,0FFH
008B EE                   OUT      DX,AL

008C BA 03C4              MOV      DX,3C4H          ;address map mask register
008F B0 02                MOV      AL,2
0091 EE                   OUT      DX,AL
0092 42                   INC      DX               ;select map mask register
0093            DISP_G1:
0093 B0 0F                MOV      AL,0FH           ;select all planes
0095 EE                   OUT      DX,AL
0096 26: 8A 05            MOV      AL,ES:[DI]       ;read memory
0099 32 C0                XOR      AL,AL            ;write black

009B AA                   STOSB
009C 4F                   DEC      DI
009D B0 03                MOV      AL,3             ;select cyan
009F EE                   OUT      DX,AL
00A0 B0 FF                MOV      AL,0FFH          ;write cyan
00A2 AA                   STOSB
00A3 E2 EE                LOOP     DISP_G1          ;repeat 38,400 times

00A5 C3                   RET

00A6           DISP_G     ENDP

00A6           MAIN       PROC     FAR

00A6 B8 ── R              MOV      AX,DATA          ;address data segment
00A9 8E D8                MOV      DS,AX

00AB B8 A000              MOV      AX,0A000H        ;address video memory
00AE 8E C0                MOV      ES,AX

00B0 B8 0012              MOV      AX,0012H         ;select VGA mode 12H
00B3 CD 10                INT      10H

00B5 E8 FF48              CALL     GETC             ;get character set
00B8 E8 FFC1              CALL     DISP_G           ;display cyan screen
00BB BE 0400 R            MOV      SI,OFFSET MES1   ;address string
00BE B6 0A                MOV      DH,10            ;line 10
00C0 B2 0F                MOV      DL,15            ;column 15
00C2 B3 0F                MOV      BL,15            ;color brightwhite
00C4 E8 FFA8              CALL     DISP_S           ;display string
00C7 BF 041D R            MOV      DI,OFFSET MES2
00CA B6 0B                MOV      DH,11            ;line 11
00CC B2 0F                MOV      DL,15            ;column 15
00CE B3 09                MOV      BL,9             ;color bright blue
00D0 E8 FF9C              CALL     DISP_S           ;display string
00D3           MAIN1:
00D3 B4 06                MOV      AH,06H           ;wait for key
00D5 B2 FF                MOV      DL,0FFH
00D7 CD 21                INT      21H
00D9 74 F8                JZ       MAIN1

00DB B8 4C00              MOV      AX,4C00H         ;return to DOS
00DE CD 21                INT      21H

00E0           MAIN       ENDP
```

00E0   CODE   ENDS

                            END   MAIN

7-6   SUMMARY

1. The assembler program assembles modules that contain PUBLIC variables and segments plus EXTRN (external) variables. The linker program links modules and library files to create a run-time program executed from the DOS command line. The run-time program usually has the extension EXE.

2. The MACRO and ENDM directives create a new opcode for use in programs. These macros are similar to procedures except there is no call or return. In place of them, the assembler inserts the code of the macro sequence into a program each time it is invoked. Macros can include variables that pass information and data to the macro sequence.

3. The DOS INT 21H function call provides a method of using the keyboard and video display. Function number 06H, placed into register AH, provides an interface to the keyboard and display. If DL = 0FFH, this function tests the keyboard for a keystroke. If no keystroke is detected, it returns equal. If a keystroke is detected, the standard ASCII character returns in AL. If an extended ASCII character is typed, it returns with AL = 00H, where the function must again be called to return with the extended ASCII character in AL. To display a character, DL is loaded with the character and AH with 06H before the INT 21H is used in a program.

4. Character strings are displayed using function number 09H. The DS:DX register combination addresses the character string, which must end with a $.

5. The INT 10H instruction accesses BIOS (basic I/O system) procedures that control the video display and keyboard. The BIOS functions are independent of DOS and function with any operating system.

6. Data conversion from binary to BCD is accomplished with the AAM instruction for numbers that are less than 100 or by repeated division by 10 for larger numbers. Once converted to BCD, a 30H is added to convert each digit to ASCII code for the video display.

7. When converting from an ASCII number to BCD, a 30H is subtracted from each digit. To obtain the binary equivalent, we multiply by 10.

8. Lookup tables are used for code conversion with the XLAT instruction if the code is an 8-bit code. If the code is wider than 8 bits, then a short procedure that accesses a lookup table provides the conversion. Lookup tables are also used to hold addresses so that different parts of a program or different procedures can be selected.

9. The video display is accessed through the video BIOS or directly through the video display memory. For a VGA display adapter, the video display memory exists at location A0000H–AFFFFH for the graphics modes 12H and 13H. Mode 12H is a 16 color, 640×480 display mode. Mode 13H is a 256 color, 320×200 display mode.

10. The 256 color mode stores one picture element per byte of memory, where-as the 16 color mode is a bit-mapped mode storing 8 PELs per memory byte.
11. The bit mask register selects the PEL to be written in the 16 color, 640 × 480 video display. The map mask register selects the bit planes (color) to be written in the 16 color, 640 × 480 video display.
12. The video BIOS ROM contains character sets that are copied to an application program to display text on the graphics mode VGA screen. This allows the display of standard characters, custom characters or character sets, or even multiple character sets.

7–7   GLOSSARY

Bit-mapped   The video display memory being organized as bits stored in bytes. Here one bit represents one picture element on the video display.

Bit plane   An area of memory that holds a portion of the color of a picture element.

CGA   Color graphics adapter.

Command file   A special execution file that requires less room to store on the disk. The command file may be 64K bytes in length and must have the extension .COM.

Echo   A term used with keyboards that indicates the typed character is sent (echoed) to the video display.

EGA   Enhanced color graphics adapter.

Execution file   A program file that is executed from the DOS command line by typing the name of the program file. Execution files always end with the file extension .EXE.

Home   The upper left-hand corner of the video display screen.

Library   A collection of procedures that, if they are stored in a library file, can be added to any program.

Linear program   A program that progresses from the top to the bottom without procedures. We most often use macro sequences instead of procedures to develop this type of program.

Linker   A program that links assembled object and library files into an execution file.

Macro   A sequence of instructions substituted in a program each time the macro is invoked by its name.

Palette   A series of registers inside the video display controller that select the color for a binary color code.

PEL   A picture element is the smallest visible dot on the video screen.

VGA   Variable graphics array.

7–8 QUESTIONS

1. The assembler converts a source file into an _____ file.
2. What files are generated from the source file TEST.ASM as it is processed by MASM?
3. The linker program links object files and _____ files to create an execution file.
4. What is the difference between an .EXE and a .COM file?
5. What does the PUBLIC directive indicate when placed in a program module?
6. What does the EXTRN directive indicate when placed in a program module?
7. What directives appear with labels defined external?
8. Describe how a library file works when it is linked to other object files by the linker program.
9. What assembler language directives delineate a macro sequence?
10. What is a macro sequence?
11. How are parameters transferred to a macro sequence?
12. Develop a macro called ADD32 that adds the 32-bit contents of DX–CX to the 32-bit contents of BX–AX.
13. Develop a macro called STUB that sign extends the 8-bit number in AL into a 64-bit sign-extended number in ECX–EBX. (Note that ECX is the most significant part of the 64-bit result.)
14. How is the LOCAL directive used within a macro sequence?
15. Develop a macro called ADDLIST PARA1,PARA2 that adds the contents of PARA1 to PARA2. Each of these parameters represents an area of memory. The number of bytes added are indicated by register CX before the macro is invoked.
16. Develop a macro that sums a list of byte-sized data invoked by the macro ADDM LIST,LENGTH. The label LIST is the starting address of the data block and length is the number of data elements added. The result must be a 16-bit sum found in AX at the end of the macro sequence.
17. What is the purpose of the INCLUDE directive?
18. Develop a procedure called RANDOM. This procedure must return an 8-bit random number in register CL at the end of the subroutine. (One way to generate a random number is to increment CL each time the DOS function 06H tests the keyboard and finds *no* keystroke. In this way a random number is generated.)
19. Modify the procedure of Question 18 so the random number ranges in value from 1 through and including 6.
20. Develop a procedure that displays a character string that ends with a 00H. Your procedure must use the DS:DX register to address the start of the character string.
21. Develop a procedure that reads a key and displays the hexadecimal value of an extended ASCII coded keyboard character if it is typed. If a normal character is typed, ignore it.
22. Use BIOS INT 10H to develop a procedure that positions the cursor at line 3, column 6.

23. When a number is converted from binary to BCD, the _____ instruction accomplishes the conversion, provided the number is less than 100 decimal.
24. How is a large number (over 100 decimal) converted from binary to BCD?
25. A BCD digit is converted to ASCII code by adding a _____.
26. An ASCII coded number is converted to BCD by subtracting _____.
27. Develop a procedure that reads an ASCII number from the keyboard and stores it as a BCD number into memory array DATA. The number ends when anything other than a number is typed.
28. Explain how a 3-digit ASCII coded number is converted to binary.
29. Develop a procedure that converts all lowercase ASCII coded letters into upper-case ASCII coded letters. Your procedure may not change any other character except the letters a–z.
30. Develop a lookup table that converts hexadecimal data 00H–0FH into the ASCII coded characters that represent the hexadecimal digits. Make sure to show the lookup table and any software required for the conversion.
31. Develop a program sequence that jumps to memory location ONE if AL = 6, TWO if AL = 7, and THREE if AL = 8.
32. Show how to use the XLAT instruction to access a lookup table called LOOK that is located in the stack segment.
33. Write a program that reads any decimal number between 0 and 65,535 and displays the 16-bit binary version on the video display.
34. Write a program that displays the binary powers of two (in decimal) on the video screen for the powers 0 through 7. Your display shows 2^n = value for each power of 2.
35. Using the technique learned in Question 18, develop a program that displays random numbers between 1 and 47 (or whatever your state's lottery uses for its range of numbers).
36. Develop a program that displays the hexadecimal contents of a block of 256 bytes of memory. Your software must be able to accept the starting address as a hexadecimal number between 00000H and FFF00H.
37. Where is the video data stored when mode 13H is selected for the VGA video display adapter?
38. How many bytes of memory are required to store a screen of information for the 256 color, 320 × 200 video display mode?
39. What is a bit plane?
40. How is the bit mask register used in the 16 color, 640 × 480 VGA video display mode?
41. How is the map mask register used in the 16 color, 640 × 480 VGA video display mode?
42. Develop a program that places the video adapter into mode 13H and then draws a green line, 2 PELS wide, from the upper leftmost corner to the lower rightmost corner.
43. Develop a program that places the video adapter into mode 13H and then displays a red letter T on the screen that is 50 PELs high and 30 PELs wide. The width of the lines in the letter T should be 3 PELs wide.

44. Develop a program that places the video adapter into mode 12H and then displays a vertical color bar pattern that displays 16 color bars (each 40 PELs wide) showing the 16 colors available in this mode.
45. Develop a program that places the video adapter into mode 12H and then displays a white screen (color 7) with a cyan horizontal bar across the top that is 10 scanning lines high.
46. Develop your own character set for mode 12H of the VGA display. Your character set must contain characters that are 16 bits wide and 16 bits high. Luckily you only need the characters N, a, m, e, space, and =.
47. Using the characters developed in Question 46, develop a program using mode 12H that displays a blue video screen that contains the red character line "Name =" at the upper left margin of the screen.

CHAPTER 8

Disk Memory Functions

INTRODUCTION

Programming requires a familiarity with disk memory systems and the disk operations performed by DOS. This chapter explains the organization of disk data and also how to use the DOS INT 21H function call to access disk memory data and program files via assembly language.

The DOS INT 21H function call is used to create, delete, read, write, and close files on the disk. A complete coverage of this function is provided, along with examples of its use. Also presented is a description of software that allows access to both random and sequential access disk data files.

8–1
CHAPTER OBJECTIVES

Upon completion of this chapter, you will be able to:

1. Explain how data are stored on a floppy and hard disk memory, including explanation of the track, sector, cylinder, cluster, boot sector, the file allocation table, root directory, and the organization of disk data.
2. Use DOS INT 21H function calls to create, open, read, write, delete, rename, and close disk files.
3. Create and use random and sequential access files.
4. Append data to a file.
5. Insert data in the middle of a file.

8–2
DISK FILES

Data are stored on the disk in files. The disk is organized into four areas: the boot sector, the file allocation table (FAT), the root directory, and the data storage area.

The first sector on the disk is the **boot sector.** The boot sector contains a program that loads the disk operating system (DOS) from the disk into the memory when power is applied to the computer. The **FAT** stores data that indicate the parts of the disk that contain data. All references to a disk file are handled through the FAT and a directory. The root directory is where all subdirectory and files are referenced. The disk files are all considered sequential access files, which means that they are accessed a byte at a time from the beginning of the file toward the end.

Disk Organization

Figure 8–1 illustrates the organization of sectors and tracks on the surface of a disk. This organization applies to both floppy and hard or fixed disk memory systems. The outer track is always track 0 and the inner track is 39 (double-density) or 79 (high-density) on floppy disks. The inner track on a hard disk is determined by the disk size and could be 10,000 or higher for very large hard disks.

The floppy disk comes in two different sizes, the 5 ¼″ **mini-floppy** and the 3 ½″ **micro-floppy.** The 5 ¼″ mini-floppy disk is currently available as a double-sided, double-density (DSDD) disk that stores 360K bytes and as a high-density (HD) disk that stores 1.2M bytes. The 3 ½″ micro-floppy is available as a double-sided, double-density (DSDD) disk that stores 720K bytes, a high-density (HD) disk that stores 1.44 M bytes, and an extra-high-density (EHD) disk that stores 2.88M bytes. In all cases each type of disk memory organizes its surface into **tracks** that contain **sectors.** Each sector on a disk usually contains 512 bytes of data. The number of sectors and tracks varies from one form to another. For example, the 5 ¼″ double-density, double-sided disk contains 40 tracks with 9 sectors per track (40 × 9 × 512 = 180K bytes) per side. A pair of tracks (top and bottom) is called a **cylinder.** We often use the term cylinder in place of track and upper or lower head in place of side. Some fixed disk drives have up to 16 heads per cylinder because they contain more than one disk platter.

FIGURE 8–1 Data organization on one side of a disk memory.

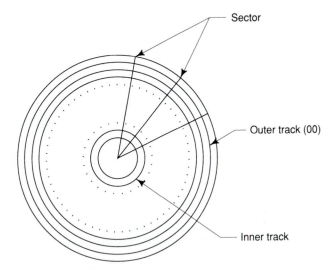

Sector

Outer track (00)

Inner track

Figure 8–2 shows the organization of data on a disk. The length of the FAT is determined by the size of the disk because each disk sector or group of sectors is represented in the FAT by a 16-bit number. Note that some extremely large fixed disks use 32-bit numbers as entries in the FAT. The length of the root directory is fixed at 4K bytes. The boot sector is always a single 512-byte-long sector located in the outer track (track 0) at sector 0, the first sector.

The boot sector contains a ***bootstrap loader*** program, which is read into RAM when the system is powered. After the bootstrap loader is read into RAM, it executes and loads the IO.SYS* and MSDOS.SYS† programs into RAM. Next, the bootstrap loader passes control to the MSDOS control program and the computer is now under the control of the DOS command processor.

The FAT indicates which sectors are free, which are corrupted (unusable), and which contain data. The FAT table is referenced each time DOS writes data to the disk, so it can find a free sector. Each free cluster is indicated by a 0000H in the FAT and each occupied cluster is indicated by the cluster number. A ***cluster*** can be anything from one sector, on a floppy disk, to any number of sectors in length. Many fixed disk memory systems use 4 or more sectors per cluster, which means the smallest file allocation unit is 512×4 or 2,048 bytes in length if a cluster contains 4 sectors. The larger the hard disk drive, the more sectors in a cluster. A 16-bit FAT table accesses $64K - 2$ clusters. One code, 0000H, indicates a free cluster, FFFFH indicates a bad cluster, and the remaining cluster numbers 0001H–FFFEH are available. If a cluster is 4 sectors of 512 bytes, a fixed disk containing a 16-bit

FIGURE 8–2 The organization of disk data beginning at track 0, sector 0.

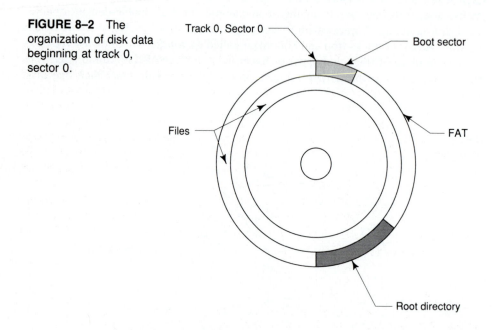

Track 0, Sector 0

Boot sector

Files

FAT

Root directory

---

*IO.SYS is an I/O control program provided by Microsoft Corporation with Microsoft DOS.
†MSDOS.SYS is the Microsoft Disk Operating System.

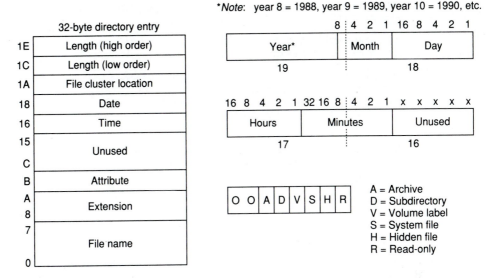

FIGURE 8–3 The format for any directory or subdirectory entry.

FAT accesses a maximum of 65,534 clusters. Because a sector is 512 bytes this hard disk drive contains a maximum of 134,213,632 bytes of data. If a larger hard disk exists, either more sectors per cluster are allocated or the FAT is increased to 32 bits.

Figure 8–3 shows the format of each directory entry in the root or any other directory or subdirectory. Each entry contains the name, extension, attribute, time, date, location, and length. The length of the file is stored as a 32-bit number. This means that a file can have a maximum length of 4G bytes. The location is the starting cluster number. The root directory has room for 128 of these 32-byte directory entries. When the root directory is full, the disk is full. This is why it is important to use subdirectories, which can contain any number of files desired. A single subdirectory requires one entry in the root directory. A sub-subdirectory entry is stored in the subdirectory and not the root directory. This allows an almost infinite combination of files stored in subdirectories.

Example 8–1 shows how part of the root directory appears in a hexadecimal dump. Try to identify the date, time, location, and length of each entry. Also identify the attribute for each entry. The listing shows hexadecimal data and also ASCII data as is customary in most computer dumps.

EXAMPLE 8–1

```
0000  49 4F 20 20 20 20 20 20 53 59 53 07 00 00 00 00    IO       SYS
0010  00 00 00 00 00 00 00 00 93 11 02 00 39 82 00 00

0020  4D 53 44 4F 53 20 20 20 53 59 53 07 00 00 00 00    MSDOS    SYS
0030  00 00 00 00 00 00 C0 44 93 12 13 00 92 00 00 00
```

```
0040  43 4F 4D 4D 41 4E 44 20 43 4F 4D 00 00 00 00 00    COMMAND  COM
0050  00 00 00 00 00 00 00 00 93 11 26 00 B5 92 00 00

0060  42 41 52 52 59 20 42 52 45 59 20 28 00 00 00 00    BARRY BREY
0070  00 00 00 00 00 00 E0 AD 6A 13 00 00 00 00 00 00

0080  50 43 54 4F 4F 4C 53 20 20 20 20 10 00 00 00 00    PCTOOLS
0090  00 00 00 00 00 00 80 AE 6A 13 5C 00 00 00 00 00

00A0  44 4F 53 20 20 20 20 20 20 20 20 10 00 00 00 00    DOS
00B0  00 00 00 00 00 00 E0 B0 6A 13 4E 00 00 00 00 00

00C0  52 55 4E 5F 46 57 20 20 42 41 54 00 00 00 00 00    FUN_FW      BAT
00D0  00 00 00 00 00 00 40 BD 6A 13 97 0F 4A 00 00 00

00E0  46 4F 4E 54 57 41 52 45 20 20 20 10 00 00 00 00    FONTWARE
00F0  00 00 00 00 00 00 60 BD 6A 13 6E 00 00 00 00 00
```

Files are usually accessed through DOS INT 21H function calls. There are two ways to access a file using an INT 21H. One method uses a file control block and the other uses a file handle. Today, all software accesses files via a file handle, so this text also uses file handles for file access. File control blocks are a carryover from an earlier operating system called CP/M* (control program/microprocessor), which was used with 8-bit computer systems based on the Z80, 8080, or 8085 microprocessor.

8–3   SEQUENTIAL ACCESS FILES

All DOS files are sequential access files. A *sequential access file* is stored and addressed from its beginning toward the end. This means that the first byte and all bytes between it and the last must be accessed to read the last byte. Fortunately, files are read and written with the DOS INT 21H function calls (refer to Appendix A), which simplifies file access and data manipulation. This section of the text describes how to create, read, write, delete, and rename a sequential access file by using the DOS INT 21H function calls.

File Creation

Before a file is used, it must exist on the disk. A file is *created* by the INT 21H function call number 3CH. To create a file, the name of the file and its extension must be stored in memory at a location addressed by DS:DX before calling the function. The CX register must also contain the attribute of the file created by function 3CH.

---

*CP/M is a registered trademark of Digital Research Corporation.

A *file name* is always stored as an ASCII-Z string and may contain the disk drive letter and directory path(s) if needed. Example 8–2 shows several ASCII-Z string file names stored in a data segment for access by the file utilities. An *ASCII-Z string* is a character string that ends with a 00H, or null character.

EXAMPLE 8–2

```
0000                    DAT    SEGMENT

0000 44 4F 47 2E 54 58  FILE1  DB    'DOG.TXT',0              ;file name DOG.TXT
     54 00
0008 43 3A 44 41 54 41  FILE2  DB    'C:DATA.DOC',0           ;file C:DATA.DOC
     2E 44 4F 43 00
0013 43 3A 5C 44 52 45  FILE3  DB    'C:\DREAD\ERROR.FIL',0   ;file C:\DREAD\ERROR.FIL
     41 44 5C 45 52 52
     4F 52 2E 46 49 4C
     00

0026                    DAT    ENDS
```

Suppose that you have filled a 256-byte memory buffer area with data that must be stored in a new file called DATA.NEW on the default disk drive. Before data can be written to this new file, it must first be created. Example 8–3 lists a short procedure that creates this new file on the disk.

EXAMPLE 8–3

```
0000                    DAT    SEGMENT

0000 44 41 54 41 2E 4E  FILE1  DB     'DATA.NEW',0      ;file name DATA.NEW
     45 57 00
0009 0100[              BUFFER DB     256 DUP (?)       ;data buffer
          ??
     ]

0109                    DAT    ENDS

0000                    COD    SEGMENT 'CODE'
                               ASSUME  CS:COD,DS:DAT

0000 B8 —— R            MOV    AX,DAT
0003 8E D8              MOV    DS,AX              ;load DS

0005 B4 3C              MOV    AH,3CH             ;load create function
0007 33 C9              XOR    CX,CX              ;00H = attribute
0009 BA 0000 R          MOV    DX,OFFSET FILE1    ;address ASCII-Z name
000C CD 21              INT    21H                ;create DATA.NEW

000E 72 20              JC     ERROR              ;on creation error
                         .      .
                         .      .
                         .      .
0030                    COD    ENDS

                               END
```

Whenever a file is created, the CX register must contain the attributes or characteristics of the file. Table 8–1 lists the attribute bit positions and defines them. A logic one in a bit selects the attribute, while a logic zero does not. For example, to create a hidden, archive file, CX is loaded with a 0022H. Note that the hidden and archives bits are set.

After returning from the INT 21H, the carry flag indicates whether or not an error occurred (C = 1) during the creation of the file. Some errors that occur during file creation, which are obtained if needed by INT 21H function call number 59H, include the following: path not found, no file handles available, or media error. If carry is cleared (C = 0), no error occurred during file creation and the AX register contains a file handle. The *file handle* is a number that refers to the file after it is created or opened. The file handle allows a file to be accessed without using the ASCII-Z string name of the file, thus speeding the operation.

Writing to a File

Now that we have created a new file in Example 8–3, called FILE.NEW, data can be written to it. Before writing to a file it must be created or opened. When a file is created or opened, the file handle returns in the AX register. The file handle is used to refer to the file whenever data are written.

Function number 40H is used to write data to an opened or newly created file. In addition to loading a 40H into AH, we must load BX with the file handle, CX with the number of bytes to be written, and DS:DX with the address of the memory buffer area to be written to the disk.

Suppose that we must write 256 bytes, from data segment memory area BUFFER, to a file. This is accomplished as illustrated in Example 8–4 using function 40H. If an error occurs during a write operation, the carry flag is set. If no error occurs, the carry flag is cleared and the number of bytes written to the file are returned in the AX register. Errors that occur during write operations usually indicate that the disk is full or that there is some type of media error. Media errors occur for a bad disk (floppy) or a problem with the disk electronics.

TABLE 8–1   File attribute definitions

Bit Position	Attribute	Function
0	Read-only	A read-only file or subdirectory
1	Hidden	Prevents the file or subdirectory name from appearing in the directory when a DIR is used from the DOS command line
2	System	Specifies a file as a system file
3	Volume	Specifies the disk volume label
4	Subdirectory	Specifies a subdirectory name
5	Archive	Indicates that a file has been changed and that it should be archived

EXAMPLE 8–4

```
                        .        .
                        .        .
                        .        .
0010  8B D8             MOV      BX,AX                ;move handle to BX
0012  B4 40             MOV      AH,40H               ;load write function
0014  B9 0100           MOV      CX,256               ;load count
0017  BA 0009 R         MOV      DX,OFFSET BUFFER     ;address BUFFER
001A  CD 21             INT      21H                  ;write 256 bytes from BUFFER

001C  72 32             JC       ERROR1               ;on write error
                        .        .
                        .        .
                        .        .
```

Opening, Reading, and Closing a File

To read an existing file, it must be opened first. When a file is opened, DOS checks the directory to determine if the file exists and returns the DOS file handle in register AX. The DOS file handle must be used for reading, writing to, and closing a file.

Example 8–5 shows a sequence of instructions that open a file, read 256 bytes from the file into memory area BUFFER, and then close the file. When a file is opened (AH = 3DH), the AL register specifies the type of operation allowed for the opened file. If AL = 00H, the file is opened for a read operation; if AL = 01H, the file is opened for a write operation; and if AL = 02H, the file is opened for a read or a write operation.

EXAMPLE 8–5

```
0000            READ    PROC     NEAR

0000  B4 3D             MOV      AH,3DH               ;load open function
0002  B0 00             MOV      AL,0                 ;select read
0004  BA 0000 R         MOV      DX,OFFSET FILE1      ;address file name
0007  CD 21             INT      21H                  ;open file

0009  72 15             JC       ERROR                ;on error

000B  8B D8             MOV      BX,AX                ;move file handle to BX

000D  B4 3F             MOV      AH,3FH               ;load read function
000F  B9 0100           MOV      CX,256               ;number of bytes
0012  BA 0009 R         MOV      DX,OFFSET BUFFER     ;address BUFFER
0015  CD 21             INT      21H                  ;read 256 bytes

0017  72 07             JC       ERROR                ;on error

0019  B4 3E             MOV      AH,3EH               ;load close function
001B  CD 21             INT      21H                  ;close file

001D  72 01             JC       ERROR                ;on error

001F  C3                RET

0020            READ    ENDP
```

Function number 3FH causes a file to be read. As with the write function, BX contains the file handle, CX contains the number of bytes to be read, and DS:DX contains the location of a memory area where the data are stored. As with all disk functions, the carry flag indicates an error when C = 1. If C = 0, the AX register indicates the number of bytes read from the file.

Closing a file is very important. If a file is left open, some serious problems occur that can actually destroy the disk and all its data. If a file is written and not closed, the FAT can become corrupted, making it difficult or impossible to retrieve data from the disk. Always be certain to close the file after it is read or written. If you suspect that a file has been written without closing, you should execute the DOS utility program CHKDSK (check disk) before any subsequent write to the disk. The CHKDSK program tests the disk and looks for lost file chains. If it detects a lost file chain (usually caused by forgetting to close a file), it can correct the problem. Run the CHKDSK/F command at the DOS prompt to fix any lost file chains.

The File Pointer

Whenever a file is opened, written, or read, the file pointer addresses the current location in the sequential file. When a file is opened, the file pointer always addresses the first byte of the file. If a file is 1,024 bytes in length, and a read function (3FH) reads 1,023 bytes, the file pointer addresses the last byte of the file, but not the end of the file.

The *file pointer* is a 32-bit number that addresses any byte in a file. Once a file is opened, the file pointer can be changed with the move file pointer function number 42H to access any location in the file. A file pointer can be moved from the start of the file (AL = 00H), from the current location (AL = 01H), or from the end of the file (AL = 02H). In practice all three directions are used to access different parts of the file. The distance moved by the file pointer is specified by registers CX and DX. The DX register holds the least significant part and CX the most significant part of the distance. Register BX must contain the file handle before using function 42H to move the file pointer.

Suppose that a file exists on the disk and that you must append the file with 256 bytes of new information. When the file is opened, the file pointer addresses the first byte of the file. If you attempt to write new data without moving the file pointer to the end of the file, the new data will overwrite the first 256 bytes of the file. Example 8–6 shows a procedure that opens a file, moves the file pointer to the end of the file, writes 256 bytes of data, and then closes the file. This *appends* the file with 256 new bytes of data.

EXAMPLE 8–6

```
0000                    APPEND  PROC    NEAR

0000 B4 3D                      MOV     AH,3DH          ;load open function
0002 B0 01                      MOV     AL,1            ;select write
0004 BA 0000 R                  MOV     DX,OFFSET FILE1 ;address file name
0007 CD 21                      INT     21H             ;open file

0009 72 21                      JC      ERROR           ;on error
```

```
000B 8B D8                   MOV    BX,AX              ;move file handle to BX

000D B4 42                   MOV    AH,42H             ;load move file pointer functioi
000F B0 02                   MOV    AL,02H             ;move from end
0011 33 C9                   XOR    CX,CX              ;move 0 bytes from end
0013 33 D2                   XOR    DX,DX
0015 CD 21                   INT    21H                ;move pointer to end

0017 72 13                   JC     ERROR              ;on error

0019 B4 40                   MOV    AH,40H             ;load write function
001B B9 0100                 MOV    CX,256             ;number of bytes
001E BA 0009 R               MOV    DX,OFFSET BUFFER   ;address BUFFER
0021 CD 21                   INT    21H                ;write 256 bytes

0023 72 07                   JC     ERROR              ;on error

0025 B4 3E                   MOV    AH,3EH             ;load close function
0027 CD 21                   INT    21H                ;close file

0029 72 01                   JC     ERROR              ;on error

002B C3                      RET

002C              APPEND     ENDP
```

One of the more difficult file maneuvers is inserting new data into the middle of a file. Figure 8–4 shows how this is accomplished by creating a second temporary file. Notice that the part of the file before the insertion point is copied into the new file. This is followed by the new information before the remainder of the file is appended to the end of the new file. Once the new file is complete, the old file is deleted and the new file is renamed by the old file name.

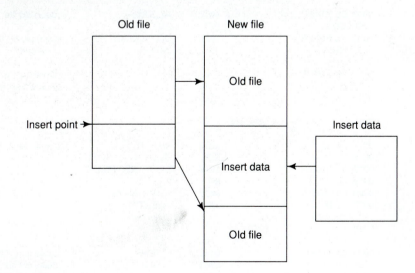

FIGURE 8–4   Inserting new data within an old file.

Example 8–7 shows a procedure that inserts new data of up to 64K bytes in length into an old file. This procedure requires some input parameters to make it a general-purpose procedure. The first parameter is the old file name that is passed to the procedure through the DS:DX register. The second parameter (register CX) is the number of new bytes of information to write to the new file. The third parameter is the location (ES:SI) of the new data to be inserted in the file. The final parameter is the insertion point located in registers BP and DI, where BP contains the most significant insertion point and DI the least.

EXAMPLE 8–7

```
0000                    DATA      SEGMENT  PUBLIC

0000 0100[                        BUFFER    DB      256 DUP (?)      ;buffer
        ??
        ]
0100 0000                         TEMP      DW      ?                ;temporary data
0102 0000                         TEMP1     DW      ?                ;temporary data
0104 0000                         OLD_HAN   DW      ?                ;old handle
0106 0000                         NEW_HAN   DW      ?                ;temp handle
0108 54 45 4D 50 2E 24            TEMPS     DB      'TEMP.$$$',0     ;temp file
     24 24 00

0111                    DATA      ENDS

0000                    CODE      SEGMENT 'CODE'

                                  ASSUME  CS:CODE,DS:DATA

0000                    INSERT    PROC    FAR

0000 89 16 0100 R                 MOV     TEMP,DX                   ;save old file address
0004 89 0E 0102 R                 MOV     TEMP1,CX                  ;save count

0008 B8 3D02                      MOV     AX,3D02H                  ;open old file
000B CD 21                        INT     21H
000D A3 0104 R                    MOV     OLD_HAN,AX                ;save old file handle
0010 73 05                        JNC     INSERT1                   ;if no error

0012 E8 009F R                    CALL    C_OLD                     ;close old file
0015 F9                           STC                               ;indicate error
0016 CB                           RET

0017                    INSERT1:

0017 B4 3C                        MOV     AH,3CH                    ;create temp file
0019 33 C9                        XOR     CX,CX
001B BA 0108 R                    MOV     DX,OFFSET TEMPS
001E CD 21                        INT     21H
0020 A3 0106 R                    MOV     NEW_HAN,AX                ;save temp file handle
0023 73 08                        JNC     INSERT2

0025                    EXIT:

0025 E8 00A8 R                    CALL    C_TEMP                    ;close temp file
0028 E8 009F R                    CALL    C_OLD                     ;close old file
002B F9                           STC                               ;indicate error
002C CB                           RET
```

```
002D                        INSERT2:

002D  81 FF 0100            CMP     DI,256              ;test insert point
0031  77 07                 JA      INSERT3             ;if greater than 256
0033  0B ED                 OR      BP,BP
0035  75 03                 JNE     INSERT3             ;if greater than 256
0037  EB 17 90              JMP     INSERT4             ;if less than 256

003A                        INSERT3:

003A  B9 0100               MOV     CX,256
003D  E8 00B1 R             CALL    R_OLD               ;read old file
0040  72 E3                 JC      EXIT                ;on error
0042  E8 00BD R             CALL    W_TEMP              ;write temp file
0045  72 DE                 JC      EXIT                ;on error

0047  81 EF 0100            SUB     DI,256              ;decrement BP-DI by 256
004B  83 DD 00              SBB     BP,0
004E  EB DD                 JMP     INSERT2

0050                        INSERT4:

0050  8B CF                 MOV     CX,DI               ;get count
0052  E8 00B1 R             CALL    R_OLD               ;read old file
0055  72 CE                 JC      EXIT                ;on error
0057  E8 00BD R             CALL    W_TEMP              ;write temp file
005A  72 C9                 JC      EXIT                ;on error

005C  8B 0E 0102 R          MOV     CX,TEMP1            ;get insert count
0060  1E                    PUSH    DS                  ;save data segment
0061  8C C0                 MOV     AX,ES
0063  8E D8                 MOV     DS,AX               ;load DS with ES

0065  8B D6                 MOV     DX,SI               ;get address
0067  B4 40                 MOV     AH,40H              ;write insert data
0069  CD 21                 INT     21H
006B  1F                    POP     DS                  ;restore DS
006C  72 B7                 JC      EXIT                ;on error

006E                        INSERT5:

006E  B9 0100               MOV     CX,256              ;write remainder of file
0071  E8 00B1 R             CALL    R_OLD               ;read old file
0074  72 AF                 JC      EXIT                ;on error
0076  0B C0                 OR      AX,AX               ;test for end of file
0078  74 05                 JE      INSERT6             ;if end file
007A  E8 00BD R             CALL    W_TEMP              ;write temp file
007D  EB EF                 JMP     INSERT5             ;repeat until end

007F                        INSERT6:

007F  E8 009F R             CALL    C_OLD               ;close old file
0082  E8 00A8 R             CALL    C_TEMP              ;close temp file
0085  8B 16 0100 R          MOV     DX,TEMP             ;delete old file
0089  B4 41                 MOV     AH,41H
008B  CD 21                 INT     21H

008D  06                    PUSH    ES                  ;save ES
008E  8C D8                 MOV     AX,DS
0090  8E C0                 MOV     ES,AX               ;load ES with DS
```

```
0092 8B 3E 0100 R              MOV      DI,TEMP              ;get old file name
0096 BA 0108 R                 MOV      DX,OFFSET TEMPS
0099 B4 56                     MOV      AH,56H               ;rename file
009B CD 21                     INT      21H
009D 07                        POP      ES
009E CB                        RET

009F            INSERT  ENDP

009F            C_OLD   PROC    NEAR

009F 8B 1E 0104 R             MOV      BX,OLD_HAN            ;close old file
00A3 B4 3E                    MOV      AH,3EH
00A5 CD 21                    INT      21H
00A7 C3                       RET

00A8            C_OLD   ENDP

00A8            C_TEMP  PROC    NEAR                         ;close temp file

00A8 8B 1E 0106 R             MOV      BX,NEW_HAN
00AC B4 3E                    MOV      AH,3EH
00AE CD 21                    INT      21H
00B0 C3                       RET

00B1            C_TEMP  ENDP

00B1            R_OLD   PROC    NEAR

00B1 8B 1E 0104 R             MOV      BX,OLD_HAN
00B5 BA 0000 R                MOV      DX,OFFSET BUFFER
00B8 B4 3F                    MOV      AH,3FH
00BA CD 21                    INT      21H
00BC C3                       RET

00BD            R_OLD   ENDP

00BD            W_TEMP  PROC    NEAR

00BD 8B C8                    MOV      CX,AX
00BF 8B 1E 0106 R            MOV      BX,NEW_HAN
00C3 BA 0000 R               MOV      DX,OFFSET BUFFER
00C6 B4 40                   MOV      AH,40H
00C8 CD 21                   INT      21H
00CA C3                      RET

00CB            W_TEMP  ENDP

00CB            CODE    ENDS

                        END
```

This procedure uses two new INT 21H function calls. The delete and rename function calls are used to delete the old file before the temporary file is renamed to the old file name. These functions, and all other DOS file functions, are listed in Appendix A for reference.

8–4 RANDOM ACCESS FILES

Random access files are developed through software using DOS sequential access files. A *random access file* is addressed by a record number rather than by passing through the file searching for data. The move pointer function is very important when random access files are created. Random access files are much easier to use for large volumes of data than sequential access files.

Creating a Random Access File

Planning is paramount in creating a random access file system. Suppose that a random access file is required for storing the names of customers. Each customer record requires 16 bytes for the last name, 16 bytes for the first name, and 1 byte for the middle initial. Each customer record contains two street address lines of 32 bytes each, a city line of 16 bytes, 2 bytes for the state code, and 9 bytes for the zip code. Just the basic customer information requires 105 bytes. Additional information expands the record to 256 bytes in length. Because the business is growing, provisions are made for 5,000 customers. This means that the total required random access file is 1,280,000 bytes in length.

Example 8–8 illustrates a short program that creates a file called CUST.FIL and inserts 5,000 blank records of 256 bytes each. A blank record contains 00H in all of its bytes. This appears to be a large file, but it fits on a single high-density 5¼″ or 3½″ floppy disk drive. In fact, this program assumes that the disk is in drive A. Note that this program takes a considerable amount of time to execute, because it writes data to virtually every byte on the floppy disk.

EXAMPLE 8–8

```
0000                    DATA    SEGMENT

0000 0100[              BUFFER  DB      256 DUP (0)    ;buffer of 00H
          00
     ]
0100 41 3A 43 55 53 54  FILE    DB      'A:CUST.FIL',0 ;file name
     2E 46 49 4C 00

010B                    DATA    ENDS

0000                    CODE    SEGMENT 'CODE'

                                ASSUME  CS:CODE,DS:DATA

0000                    MAKE    PROC    FAR

0000 B8 —— R                    MOV     AX,DATA        ;load DS
0003 8E D8                      MOV     DS,AX

0005 B4 3C                      MOV     AH,3CH         ;create CUST.FIL
0007 33 C9                      XOR     CX,CX
```

```
0009 BA 0100 R              MOV      DX,OFFSET FILE
000C CD 21                  INT      21H

000E 8B D8                  MOV      BX,AX                    ;handle to BX

0010 BF 1388                MOV      DI,5000                  ;load record count

0013              MAKE1:

0013 B4 40                  MOV      AH,40H                   ;write a record of 00H
0015 B9 0100                MOV      CX,256
0018 BA 0000 R              MOV      DX,OFFSET BUFFER
001B CD 21                  INT      21H
001D 4F                     DEC      DI
001E 75 F3                  JNZ      MAKE1                    ;repeat 5000 times

0020 B4 3E                  MOV      AH,3EH                   ;close file
0022 CD 21                  INT      21H

0024 B4 4C                  MOV      AH,4CH                   ;exit to DOS
0026 CD 21                  INT      21H

0028              MAKE      ENDP

0028              CODE      ENDS

                            END      MAKE
```

Reading and Writing a Record

Whenever a record is read, the record number is loaded into the BP register and the procedure listed in Example 8–9 is called. This procedure assumes that FIL contains the file handle number and that the CUST.FIL remains open at all times.

EXAMPLE 8–9

```
0000              READ      PROC     FAR

0000 8B 1E 0100 R           MOV      BX,FIL                   ;get handle
0004 B8 0100                MOV      AX,256                   ;multiply by 256
0007 F7 E5                  MUL      BP
0009 8B CA                  MOV      CX,DX
000B 8B D0                  MOV      DX,AX
000D B8 4200                MOV      AX,4200H                 ;move pointer
0010 CD 21                  INT      21H

0012 B4 3F                  MOV      AH,3FH                   ;read record
0014 B9 0100                MOV      CX,256
0017 BA 0000 R              MOV      DX,OFFSET BUFFER
001A CD 21                  INT      21H
001C CB                     RET

001D              READ      ENDP
```

Notice how the record number is multiplied by 256 to obtain a count for the move pointer function. In each case the file pointer is moved from the start of the file to the desired record before it is read into memory area BUFFER. Although not shown, writing a record is performed in the same manner as reading.

Creating a Simple Database

Although many good database programs are available, we can use assembly language and a random access file to create a tailored and highly efficient database. This section shows a simple database system that catalogs video tapes for a small home videotape library. Before the program is created, the size of each record is determined. The record size and allocation of various data fields are listed in Table 8–2. The entire size of each record is 1K byte. Because this is designed for a home video library, we limit the number of records to 1,000. This can be modified, if needed, by changing the program.

This program uses many of the techniques presented in this chapter and the last one. These techniques use the keyboard, disk random access files, and the graphics display mode to display a pleasing image of each videotape in the library. The display used with this program is a 16 color, 640 × 480 VGA display using graphics mode 12H. The contents of one screen of information is the display of one record. The program can also print data on the printer connected to LPT1, the parallel printer port. This program does not allow changes in the display form or in the printed output. The partial program listing (FILL and EDIT are not shown) is found in Example 8–10. The complete program listing is too long to include in the text. Some things to note are the way the VID.DAT file is created, and if needed, how the VIDLIB subdirectory is created.

EXAMPLE 8–10

```
STACK           SEGMENT STACK
                DW          1024 DUP (?)
STACK           ENDS

DATA            SEGMENT

NUMB            DW      0                               ;tape number
RUN             DW      0                               ;running time
TITL            DB      128 DUP (0)                     ;title
```

TABLE 8–2 Record for a videotape cataloging system

Field	Size	Contents
0	128	Title
1	1	Year
2	2	Running time
3	32	Actor 1
4	32	Actor 2
5	32	Actor 3
6	1	Rating
7	2	Tape number
8	794	Description

```
YEAR        DB      0                                   ;year
ACT1        DB      32 DUP (0)                          ;actor 1
ACT2        DB      32 DUP (0)                          ;actor 2
ACT3        DB      32 DUP (0)                          ;actor 3
RATE        DB      0                                   ;rating
NOTE        DB      794 DUP (0)                         ;notes

CHAR        DB      1024 DUP (?)                        ;character lookup table

VDAT        DW      80*9,7                              ;top white bar
            DW      160*9+10,3                          ;2 lines of cyan
            DW      64,8,16,3,64,8,16,3,64,8,16,3,64,8,16,3
            DW      64,8,16,3,64,8,16,3,64,8,16,3,64,8,16,3,64,8,16,3
            DW      64,8,16,3,64,8,16,3,64,8,16,3,64,8,16,3
            DW      64,8,16,3,64,8,16,3,64,8,16,3,64,8,16,3,64,8,16,3
            DW      160*9,3                             ;end title
            DW      3,8,77,3,3,8,77,3,3,8,77,3,3,8,77,3
            DW      3,8,77,3,3,8,77,3,3,8,77,3,3,8,77,3,3,8,77,3
            DW      160*9,3                             ;end tape number
            DW      4,8,76,3,4,8,76,3,4,8,76,3,4,8,76,3
            DW      4,8,76,3,4,8,76,3,4,8,76,3,4,8,76,3,4,8,76,3
            DW      160*9,3                             ;end year
            DW      5,8,75,3,5,8,75,3,5,8,75,3,5,8,75,3
            DW      5,8,75,3,5,8,75,3,5,8,75,3,5,8,75,3,5,8,75,3
            DW      160*9,3                             ;end running time
            DW      1,8,79,3,1,8,79,3,1,8,79,3,1,8,79,3
            DW      1,8,79,3,1,8,79,3,1,8,79,3,1,8,79,3,1,8,79,3
            DW      160*9,3                             ;end rating
            DW      32,8,48,3,32,8,48,3,32,8,48,3,32,8,48,3
            DW      32,8,48,3,32,8,48,3,32,8,48,3,32,8,48,3
            DW      160*9,3                             ;end actor 1
            DW      32,8,48,3,32,8,48,3,32,8,48,3,32,8,48,3
            DW      32,8,48,3,32,8,48,3,32,8,48,3,32,8,48,3,32,8,48,3
            DW      160*9,3                             ;end actor 2
            DW      32,8,48,3,32,8,48,3,32,8,48,3,32,8,48,3
            DW      32,8,48,3,32,8,48,3,32,8,48,3,32,8,48,3
            DW      160*9,3                             ;end actor 3
            DW      64,8,16,3,64,8,16,3,64,8,16,3,64,8,16,3
            DW      64,8,16,3,64,8,16,3,64,8,16,3,64,8,16,3,64,8,16,3
            DW      64,8,16,3,64,8,16,3,64,8,16,3,64,8,16,3
            DW      64,8,16,3,64,8,16,3,64,8,16,3,64,8,16,3,64,8,16,3
            DW      64,8,16,3,64,8,16,3,64,8,16,3,64,8,16,3
            DW      64,8,16,3,64,8,16,3,64,8,16,3,64,8,16,3,64,8,16,3
            DW      64,8,16,3,64,8,16,3,64,8,16,3,64,8,16,3
            DW      64,8,16,3,64,8,16,3,64,8,16,3,64,8,16,3,64,8,16,3
            DW      64,8,16,3,64,8,16,3,64,8,16,3,64,8,16,3
            DW      64,8,16,3,64,8,16,3,64,8,16,3,64,8,16,3,64,8,16,3
            DW      64,8,16,3,64,8,16,3,64,8,16,3,64,8,16,3
            DW      64,8,16,3,64,8,16,3,64,8,16,3,64,8,16,3,64,8,16,3
            DW      64,8,16,3,64,8,16,3,64,8,16,3,64,8,16,3
            DW      64,8,16,3,64,8,16,3,64,8,16,3,64,8,16,3,64,8,16,3
            DW      64,8,16,3,64,8,16,3,64,8,16,3,64,8,16,3
            DW      64,8,16,3,64,8,16,3,64,8,16,3,64,8,16,3,64,8,16,3
            DW      64,8,16,3,64,8,16,3,64,8,16,3,64,8,16,3
            DW      64,8,16,3,64,8,16,3,64,8,16,3,64,8,16,3,64,8,16,3
            DW      64,8,16,3,64,8,16,3,64,8,16,3,64,8,16,3
            DW      64,8,16,3,64,8,16,3,64,8,16,3,64,8,16,3,64,8,16,3
            DW      64,8,16,3,64,8,16,3,64,8,16,3,64,8,16,3
            DW      64,8,16,3,64,8,16,3,64,8,16,3,64,8,16,3,64,8,16,3
```

```
                        DW          26,8,54,3,26,8,54,3,26,8,54,3,26,8,54,3
                        DW          26,8,54,3,26,8,54,3,26,8,54,3,26,8,54,3,26,8,54,3
                        DW          480*9-10,3
                        DW          400*9,11
                        DW          0                                           ;end video data
        MLST            DW          M1,M1A,M2,M3,M4,M5,M6,M7,M8,M9,M10,M11,M12,M13,0
        M1              DB          13,0,20,4,'VIDEO LIBRARY'
        M1A             DB          25,0,34,0,'(c) 1992 by Barry B. Brey'
        M2              DB          5,3,4,0,'TITLE'
        M3              DB          6,7,3,0,'NUMBER'
        M4              DB          4,10,5,0,'YEAR'
        M5              DB          4,13,5,0,'TIME'
        M6              DB          6,16,3,0,'RATING'
        M7              DB          7,19,2,0,'ACTOR 1'
        M8              DB          7,22,2,0,'ACTOR 2'
        M9              DB          7,25,2,0,'ACTOR 3'
        M10             DB          5,28,4,0,'NOTES'
        M11             DB          25,48,2,0,'F9 = Search   F10 = Print'
        M12             DB          25,49,2,0,'F7 = Erase   F8 = Quit '
        M13             DB          25,50,2,0,'F5 = Append  F6 = Redo '
        M14             DB          24,49,40,4,'***ERROR*** disk problem'
        M15             DB          24,49,40,4,'***PLEASE WAIT***      '
        M16             DB          24,49,40,11,'***PLEASE WAIT***      '

        DIRN            DB          'C:\VIDLIB',0
        FILEN           DB          'C:\VIDLIB\VID.DAT',0
        FILEH           DW          ?

        DATA            ENDS


        CODE            SEGMENT  'CODE'
                        ASSUME      CS:CODE,DS:DATA,SS:STACK

        MAIN            PROC        FAR

                        CLD                                                     ;select auto-increment
                        MOV         AX,DATA                                     ;address DATA segment
                        MOV         DS,AX

                        MOV         AX,0012H                                    ;select video mode
                        INT         10H

                        MOV         AX,1130H                                    ;get 8 x 8 character set
                        MOV         BH,3
                        INT         10H

                        MOV         CX,1024
                        MOV         DI,OFFSET CHAR
                        MOV         SI,BP
        MAIN1:                                                                  ;get character table
                        MOV         AL,ES:[SI]
                        MOV         [DI],AL
                        INC         SI
                        INC         DI
                        LOOP        MAIN1

                        MOV         AX,0A000H                                   ;address video memory
                        MOV         ES,AX

                        CALL        PAINT                                       ;paint graphics screen and lables
```

```
              MOV     AX,3D02H                        ;open file
              MOV     DX,OFFSET FILEN
              INT     21H
              JNC     MAIN4
              CMP     AX,2                            ;file not found
              JE      MAIN3
              CMP     AX,3                   ;path not found
              JE      MAIN2
MAIN1A:                                               ;if disk error
              MOV     SI,OFFSET M14
              CALL    DMES
TRAP:                                                 ;if disk error trap
              JMP     TRAP
MAIN2:                                                ;make subdirectory VIDLIB
              MOV     AH,39H
              MOV     DX,OFFSET DIRN
              INT     21H
              JC      MAIN1A

MAIN3:                                                ;create file
              MOV     AH,3CH
              XOR     CX,CX
              MOV     DX,OFFSET FILEN
              INT     21H
              JC      MAIN1A                          ;if error
              MOV     FILEH,AX                        ;save handle

              MOV     SI,OFFSET M15                   ;show program busy
              CALL    DMES
              MOV     BP,1000                         ;initialize record count
              MOV     NUMB,1                          ;set record number
MAIN3A:                                               ;initialize data base
              MOV     AH,40H                          ;write record
              MOV     BX,FILEH                        ;get handle
              MOV     CX,1024                         ;byte count
              MOV     DX,OFFSET NUMB
              INT     21H
              JC      MAIN1A                          ;if error
              INC     NUMB                            ;increment record number
              DEC     BP                              ;decrement count
              JNZ     MAIN3A                          ;repeat 1,000 times

              MOV     SI,OFFSET M16                   ;clear busy
              CALL    DMES
              MOV     AX,FILEH
MAIN4:                                                ;get first record
              MOV     FILEH,AX                        ;save file handle
              MOV     BX,AX
              MOV     AX,4200H                        ;front of file
              XOR     DX,DX
              XOR     CX,CX
              INT     21H
              MOV     AH,3FH                          ;read first record
              MOV     CX,1024
              MOV     DX,OFFSET NUMB
              INT     21H
              JC      MAIN1A                          ;if error
              CALL    FILL                            ;display record
              CALL    EDIT                            ;enter editor

              MOV     AH,3EH                          ;close file
```

```
                        MOV         BX,FILEH
                        INT         21H

                        MOV         AX,4C00H                    ;return to DOS
                        INT         21H

MAIN                    ENDP

PAINT                   PROC        NEAR                        ;paint main screen

                        MOV         DI,0                        ;address video RAM
                        MOV         SI,OFFSET VDAT              ;address video data

                        MOV         DX,3CEH                     ;address bit mask register
                        MOV         AL,8
                        OUT         DX,AL
                        INC         DX
                        MOV         AL,0FFH                     ;select all PELs
                        OUT         DX,AL
                        MOV         DX,3C4H                     ;address map mask register
                        MOV         AL,2
                        OUT         DX,AL
                        INC         DX
PAINT1:
                        LODSW
                        OR          AX,AX
                        JZ          PAINT3                      ;if end video data
                        MOV         CX,AX                       ;get byte count
                        LODSW                                   ;get color
                        MOV         BL,AL
PAINT2:
                        MOV         AL,0FH
                        OUT         DX,AL                       ;enable all bit planes
                        MOV         AL,ES:[DI]                  ;load latch
                        MOV         BYTE PTR ES:[DI],0          ;black
                        MOV         AL,BL
                        OUT         DX,AL                       ;select color
                        MOV         AL,0FFH
                        STOSB
                        LOOP        PAINT2
                        JMP         PAINT1                      ;get next
PAINT3:                                                         ;display text data
                        MOV         BP,OFFSET MLST             ;address text data
PAINT4:
                        MOV         SI,DS:[BP]
                        ADD         BP,2
                        OR          SI,SI                       ;if end of text data
                        JZ          PAINT5
                        CALL        DMES                        ;display string
                        JMP         PAINT4                      ;get next text string
PAINT5:
                        RET

PAINT                   ENDP

DMES                    PROC        NEAR                        ;display message

                        LODSB                                   ;DS:SI = address of string
                        MOV         CL,AL                       ;get count
                        XOR         CH,CH
                        LODSB                                   ;get line number
```

```
                MOV         DH,AL
                LODSB                                   ;get column number
                MOV         DL,AL
                LODSB                                   ;get text color
                MOV         BL,AL
DMES1:
                LODSB                                   ;get character
                CALL        DCHAR                       ;display character
                INC         DL                          ;address next column
                LOOP        DMES1                       ;repeat
                RET

DMES            ENDP

DCHAR           PROC        NEAR                        ;display character

                PUSH        SI                          ;save registers
                PUSH        DX
                PUSH        CX
                PUSH        AX                          ;AL = ASCII character
                MOV         CX,720                      ;BL = character color
                MOV         AL,DH                       ;DH = line number
                XOR         AH,AH                       ;DL = column number
                PUSH        DX
                MUL         CX
                POP         DX
                MOV         DI,AX
                XOR         DH,DH
                ADD         DI,DX
                MOV         SI,OFFSET CHAR              ;address character
                POP         AX
                XOR         AH,AH
                SHL         AX,1
                SHL         AX,1
                SHL         AX,1
                ADD         SI,AX
                MOV         CX,8
DCHAR1:
                MOV         DX,3CEH                      ;address bit mask register
                MOV         AL,8
                OUT         DX,AL
                INC         DX
                LODSB                                    ;get scan line
                OUT         DX,AL                        ;set mask
                MOV         DX,3C4H                      ;address map mask register
                MOV         AL,2
                OUT         DX,AL
                INC         DX
                MOV         AL,0FH                       ;enable all bit planes
                OUT         DX,AL
                MOV         AL,ES:[DI]                   ;load latch
                MOV         BYTE PTR ES:[DI],0           ;black
                MOV         AL,BL
                OUT         DX,AL                        ;set color
                MOV         BYTE PTR ES:[DI],0FFH
                ADD         DI,80
                LOOP        DCHAR1
                POP         CX
                POP         DX
                POP         SI
                RET
```

```
DCHAR          ENDP

FILL           PROC         NEAR                              ;not shown

;This procedure displays the data for the current record

FILL           ENDP

EDIT           PROC         NEAR                              ;not shown

;This procedure edits the screen and responds to the function keys

EDIT           ENDP

CODE           ENDS

               END          MAIN
```

8–5   SUMMARY

1. Two types of floppy disks are found in application today: (a) the mini-floppy 5¼″ disk stores 360K bytes (DSDD) or 1.2M bytes (HD), and (b) the micro-floppy 3½″ disk stores 720K bytes (DSDD), 1.44M bytes (HD), or 2.88M bytes (EHD).
2. Disk data are stored in sectors that often contain 512 bytes of data. A track is a concentric ring that holds 9 or more sectors of data. A cylinder contains two or more tracks. A cluster often contains 4 or more sectors.
3. Data are stored on a disk starting with the boot sector located in the outermost track at its first sector. Following the boot sector, which contains the bootstrap loader program, is the FAT (file allocation table) that shows which sectors (clusters) are occupied and which are free. Located after the FAT is the root directory, which contains the system programs entries and root directory. The remaining part of the disk contains files and subdirectories.
4. Files are accessed by first opening or creating them using DOS function 21H calls. Once a file is opened or created, it can be read or written to using other DOS function 21H calls. Before a program that uses a file ends, the file must be closed or problems can result. Files are accessed by the use of a file name stored in an ASCII-Z string. The ASCII-Z string is an ASCII character string that ends with a NUL (00H).
5. Once a file is opened via an ASCII-Z string, it is referred to by a file handle. The file handle allows DOS to process file references more efficiently.
6. The file pointer is a 32-bit number used to refer to any byte in a file. A DOS function 21H call is provided that allows the file pointer to be moved any number of bytes from the start of the file, the current location, or the end of the file.
7. Random access files are created by using sequential access files and the file pointer. The file pointer can be used to access, randomly, any portion of the sequential file.

8–6   **GLOSSARY**

Boot sector   The first sector in the outermost track of a disk, which contains a program (bootstrap loader) that loads the disk operating system into the computer memory.

Bootstrap loader   A program that is used to load the disk operating system into the memory from the disk.

Cluster   A grouping of sectors. A cluster on many hard disk systems is a group of four sectors.

Cylinder   A grouping of two more more tracks on a disk. On a floppy disk a cylinder is composed of a top and bottom track.

FAT   A file allocation table indicates which sectors (clusters) on the disk contain data and which are free.

File pointer   A mechanism that addresses a section of a sequential access file that allows it to perform as a random access file.

Handle   A number used to refer to a file in a DOS based computer system.

Micro-floppy disk   A 3 ½″ disk used with a computer system. Micro-floppy disks are double-sided, double-density that store 720K bytes, high-density that store 1.44M bytes, or extra-high-density that store 2.88M bytes.

Mini-floppy disk   A 5 ¼″ disk used with a computer system. Mini-floppy disks are either double-sided, double-density that store 360K bytes or high-density that store 1.2M bytes.

Random access file   A file that is accessed at any point rather than through each byte as a sequential access file.

Sector   A section of disk track that usually contains 512 bytes of data.

Sequential access file   A file that is accessed sequentially from the first byte through the last byte.

Track   A concentric ring on a disk that contains sectors of data.

8–7   **QUESTIONS**

1. A 5 ¼″ floppy disk is called a _____ floppy disk.
2. A 3 ½″ floppy disk is called a _____ floppy disk.
3. How much data is stored on a 5 ¼″ double-sided, double-density disk?
4. The high-density 3 ½″ floppy disk stores _____ bytes of data.
5. How much information is stored on a 3 ½″ extra-high-density disk?
6. Define the following terms:
   a. Track
   b. Sector

   c. Cluster

   d. Cylinder

7. What information is stored in the file allocation table?

8. Detail the purpose of the boot sector.

9. The root directory contains _____ entries.

10. How many directory entries can be placed in a subdirectory?

11. How many characters can be used for a file name?

12. How many characters can be used in a file extension?

13. How many bytes are used to store a directory name?

14. Develop a short sequence of instructions that opens a file whose name is stored at data offset address FILEN as a read-only file.

15. Develop a short sequence of instructions that creates a hidden file whose name is stored at data segment offset address FILES.

16. Develop a procedure that creates a file whose name is FROG.LST and fills it with 512 bytes of 00H. (Don't forget to close the file.)

17. Write a macro instruction sequence called READ. The syntax for this macro is READ BUFFER,COUNT,HANDLE where BUFFER is the offset address of a file buffer, COUNT is the number of bytes read, and HANDLE is the address of the file handle.

18. Develop a macro sequence called APPEND. The syntax for this macro is AP-PEND BUFFER,COUNT,HANDLE where BUFFER is the offset address of the file buffer, COUNT is the number of bytes to append to the file, and HAN-DLE is the address of the file handle. (Don't forget to use the move file pointer function to append data to the end of the file.)

19. Create a random access file (RAN.FIL) containing 200 records with a record length of 100 bytes. To initialize this random access file, place a 00H in every byte.

20. Using the random access file developed in Question 19, develop a macro sequence that accesses and reads any record. The syntax for this macro is AC-CESS BUFFER,RECORD,HANDLE where BUFFER is the offset address of the record buffer, RECORD is the record number, and HANDLE is the address of the file handle.

21. Modify the macro of Question 20 so it can read or write a record. The syntax is ACCESS T,BUFFER,RECORD,HANDLE where T is added to specify a read 'R' or a write 'W' using the ASCII characters for the letters R and W.

22. Develop a short program that displays the first 256 bytes of a file on the video screen. Your display must appear as in Example 8–11, showing the actual data in any file. It must ask for the name of the file to be displayed, as in the example, and then display the first 256 bytes. (Use text mode.)

EXAMPLE 8–11

FILENAME = TEST.TXT

```
00: 00 00 00 00 00 00 00 00 00 00 00 00 00 00 00 00
10: 00 00 00 00 00 00 00 00 00 00 00 00 00 00 00 00
20: 00 00 00 00 00 00 00 00 00 00 00 00 00 00 00 00
```

```
30: 00 00 00 00 00 00 00 00 00 00 00 00 00 00 00 00
40: 00 00 00 00 00 00 00 00 00 00 00 00 00 00 00 00
50: 00 00 00 00 00 00 00 00 00 00 00 00 00 00 00 00
60: 00 00 00 00 00 00 00 00 00 00 00 00 00 00 00 00
70: 00 00 00 00 00 00 00 00 00 00 00 00 00 00 00 00
80: 00 00 00 00 00 00 00 00 00 00 00 00 00 00 00 00
90: 00 00 00 00 00 00 00 00 00 00 00 00 00 00 00 00
A0: 00 00 00 00 00 00 00 00 00 00 00 00 00 00 00 00
B0: 00 00 00 00 00 00 00 00 00 00 00 00 00 00 00 00
C0: 00 00 00 00 00 00 00 00 00 00 00 00 00 00 00 00
D0: 00 00 00 00 00 00 00 00 00 00 00 00 00 00 00 00
E0: 00 00 00 00 00 00 00 00 00 00 00 00 00 00 00 00
F0: 00 00 00 00 00 00 00 00 00 00 00 00 00 00 00 00
```

23. Repeat Question 22 using graphics mode to display black numerals on a cyan video screen.

CHAPTER 9

Interrupt Hooks and the TSR

INTRODUCTION

In order to use the personal computer in many applications a few additional software techniques are required. This chapter presents a technique called an interrupt hook that allows a program to hook into the interrupt structure of the computer system. Interrupts occur for almost every I/O event in a computer. An interrupt hook allows assembly language access to these systems events.

A TSR is a terminate and stay resident program that remains in the memory, dormant, until activated by some event. A TSR is often activated by a special key called a hot-key. A hot-key might be a shift F2, or an alternate C, or any other non-standard ASCII key code on the keyboard. A program accessed via a hot-key is often called a pop-up program.

9-1 CHAPTER OBJECTIVES

Upon completion of this chapter you will be able to:

1. Develop an understanding of the function and application of the interrupt structure of the personal computer.
2. Hook into the interrupts for the keyboard and clock tick to perform tasks on the computer.
3. Develop terminate and stay resident software using a hot-key to access a program.
4. Control the timer and speaker in the personal computer.
5. Install and remove TSR programs.

9–2   INTERRUPT HOOKS

Interrupt **hooks** are used to tap into the interrupt structure of the microprocessor. For example, we might hook into the keyboard interrupt so we can detect a special keystroke called a hot-key. Whenever the hot-key is typed, we access a terminate and stay resident (TSR) program that performs a special task. Some examples of hot-key software are pop-up calculators, pop-up clocks, and so forth. We might also hook into the clock tick interrupt to time events.

Linking to an Interrupt

In order to link to or hook into an interrupt, we use a DOS function call that reads the current address from the interrupt vector. The DOS INT 21H function call number 35H reads the current interrupt vector and DOS INT 21H function call number 25H changes the address of the current vector. With both DOS function calls, AL indicates the vector type number (00H–FFH) and AH indicates the DOS function call number.

When the vector is read using function 35H, the offset address is returned in register BX and the segment address is in register ES. These two registers' contents are saved so they can be restored when the interrupt hook is removed from memory. When the vector is changed, it changes to the address stored at the memory location addressed by DS:DX using function number 25H.

The process of installing an interrupt handler through a hook is illustrated in Example 9–1. This procedure reads the current interrupt vector's address (DOS INT 21H function 35H) and stores it into a double-word memory location for access by the new interrupt service procedure. Please note how this double word is addressed in this procedure. Next, the address of the new interrupt service procedure, located in DS:DX is placed into the vector using the DOS function call number 25H. Notice that this sequence of instructions is organized as a .COM file, which has an origin of 100H. As mentioned, a .COM file is created from an .EXE file by using EXE2BIN with version 5.1 of the MASM assembler or by using an option selected in version 6.0.

EXAMPLE 9–1

```
                        ORG     100H                            ;origin for a COM program

0100                    START:

0100 EB 04                      JMP     MAIN

0102 00000000           ADDRESS DD      ?                       ;old interrupt vector

0106                    MAIN:

0106 8C C8                      MOV     AX,CS                   ;address CS with DS
0108 8E D8                      MOV     DS,AX

                ;get vector 0 address

010A B8 3500                    MOV     AX,3500H
010D CD 21                      INT     21H
```

```
                              ;save vector address

010F  2E: 89 1E 0102 R        MOV       WORD PTR ADDRESS,BX
0114  2E: 8C 06 0104 R        MOV       WORD PTR ADDRESS+2,ES

                              ;install new interrupt vector 0 address

0119  B8 2500                 MOV       AX,2500H
011C  BA 0300 R               MOV       DX,OFFSET NEW
011F  CD 21                   INT       21H
```

Hooking into the Clock Tick Interrupt

The *clock tick interrupt* is constant on all personal computer systems and is an excellent way to time events and programs. The clock tick interrupt occurs about 18.2 times a second (the actual number is 18.2064819336). The clock tick interrupt uses vector 8 in the personal computer. When an interrupt occurs it pauses the program that is currently executing in order to execute the interrupt service procedure addressed by the interrupt vector. After the interrupt service procedure executes, the program that was paused resumes execution.

Suppose we need a program that beeps the speaker once on each half hour and twice on the hour. This is accomplished by hooking into interrupt vector 8, which occurs 18.2 times per second. As each clock tick interrupt occurs, the time is tested to see if it is at the half-hour or hour mark. In either of these cases, our interrupt service procedure beeps the speaker the correct number of times.

Before we can write this program, we need to learn how to control the speaker in the computer. Figure 9–1 illustrates the internal organization of the timer (8253) and speaker. The 8253's timer 0 provides the clock tick interrupt, timer 1 provides an interrupt that controls DRAM refresh in some personal computers, and timer 2 provides a signal to the speaker through a NAND gate.

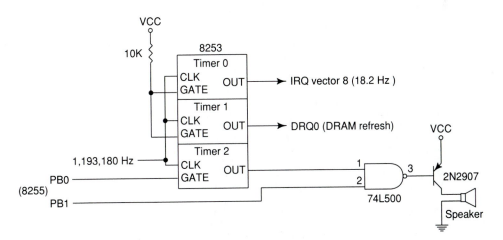

FIGURE 9–1 The speaker circuit connected to the timer inside the personal computer. (The 8255 is at I/O ports 60H–63H, and the 8253 timer is at I/O ports 40H–43H.)

15							0
TAG (7)	TAG (6)	TAG (5)	TAG (4)	TAG (3)	TAG (2)	TAG (1)	TAG (0)

Note:
The index i of tag(i) is not top-relative. A program typically uses the "top" field of Status Word to determine which tag(i) field refers to logical top of stack.

Tag values:
00 = valid
01 = zero
10 = invalid or infinity
11 = empty

FIGURE 10–7 The TAG register and conditions held in each TAG. (Courtesy of Intel Corporation)

4. *Exception Masks*—determine whether the error indicated by the exception affects the error bit in the status register. If a logic 1 is placed in one of the exception control bits, the corresponding status register bit is masked off.

Tag Register. The tag register marks the contents of each register in the 80X87 stack. Figure 10–7 illustrates the tag register and the status indicated by each tag. The tag indicates whether a register is: valid; zero; invalid or infinity; or empty. This register is set to a FFFFH when the coprocessor is reset or initialized. The only way that a program can view the tag register is by storing the coprocessor environment using the FSTENV, FSAVE, or FRSTOR instructions. Each of these instructions stores the tag register along with other coprocessor data.

The RESET pin initializes the 80X87 whenever the microprocessor is reset. The 80X87 responds to a reset input or to the software reset instruction, FINIT, in almost the same manner. The hardware reset forces the 80X87 to operate in the real mode and the FINIT instruction does not change the mode. Table 10–2 shows how the various internal sections of the 80X87 respond to a reset. In all but special cases we normally operate the 80X87 in the default reset mode of operation.

TABLE 10–2 Internal state of the 80X87 after a RESET or FINIT

Field	Value	Condition
Infinity	0	Projective
Rounding	00	Round-to-nearest
Precision	11	64-bit
Error Masks	11111	Error bits disabled
Busy	0	Not busy
C_3–C_0	????	Unknown
TOP	000	Register 000
ES	0	No error
Error Bits	00000	No errors
Tags	11	Empty
Registers	—	Not changed

10–4   THE INSTRUCTION SET

The 80X87 arithmetic coprocessor is able to execute 68 different instructions. Whenever a coprocessor instruction references the memory, the microprocessor automatically generates the memory address for the instruction. The 80X87 uses the data bus for data transfers during coprocessor instructions and the microprocessor uses it during normal microprocessor instructions.

This section of the text describes the function of each instruction and lists its assembly language form. Because the coprocessor uses all of the microprocessor memory-addressing modes, not all possible forms of each instruction are illustrated. Each time that the assembler encounters one of the 80X87 mnemonic opcodes, it converts it into a machine language ESC instruction. The ESC instruction represents an opcode to the 80X87.

Data Transfer Instructions

There are three basic data transfers: floating-point, signed-integer, and BCD. The only time that data ever appear in signed-integer or BCD form is in the memory. Inside the coprocessor, data are always stored as 80-bit extended-precision floating-point data.

Floating-Point Data Transfers.   There are four floating-point data transfer instructions in the 80287 instruction set: FLD (load real), FST (store real), FSTP (store real and pop), and FXCH (exchange).

The FLD instruction loads memory data to the top of the internal stack. This instruction stores the data on the top of the stack and then decrements the stack pointer by one. Data loaded to the top of the stack are from any memory location or from another register. For example, an FLD ST(2) instruction copies the contents of register 2 to the stack top, which is ST. The top of the stack is register 0 when the coprocessor is reset or initialized. Another example is the FLD DATA7 instruction that copies the contents of memory location DATA7 to the top of the stack. The size of the transfer is automatically determined by the assembler through the directives DD for single-precision, DQ for double-precision, and DT for extended-precision.

The FST instruction stores a copy of the top of the stack into the memory location or coprocessor register indicated by the operand. At the time of storage, the internal, extended-precision floating-point number is rounded to the size of the floating point number indicated by the control register.

The FSTP instruction stores a copy of the top of the stack into memory or any coprocessor register and then pops the data from the top of the stack. You might think of FST as a *copy* instruction and FSTP as a *copy and remove* instruction.

The FXCH instruction exchanges the register indicated by the operand with the top of the stack. For example, the FXCH ST(2) instruction exchanges the top of the stack with register 2.

Integer Data Transfer Instructions. The 80X87 supports three integer data transfer instructions: FILD (load integer), FIST (store integer), and FISTP (store integer and pop). These three instructions function as did FLD, FST, and FSTP except the data transferred are integer data. The 80X87 automatically converts the internal extended-precision data to integer data. The size of the data is determined by the way the label is defined with DW, DD, or DQ.

BCD Data Transfer Instructions. Two instructions are used to load or store BCD data. The FBLD instruction loads the top of the stack with BCD memory data and the FBSTP stores the top of the stack and does a pop.

Example 10-6 shows how the assembler automatically adjusts the FLD, FILD, and FBLD instructions for different size operands. (Look closely at the machine coded forms of the instructions.) Note in this example that it begins with the .286 and .287 directives that identify the microprocessor as an 80286 and the coprocessor as an 80287. The assembler by default assumes that the software is assembled for an 8086/8088 with an 8087 coprocessor. If the 80387 is in use, we use the .387 directive.

EXAMPLE 10-6

```
                        .286
                        .287

0000                    DATAS     SEGMENT

0000 41F00000           DATA1     DD    30.0        ;single-precision
0004 4059100000000000   DATA1     DQ    100.25      ;double-precision
000C 400487F34D6A161E4F76  DATA3  DT    33.9876     ;extended-precision

0016 001E               DATA4     DW    30          ;16-bit integer
0018 0000001E           DATA5     DD    30          ;32-bit integer
001C 000000000000001E   DATA6     DQ    30          ;64-bit integer

0024 00000000000000000030  DATA7  DT    30H         ;BCD 30

002E                    DATAS     ENDS

0000                    CODE      SEGMENT 'CODE'

                                  ASSUME  CS:CODE,DS:DATAS

0000 D9 06 0000 R                 FLD     DATA1
0004 DD 06 0004 R                 FLD     DATA2
0008 DB 2E 000C R                 FLD     DATA3

000C DF 06 0016 R                 FILD    DATA4
0010 DB 06 0018 R                 FILD    DATA5
0014 DF 2E 001C R                 FILD    DATA6

0018 DF 26 0024 R                 FBLD    DATA7

001C                    CODE      ENDS
                                  END
```

Arithmetic Instructions

The coprocessor arithmetic instructions include addition, subtraction, multiplication, division, and square root. Arithmetic-related instructions include scaling, rounding, absolute value, and changing the sign.

Table 10–3 shows the basic addressing modes allowed for the arithmetic operations. Each addressing mode is shown with an example using the FADD (real addition) instruction. All arithmetic operations are floating-point except in some cases when memory data are referenced as an operand.

The classic stack form of addressing operand data (stack addressing) uses the top of the stack as a source operand and the next to the top of the stack as a destination operand. Afterwards, the two original datum are removed from the stack and only the result remains at the top of the stack. To use this addressing form the instruction is placed in the program without any operands such as FADD or FSUB. The FADD instruction adds ST and ST(1) and stores the answer at the top of the stack; it also removes the original two datum from the stack by popping. This addressing mode is identical to the one used in a reverse polish entry calculator such as an HP scientific calculator.

The register-addressing mode uses ST for the top of the stack and ST(n) for another location, where n is the register number. With this form, one operand must be ST and the other is ST(n). Note that to double the top of the stack we use the FADD ST,ST(0) instruction, where ST(0) addresses the top of the stack. One of the two operands in the register-addressing mode must be ST, while the other must be in the form ST(n), where n is a stack register 0–7.

The memory addressing mode always uses the top of the stack as a destination because the 80X87 is a stack-oriented machine. For example, the FADD DATA instruction adds the real number contents of memory location data to the top of the stack. The result is also found at the top of the stack after the addition.

Arithmetic Operations. The letter P in an opcode specifies a register pop after the operation (FADDP compared to FADD). The letter R in an opcode (subtraction and division only) indicates reverse mode. The reverse mode is useful for memory data because normally memory data subtracts from the top of the stack. A reversed subtract instruction subtracts the top of the stack from memory and stores the result in the top of the stack. For example, if the top of the stack contains a 10 and memory

TABLE 10–3 Coprocessor addressing modes

Mode	Form	Examples
Stack	ST,ST(1)	FADD
Register	ST,ST(n)	FADD ST,ST(2)
	ST(n),ST	FADD ST(6),ST
Register pop	ST(n),ST	FADDP ST(3),ST
Memory	operand	FADD DATA

Note: Stack addressing is fixed as ST,ST(1) and n = register number 0–7

location DATA1 contains a 1, then the FSUB DATA1 instruction results in a +9 on the stack top. The FSUB DATA1 instruction subtracts the 1, in memory, from the top of the stack. The FSUBR instruction results in a −9 because it subtracts the 10 on the top of the stack from the 1 in memory.

The letter I as a second letter in an opcode indicates that the memory operand is an integer. For example, the FADD DATA instruction is a floating-point addition, while the FIADD DATA is an integer addition. The FIADD DATA instruction adds the integer at memory location DATA to the floating-point number at the top of the stack. The same rules apply to FADD, FSUB, FMUL, and FDIV instructions.

Arithmetic-Related Operations. Other operations that are arithmetic in nature include: FSQRT (square root), FSCALE (scale a number), FPREM (find partial remainder), FRNDINT (round to integer), FXTRACT (extract exponent and significand), FABS (find absolute value), and FCHG (change sign). These instructions and the functions they perform follow:

1. FSQRT—finds the square root of the top of the stack and leaves the resultant square root on the top of the stack. An invalid error occurs for the square root of a negative number. The IE bit in the status register, which indicates an invalid error, should be tested whenever an invalid result can occur.
2. FSCALE—adds the contents of ST(1) (interpreted as an integer) to the exponent at the top of the stack. FSCALE can multiply or divide rapidly by powers of two. The value in ST(1) must be between 2^{-15} and 2^{+15}.
3. FPREM—performs modulo division on ST by ST(1). The resultant modulo remainder is found in the top of the stack and has the same sign as the original dividend.
4. FRNDINT—rounds the top of the stack to an integer.
5. FXTRACT—decomposes the number at the top of the stack into two separate parts that represent the value of the exponent and the value of the significand. The extracted significand is found at the top of the stack and the exponent at ST(1). We often use this instruction to convert a floating-point number into a form that can be printed.
6. FABS—changes the sign of the top of the stack to positive.
7. FCHS—changes the sign from positive to negative or negative to positive.

Comparison Instructions

The comparison instructions examine data at the top of the stack in relation to another element and return the result of the comparison in status register condition code bits C_3–C_0. Comparisons that are allowed by the coprocessor are FCOM (floating-point compare), FCOMP (floating-point compare with a pop), FCOMPP (floating-point compare with 2 pops), FICOM (integer compare), FICOMP (integer compare and pop), FSTS (test), and FXAM (examine). Following is a list of these instructions with a description of their function:

1. FCOM—compares the floating-point data at the top of the stack with an operand, which may be any register or any memory operand. If the operand is not coded

with the instruction, the next stack element ST(1) is compared with the stack top ST.

2. FCOMP and FCOMPP—both instructions perform as FCOM, but they also pop one or two register from the stack. We often use these forms to clear data from the stack at the end of a program or procedure that uses the arithmetic co-processor.

3. FICOM and FICOMP—the top of the stack is compared with the integer stored at a memory operand. In addition to the compare, FICOMP also pops the top of the stack.

4. FTST—tests the contents of the top of the stack against a zero. The result of the comparison is coded in the status register condition code bits as illustrated in Table 10–1 with the status register.

5. FXAM—examines the stack top and modifies the condition code bits to indicate whether the contents are positive, negative, normalized, and the like. Refer back to the status register in Table 10–1.

Transcendental Operations

The transcendental instructions include: FPTAN (partial tangent), FPATAM (partial arctangent), F2XM1 ($2^x - 1$), FYL2X ($Y \log_2 X$), and FYL2XP1 ($Y \log_2(X + 1)$). The 80387 and 80486 also contain FSIN, FCOS, and FSINCOS instructions not found in earlier versions of the arithmetic coprocessor. A list of the transcendental operations follows with a description of each transcendental operation:

1. FPTAN—finds the partial tangent of $Y/X = \tan \theta$. The value of θ is at the top of the stack and must be between 0 and $\pi/4$ (45°). Note that the angle is measured in radians where 2π radians equals 360°. The result is a ratio that is found as ST = X and ST(1) = Y. If the value is outside of the allowable range, an invalid error occurs as indicated by the status register.

2. FPATAN—finds the partial tangent at θ = ARCTAN X/Y. The value of X is at the top of the stack and Y is at ST(1). The values of X and Y must be as follows: $0 \le Y < X < \infty$. The instruction pops the stack and leaves θ at the top of the stack.

3. F2AM1—finds the function $2^X - 1$. The value of X is taken from the top of the stack and the result is returned to the top of the stack. To obtain 2^X, add one to the result at the top of the stack. This function can be used to derive the functions listed in Table 10–4. Note that the constants $\log_2 10$ and $\log 2^e$ are built in as standard values for the 80X87.

4. FYL2X—finds $Y \log_2 X$. The value X is taken from the stack top and Y is taken from ST(1). The result is found at the top of the stack after a pop. The

TABLE 10–4 Exponential functions

Function	Equation
10^x	$2^x \log_2 10$
e^x	$2^x \log_2 e$
y^x	$2^x \log_2 y$

value of X must range between 0 and ∞ and the value of Y must be between $-\infty$ and $+\infty$.

5. FYL2XP1—finds $Y\log_2(X+1)$. The value of X is taken from the stack top and Y is taken from ST(1). The result is found at the top of the stack after a pop. The value of X must range between 0 and $1 - \sqrt{2}/2$ and the value of Y must be between $-\infty$ and $+\infty$.

6. FSIN, FCOS—finds the sine or cosine of ST and leaves the result at the top of the stack. The range of the values at the ST must be between 0 and 2^{63} for the FSIN or FCOS instruction to function properly. As with FPTAN, FSIN and FCOS assume that the ST contains an angle in radians.

7. FSINCOS—finds both the sine and cosine of the ST. The result is found in ST (cosine) and ST(1) (sine) after this instruction executes.

Constant Operations

The 80X87 instruction set includes opcodes that return constants to the top of the stack. A list of these instructions appears in Table 10–5.

80X87 Control Instructions

The 80X87 coprocessor has control instructions for initialization, exception handling, and task switching. The control instructions have two forms. For example, FINIT initializes the coprocessor and so does FNINIT. The difference is that FNINIT does not cause any wait states, while FINIT does cause waits. The microprocessor will wait for the FINIT instruction by testing the $\overline{\text{BUSY}}$ pin on the coprocessor. All control instructions have these two forms. Following is a list of each control instruction with its function:

1. FINIT/FNINIT—this instruction performs the same basic function as reset as described in Section 10–3. The 80X87 operates with a closure of projective, rounds to the nearest, and uses a precision of 64 bits when reset or initialized.

2. FSETPM—changes the addressing mode of the 80X87 to the protected addressing mode. This mode is used when the microprocessor is operated in the protected mode.

3. FLDCW—loads the control register with the word addressed by the operand.

4. FSTCW/FNSTCW—stores the control register into the word-sized memory operand.

TABLE 10–5 Constant operations

Instruction	Constant pushed to ST
FLDZ	+0.0
FLD1	+1.0
FLDPI	π
FLDL2T	$\log_2 10$
FLDL2E	$\log_2 e$
FLDLG2	$\log_{10} 2$
FLDLN2	$\log_e 2$

FIGURE 10–8 Memory format when the 80X87 registers are stored by the FSAVE instruction. (Courtesy of Intel Corporation)

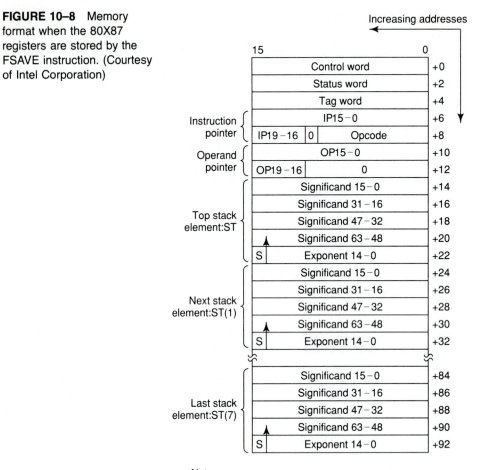

Notes:
S = sign
Bit 0 of each field is rightmost, least significant bit of corresponding register field.
Bit 63 of significand is integer bit (assumed binary point is immediately to the right).

5. FSTSW AX/FNSTSW AX—copies the contents of the control register to the AX register.

6. FCLEX/FNCLEX—clears the error flags in the status register and also the busy flag.

7. FSAVE/FNSAVE—writes the entire state of the machine to memory. Figure 10–8 shows the memory layout for this instruction.

8. FRSTOR—restores the state of the machine from memory. This instruction is used to restore the information saved by FSAVE/FNSAVE.

9. FSTENV/FNSTENV—stores the environment of the 80X87 as shown in Figure 10–9.

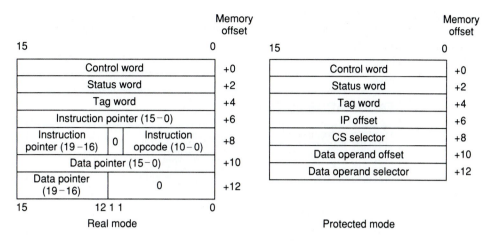

FIGURE 10–9 The memory format for the FSTENV instruction. (Courtesy of Intel Corporation)

10. FLDENV—reloads the environment saved by FSTENV/FNSTENV.
11. FINCST—increments the stack pointer.
12. FDECSTP—decrements the stack pointer.
13. FFREE—frees a register by changing the destination register's tag to empty. It does not affect the contents of the register.
14. FNOP—floating-point coprocessor NOP.
15. FWAIT—causes the microprocessor to wait for the coprocessor to finish an operation. FWAIT should be used before the microprocessor accesses memory data that is affected by the coprocessor.

The 80X87 Instruction Set Summary

Table 10–6 uses some shorthand notations to represent the displacement that may or may not be required for an instruction that uses a memory-addressing mode. It also uses the abbreviation oo to represent the mode, mmm to represent a register/memory addressing mode, and rrr to represent one of the floating-point coprocessor registers ST(0)–ST(7). The d-bit that appears in some instruction opcodes defines the direction of the data flow as in FADD ST,ST(2) or FADD ST(2),ST. The d-bit is a logic 0 for flow towards ST as in FADD ST,ST(2) and a logic 1 for FADD ST(2),ST.

TABLE 10–6 The instruction set of the arithmetic coprocessor

F2XM1	$2^{ST} - 1$
11011001 11110000	
Example	
F2XM1	

FABS	Absolute value of ST

TABLE 10–6   *(Continued)*

11011001 11100001
Example
FABS

FADD/FADDP/FIADD   Addition

11011000 oo000mmm disp	32-bit memory (FADD)
11011100 oo000mmm disp	64-bit memory (FADD)
11011d00 11000rrr	FADD ST,ST(rrr)
11011110 11000rrr	FADDP ST,ST(rrr)
11011110 oo000mmm disp	16-bit memory (FIADD)
11011010 oo000mmm disp	32-bit memory (FIADD)

Format	Examples
FADD	FADD DATA
FADDP	FADD ST,ST(1)
FIADD	FADDP
	FIADD NUMBER
	FADD ST,ST(3)

FCLEX/FNCLEX   Clear errors

11011011 11100010
Examples
FCLEX FNCLEX

FCOM/FCOMP/FCOMPP/FICOM/FICOMP   Compare

11011000 oo010mmm disp	32-bit memory (FCOM)
11011100 oo010mmm disp	64-bit memory (FCOM)
11011000 11010rrr	FCOM ST(rrr)
11011000 oo011mmm disp	32-bit memory (FCOMP)
11011100 oo011mmm disp	64-bit memory (FCOMP)
11011000 11011rrr	FCOMP ST(rrr)
11011110 11011001	FCOMPP
11011110 oo010mmm disp	16-bit memory (FICOM)
11011010 oo010mmm disp	32-bit memory (FICOM)
11011110 oo011mmm disp	16-bit memory (FICOMP)
11011010 oo011mmm disp	32-bit memory (FICOMP)

Format	Examples
FCOM	FCOM ST(2)
FCOMP	FCOMP DATA
FCOMPP	FCOMPP
FICOM	FICOM NUMBER
FICOMP	FICOMP DATA3

FCOS   Cosine of ST   (80387/80486/7)

11011001 11111111
Example
FCOS

FDECSTP   Decrement stack pointer

11011001 11110110

TABLE 10–6 *(Continued)*

Example
FDECSTP

FDISI/FNDISI Disable interrupts

11011011 11100001
(ignored on the 80287, 80387, and 80486/7)
Examples
FDISI FNDISI

FDIV/FDIVP/FIDIV Divison

11011000 oo110mmm disp	32-bit memory (FDIV)
11011100 oo100mmm disp	64-bit memory (FDIV)
11011d00 11111rrr	FDIV ST,ST(rrr)
11011110 11111rrr	FDIVP ST,ST(rrr)
11011110 oo110mmm disp	16-bit memory (FIDIV)
11011010 oo110mmm disp	32-bit memory (FIDIV)

Format	Examples
FDIV FDIVP FIDIV	FDIV DATA FDIV ST,ST(3) FDIVP FIDIV NUMBER

FDIVR/FDIVRP/FIDIVR Divison reversed

11011000 oo111mmm disp	32-bit memory (FDIVR)
11011100 oo111mmm disp	64-bit memory (FDIVR)
11011d00 11110rrr	FDIVR ST,ST(rrr)
11011110 11110rrr	FDIVRP ST,ST(rrr)
11011110 oo111mmm disp	16-bit memory (FIDIVR)
11011010 oo111mmm disp	32-bit memory (FIDIVR)

Format	Examples
FDIVR FDIVRP FIDIVR	FDIVR DATA FDIVR ST,ST(3) FDIVRP FIDIVR NUMBER

FENI/FNENI Disable interrupts

11011011 11100000
(ignored on the 80287, 80387, and 80486/7)
Examples
FENI FNENI

FFREE Free register

11011101 11000rrr

Format	Examples
FFREE	FFREE FFREE ST(1) FFREE ST(2)

TABLE 10–6   *(Continued)*

FINCSTP   Increment stack pointer
11011001 11110111
Example
FINCSTP

FINIT/FNINIT   Initialize coprocessor
11011001 11110110
Examples
FINIT FNINIT

FLD/FILD/FBLD   Load data to ST(0)

11011001 oo000mmm disp	32-bit memory (FLD)
11011101 oo000mmm disp	64-bit memory (FLD)
11011011 oo101mmm disp	80-bit memory (FLD)
11011111 oo000mmm disp	16-bit memory (FILD)
11011011 oo000mmm disp	32-bit memory (FILD)
11011111 oo101mmm disp	64-bit memory (FILD)
11011111 oo100mmm disp	80-bit memory (FBLD)

Format	Examples
FLD	FLD DATA
FILD	FILD DATA1
FBLD	FBLD DEC_DATA

FLD1   Load + 1.0 to ST(0)
11011001 11101000
Example
FLD1

FLDZ   Load + 0.0 to ST(0)
11011001 11101110
Example
FLDZ

FLDPI   Load π to ST(0)
11011001 11101011
Example
FLDPI

FLDL2E   Load $\log_2$ e to ST(0)

11011001 11101010
Example

TABLE 10-6 *(Continued)*

FLDL2E

FLDL2T   Load $\log_2$ 10 to ST(0)

11011001 11101001

Example

FLDL2T

FLDLG2   Load $\log_{10}$ 2 to ST(0)

11011001 11101000

Example

FLDLG2

FLDLN2   Load $\log_e$ 2 to ST(0)

11011001 11101101

Example

FLDLN2

FLDCW   Load control register

11011001 oo101mmm disp

Format	Examples
FLDCW	FLDCW DATA FLDCW STATUS

FLDENV   Load environment

11011001 oo100mmm disp

Format	Examples
FLDENV	FLDENV ENVIRON FLDENV DATA

FMUL/FMULP/FIMUL   Multiplication

11011000 oo001mmm disp	32-bit memory (FMUL)
11011100 oo001mmm disp	64-bit memory (FMUL)
11011d00 11001rrr	FMUL ST,ST(rrr)
11011110 11001rrr	FMULP ST,ST(rrr)
11011110 oo001mmm disp	16-bit memory (FIMUL)
11011010 oo001mmm disp	32-bit memory (FIMUL)

Format	Examples
FMUL FMULP FIMUL	FMUL DATA FMUL ST,ST(2) FMUL ST(2),ST FMULP

FNOP   No operation

11011001 11010000

TABLE 10–6   *(Continued)*

Example
FNOP

FPATAN   Partial arctangent of ST(0)
11011001 11110011
Example
FPATAN

FPREM   Partial remainder
11011001 11111000
Example
FPREM

FPREM1   Partial remainder (IEEE)   (80387/80486/7)
11011001 11110101
Example
FPREM1

FPTAN   Partial tangent of ST(0)
11011001 11110010
Example
FPTAN

FRNDINT   Round ST(0) to an integer
11011001 11111100
Example
FRNDINT

FRSTOR Restore state	
11011101 oo110mmm disp	
Format	Examples
FRSTOR	FRSTOR DATA FRSTOR STATE FRSTOR NACHINE

FSAVE/FNSAVE Save machine state	
11011101 oo110mmm disp	
Format	Examples
FSAVE FNSAVE	FSAVE STATE FNSAVE STATUS FSAVE MACHINE

TABLE 10–6 *(Continued)*

FSCALE   Scale ST(0) by ST(1)
11011001 11111101
Example
FSCALE

FSETPM   Set protected mode   (80287/80387/80486/7)
11011011 11100100
Example
FSETPM

FSIN   Sine of ST(0)   (80387/80486/7)
11011001 11111110
Example
FSIN

FSINCOS   Find sine and cosine of ST(0)   (80387/80486/7)
11011001 11111011
Example
FSINCOS

FSQRT   Square root of ST(0)
11011001 11111010
Example
FSQRT

FST/FSTP/FIST/FISTP/FBSTP   Store

11011001 oo010mmm disp	32-bit memory (FST)
11011101 oo010mmm disp	64-bit memory (FST)
11011101 11010rrr	FST ST(rrr)
11011011 oo011mmm disp	32-bit memory (FSTP)
11011101 oo011mmm disp	64-bit memory (FSTP)
11011011 oo111mmm disp	80-bit memory (FSTP)
11011101 11001rrr	FSTP ST(rrr)
11011111 oo010mmm disp	16-bit memory (FIST)
11011011 oo010mmm disp	32-bit memory (FIST)
11011111 oo011mmm disp	16-bit memory (FISTP)
11011011 oo011mmm disp	32-bit memory (FISTP)
11011111 oo111mmm disp	64-bit memory (FISTP)
11011111 oo110mmm disp	80-bit memory (FBSTP)

Format	Examples
FST	FST DATA
FSTP	FST ST(3)
FIST	FST
FISTP	FSTP
FBSTP	FIST DATA2
	FBSTP DATA6

TABLE 10–6   *(Continued)*

FSTCW/FNSTCW   Store control register

11011001 oo111mmm disp	
Format	Examples
FSTCW FNSTCW	FSTCW CONTROL FNSTCW STATUS FSTCW MACHINE

FSTENV/FNSTENV   Store environment

11011001 oo110mmm disp	
Format	Examples
FSTENV FNSTENV	FSTENV CONTROL FNSTENV STATUS FSTENV MACHINE

FSTSW/FNSTSW   Store status register

11011101 oo111mmm disp	
Format	Examples
FSTSW FNSTSW	FSTSW CONTROL FNSTSW STATUS FSTSW MACHINE

FSUB/FSUBP/FISUB   Subtraction

11011000 oo100mmm disp	32-bit memory (FSUB)
11011100 oo100mmm disp	64-bit memory (FSUB)
11011d00 11101rrr	FSUB ST,ST(rrr)
11011110 11101rrr	FSUBP ST,ST(rrr)
11011110 oo100mmm disp	16-bit memory (FISUB)
11011010 oo100mmm disp	32-bit memory (FISUB)

Format	Examples
FSUB FSUBP FISUB	FSUB DATA FSUB ST,ST(2) FSUB ST(2),ST FSUBP FISUB DATA3

FSUBR/FSUBRP/FISUBR   Reverse subtraction

11011000 oo101mmm disp	32-bit memory (FSUBR)
11011100 oo101mmm disp	64-bit memory (FSUBR)
11011d00 11100rrr	FSUBR ST,ST(rrr)
11011110 11100rrr	FSUBRP ST,ST(rrr)
11011110 oo101mmm disp	16-bit memory (FISUBR)
11011010 oo101mmm disp	32-bit memory (FISUBR)

Format	Examples
FSUBR FSUBRP FISUBR	FSUBR DATA FSUBR ST,ST(2) FSUBR ST(2),ST FSUBRP FISUBR DATA3

FTST   Compare ST(0) with + 0.0

TABLE 10–6 *(Continued)*

11011001 11100100
Example
FTST

FUCOM/FUCOMP/FUCOMPP Unordered compare (80837/80486/7)

11011101 11100rrr	FUCOM ST,ST(rrr)
11011101 11101rrr	FUCOMP ST,ST(rrr)
11011101 11101001	FUCOMPP

Format	Examples
FUCOM	FUCOM ST,ST(2)
FUCOMP	FUCOM
FUCOMPP	FUCOMP ST,ST(3)
	FUCOMP
	FUCOMPP

FWAIT Wait

10011011
Example
FWAIT

FXAM Examine ST(0)

11011001 11100101
Example
FXAM

FXCH Exchange ST(0) with another register

11011001 11001rrr	FXCH ST,ST(rrr)

Format	Examples
FXCH	FXCH ST,ST(1)
	FXCH
	FXCH ST,ST(4)

FXTRACT Extract components of ST(0)

11011001 11110100
Example
FXTRACT

FYL2X $ST(1) \times \log_2 ST(0)$

11011001 11110001
Example
FYL2X

FXL2XP1 $ST(1) \times \log_2 [ST(0) + 1.0]$

TABLE 10–6   *(Continued)*

11011001 11111001
Example
FXL2XP1

Note: d = direction, where d = 0 for ST as the destination and d = 1 for ST as the source; rrr = floating-point register number; oo = mode; mmm = r/m field; and disp = displacement.

10–5   PROGRAMMING THE ARITHMETIC COPROCESSOR

This section of the chapter provides programming examples for the arithmetic co-processor. Each example is chosen to illustrate a programming technique for the coprocessor. All the example programs use the 80387/80486/7 coprocessor.

Calculating the Area of a Circle

This programming example provides a simple illustration of a method of addressing the 80X87 stack. First recall that the equation for calculating the area of a circle is $A = \pi R^2$. A procedure that performs this calculation is listed in Example 10–7.

EXAMPLE 10–7

```
                    ;Procedure that calculates the area of a circle.
                    ;
                    ;The radius must be stored at memory location RADIUS
                    ;before calling this procedure.  The result is found
                    ;in memory location AREA after the procedure.
                    ;
0000                AREAS    PROC    FAR

0000 D9 06 0004 R            FLD     RADIUS        ;radius to ST
0004 D8 C8                   FMUL    ST,ST(0)      ;square radius
0006 D9 EB                   FLDPI                 ;π to ST
0008 DE C9                   FMUL                  ;multiply ST = ST x ST(1)
000A D9 1E 0000 R            FSTP    AREA          ;save area
000E 9B                      FWAIT                 ;wait for coprocessor
000F CB                      RET

0010                AREAS    ENDP
```

This is a rather simple procedure, yet it does illustrate the operation of the stack. To provide a better understanding of the operation of the stack, Figure 10–10 shows the contents of the stack after each instruction of Example 10–7 executes.

The first instruction loads the contents of memory location RADIUS to the top of the stack. Next the FMUL ST,ST(0) instruction squares the radius on the top of the stack. The FLDPI instruction loads π to the stack top. The FMUL instruction uses

FIGURE 10–10 Operation of the 80X87 stack with the procedure of Example 10–7. (*Note:* Stack shown after the execution of the indicated instruction.)

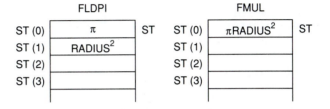

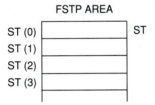

the classic stack addressing mode to multiply ST by ST(1). After the multiplication, both prior values are removed from the stack and the product replaces them at the top of the stack. Finally the FSTP instruction copies the top of the stack, which is the area, to memory location AREA and clears the stack.

The FWAIT instruction appears just before the return instruction in this example. The reason that we use the FWAIT instruction is to wait for the coprocessor to finish finding the area before returning. If we were not to wait, the main program might access memory location AREA before the coprocessor stores the result into location AREA.

Finding the Resonant Frequency

An equation commonly used in electronics is the formula for determining the resonant frequency of an *LC* circuit. The equation solved by the procedure illustrated in Example 10–8 is $Fr = 1/(2\pi \sqrt{LC})$.

EXAMPLE 10–8

```
0000                DATAS    SEGMENT

0000 00000000       RESO     DD      ?                        ;resonant frequency
```

```
0004 358637BD        L       DD      .000001         ;inductance
0008 358637BD        C       DD      .000001         ;capacitance
000C 40000000        TWO     DD      2.0             ;constant

0010                 DATAS   ENDS

0000                 CODE    SEGMENT USE16 'CODE'

                             ASSUME  CS:CODE,DS:DATAS

;Procedure that finds the resonant frequency.

0000                 FREQ    PROC    FAR

0000 D9 06 0004 R            FLD     L               ;get L
0004 D8 0E 0008 R            FMUL    C               ;find LC

0008 D9 FA                   FSQRT                   ;find √LC

000A D8 0E 000C R            FMUL    TWO             ;find 2√LC

000E D9 EB                   FLDPI                   ;get π
0010 DE C9                   FMUL                    ;get 2π√LC

0012 D9 E8                   FLD1                    ;get 1
0014 DE F1                   FDIVR                   ;form 1/(2π√LC)

0016 D9 1E 0000 R            FSTP    RESO            ;save frequency

001A 9B                      FWAIT
001B CB                      RET

001C                 FREQ    ENDP

001C                 CODE    ENDS

                             END
```

Notice the straightforward manner in which the procedure solves this equation. Very little extra data manipulation is required because of the stack inside the 80X87. Also notice how the constant TWO is defined for the program and how the DIVRP, using classic stack addressing, is used to form the reciprocal.

Finding the Roots Using the Quadratic Equation

This example illustrates how to find the roots of a polynomial expression ($ax^2 + bx + c = 0$) using the quadratic equation. The quadratic equation is ($b \pm \sqrt{b^2 - 4ac}/2a$). Example 10–9 illustrates a procedure that finds the roots (R1 and R2) for the quadratic equation. The constants are stored in memory locations A, B, and C.

EXAMPLE 10–9

```
0000                 DATAS   SEGMENT
0000 40000000        TWO     DD      2.0
0004 40800000        FOUR    DD      4.0
0008 3F800000        A       DD      1.0
```

```
000C C1800000      B       DD      -16.0
0010 421C0000      C       DD      +39.0
0014 00000000      R1      DD      ?
0018 00000000      R2      DD      ?
001C               DATAS   ENDS

0000               CODE    SEGMENT USE16 'CODE'

                           ASSUME  CS:CODE,DS:DATAS

                   ;Procedure that solves the quadratic equation.
                   ;Note: this procedure does not detect an error that
                   ;occurs for a negative square root.

0000               ROOTS   PROC    FAR

0000 D9 06 0000 R          FLD     TWO
0004 D8 0E 0008 R          FMUL    A                       ;form 2a
0008 D9 06 0004 R          FLD     FOUR
000C D8 0E 0008 R          FMUL    A
0010 D8 0E 0010 R          FMUL    C                       ;form 4ac
0014 D9 06 000C R          FLD     B
0018 D8 0E 000C R          FMUL    B                       ;form b²
001C DE E1                 FSUBR                           ;form b² - 4ac
001E D9 FA                 FSQRT                           ;form square root of b² - 4ac

0020 D9 06 000C R          FLD     B
0024 D8 E1                 FSUB    ST,ST(1)
0026 D8 F2                 FDIV    ST,ST(2)
0028 D9 1E 0014 R          FSTP    R1                      ;save root1

002C D9 06 000C R          FLD     B
0030 DE C1                 FADD
0032 DE F1                 FDIVR
0034 D9 1E 0018 R          FSTP    R2                      ;save root2

0038 9B                    FWAIT
0039 CB                    RET

003A               ROOTS   ENDP

003A               CODE    ENDS

                           END
```

Displaying a Single-Precision Floating-Point Number

This section of the text shows how to take the floating-point contents of a 32-bit single-precision floating-point number and display it on the video display. The procedure displays the floating-point number as a mixed number with an integer part and a fractional part separated by a decimal point. In order to simplify the procedure we have placed a limit on the display size of the mixed number so that the integer portion is a 32-bit binary number and the fraction is a 24-bit binary number. The procedure will not function properly for larger or smaller numbers.

Example 10–10 lists the procedure for displaying the contents of memory location NUMB on the video display at the current cursor position. The procedure first tests the sign of the number and displays a space for positive and a minus sign for a negative number. After displaying the sign, the number is made positive by the FABS instruction. Next we divide it into integer and fractional parts and store them at WHOLE and FRACT.

EXAMPLE 10–10

```
0000                    DATAS    SEGMENT

0000 C50B0C00           NUMB     DD       -2224.75
0004 0000               TEMP     DW       ?
0006 0000               WHOLE    DW       ?
0008 00000000           FRACT    DD       ?

000C                    DATAS    ENDS

0000                    CODE     SEGMENT USE16 'CODE'

                                 ASSUME  CS:CODE,DS:DATAS

                        ;Main program that displays NUMB
                        ;
0000                    MAIN     PROC     FAR

0000 B8 —— R                     MOV      AX,DATAS
0003 8E D8                       MOV      DS,AX
0005 E8 0013 R                   CALL     DISP              ;display NUMB
0008 B4 4C                       MOV      AH,4CH            ;exit to DOS
000A CD 21                       INT      21H

000C                    MAIN     ENDP

000C                    DISPS    PROC     NEAR

000C B4 06                       MOV      AH,6              ;display AL
000E 8A D0                       MOV      DL,AL
0010 CD 21                       INT      21H
0012 C3                          RET

0013                    DISPS    ENDP

0013                    DISP     PROC     NEAR

0013 9B D9 3E 0004 R             FSTCW    TEMP              ;set rounding to chop
0018 81 0E 0004 R 0C00           OR       TEMP,0C00H
001E 9B D9 2E 0004 R             FLDCW    TEMP

0023 D9 06 0000 R                FLD      NUMB              ;get NUMB
0027 D9 E4                       FTST                       ;test NUMB
0029 9B DF E0                    FSTSW    AX                ;status to AX
002C 25 4500                     AND      AX,4500H          ;get C3, C2, and C0
002F 3D 0100                     CMP      AX,0100H          ;test for -
0032 75 05                       JNE      DISP1             ;if positive
0034 B0 2D                       MOV      AL,'-'
0036 E8 000C R                   CALL     DISPS             ;display minus

0039                    DISP1:

0039 D9 E1                       FABS                       ;make ST positive
```

```
003B  D9 FC              FRNDINT                         ;get integer
003D  DF 16 0006 R       FIST      WHOLE                 ;store integer
0041  D9 06 0000 R       FLD       NUMB
0045  D9 E1              FABS
0047  DE E9              FSUB                            ;get fraction
0049  D9 E1              FABS
004B  D9 1E 0008 R       FSTP      FRACT                 ;save fraction
004F  9B                 FWAIT

                         ;display integer part

0050  A1 0006 R          MOV       AX,WHOLE
0053  B9 0000            MOV       CX,0
0056  BB 000A            MOV       BX,10

0059             DISP2:

0059  41                 INC       CX
005A  33 D2              XOR       DX,DX
005C  F7 F3              DIV       BX
005E  83 C2 30           ADD       DX,'0'                ;convert to ASCII
0061  52                 PUSH      DX
0062  0B C0              OR        AX,AX
0064  75 F3              JNE       DISP2                 ;if not zero

0066             DISP3:

0066  58                 POP       AX
0067  E8 000C R          CALL      DISPS                 ;display it
006A  E2 FA              LOOP      DISP3
006C  B0 2E              MOV       AL,'.'                ;display decimal point
006E  E8 000C R          CALL      DISPS

                         ;display fractional part

0071  A1 0008 R          MOV       AX,WORD PTR FRACT
0074  8B 16 000A R       MOV       DX,WORD PTR FRACT+2
0078  B9 0008            MOV       CX,8

007B             DISP4:

007B  D1 E0              SHL       AX,1
007D  D1 D2              RCL       DX,1
007F  E2 FA              LOOP      DISP4
0081  81 CA 8000         OR        DX,8000H              ;set implied bit

0085  92                 XCHG      AX,DX
0086  BB 000A            MOV       BX,10

0089             DISP5:

0089  F7 E3              MUL       BX
008B  50                 PUSH      AX
008C  92                 XCHG      DX,AX
008D  04 30              ADD       AL,'0'
008F  E8 000C R          CALL      DISPS                 ;display digit
0092  58                 POP       AX
0093  0B C0              OR        AX,AX
0095  75 F2              JNZ       DISP5
0097  C3                 RET

0098             DISP      ENDP
```

```
0098                          CODE    ENDS

                              END     MAIN
```

The last part of the procedure displays the whole number part followed by the fractional part. Note that the fractional part may contain a rounding error for certain values. We have not adjusted the number to remove the rounding error that is inherent in floating-point fractional numbers.

Reading a Mixed Number from the Keyboard

If floating-point arithmetic is used in a program we must have a method of reading the number from the keyboard and converting it to floating-point. The procedure listed in Example 10–11 reads a signed mixed number from the keyboard and converts it to a floating-point number located at the top of the stack inside the coprocessor.

EXAMPLE 10–11

```
0000                          DATA    SEGMENT

0000 00                       SIGN    DB      ?
0001 0000                     TEMP1   DW      ?
0003 41200000                 TEN     DD      10.0

0007                          DATA    ENDS

0000                          CODE    SEGMENT USE16 'CODE'

                              ASSUME CS:CODE,DS:DATA

                     ;procedure that reads a mixed number from the keyboard
                     ;and leaves it at the top of the coprocessor stack.

0000                          READ    PROC    FAR

0000 B8 ---- R                        MOV     AX,DATA         ;address data segment
0003 8E D8                            MOV     DS,AX
0005 9B D9 EE                         FLDZ                    ;clear ST
0008 C6 06 0000 R 00                  MOV     SIGN,0          ;clear sign
000D E8 007C                          CALL    GET             ;read a character
0010 3C 2D                            CMP     AL,'-'          ;test for minus
0012 75 07                            JNE     READ1           ;if not minus
0014 C6 06 0000 R FF                  MOV     SIGN,0FFH       ;set sign for minus
0019 EB 0F                            JMP     READ3           ;get integer part

001B                          READ1:

001B 3C 2B                            CMP     AL,'+'          ;test for plus
001D 74 0B                            JE      READ3           ;get integer part
001F 3C 30                            CMP     AL,'0'          ;test for number
0021 72 06                            JB      READ2
0023 3C 39                            CMP     AL,'9'
0025 77 02                            JA      READ2
0027 EB 04                            JMP     READ4           ;if a number

0029                          READ2:
```

```
0029 CB                          RET

002A                   READ3:

002A E8 005F                     CALL     GET               ;read integer part

002D                   READ4:

002D 3C 2E                       CMP      AL,'.'            ;test for fraction
002F 74 27                       JE       READ7             ;if fraction
0031 3C 30                       CMP      AL,'0'            ;test for number
0033 72 17                       JB       READ5
0035 3C 39                       CMP      AL,'9'
0037 77 13                       JA       READ5
0039 9B D8 0E 0003 R             FMUL     TEN               ;form integer
003E 32 E4                       XOR      AH,AH
0040 2C 30                       SUB      AL,'0'
0042 A3 0001 R                   MOV      TEMP1,AX
0045 9B DE 06 0001 R             FIADD    TEMP1
004A EB DE                       JMP      READ3

004C                   READ5:

004C 80 3E 0000 R 00             CMP      SIGN,0            ;adjust sign
0051 75 01                       JNE      READ6
0053 CB                          RET

0054                   READ6:

0054 9B D9 E0                    FCHS
0057 CB                          RET

0058                   READ7:

0058 9B D9 E8                    FLD1                       ;from fraction
005B 9B D8 36 0003 R             FDIV     TEN

0060                   READ8:

0060 E8 0029                     CALL     GET               ;read character
0063 3C 30                       CMP      AL,'0'            ;test for number
0065 72 20                       JB       READ9
0067 3C 39                       CMP      AL,'9'
0069 77 1C                       JA       READ9
006B 32 E4                       XOR      AH,AH
006D 2C 30                       SUB      AL,'0'
006F A3 0001 R                   MOV      TEMP1,AX
0072 9B DF 06 0001 R             FILD     TEMP1             ;load number
0077 9B D8 C9                    FMUL     ST,ST(1)          ;form fraction
007A 9B DC C2                    FADD     ST(2),ST
007D 9B D8 D9                    FCOMP
0080 9B D8 36 0003 R             FDIV     TEN
0085 EB D9                       JMP      READ8

0087                   READ9:

0087 9B D8 D9                    FCOMP                      ;clear stack
008A EB C0                       JMP      READ5

008C          READ     ENDP

008C          GET      PROC     NEAR
```

```
008C  B4 06              MOV    AH,6              ;read character
008E  B2 FF              MOV    DL,OFFH
0090  CD 21              INT    21H
0092  74 F8              JZ     GET
0094  C3                 RET

0095            GET      ENDP

0095            CODE     ENDS

                         END
```

Here the sign is first read from the keyboard, if present, and saved for later use in adjusting the sign of the resultant floating-point number. Next, the integer portion of the number is read. This portion terminates with a period, space, or carriage return. If a period is typed, then the procedure continues and reads a fractional part. If a space or carriage return is entered, the number is converted to floating-point form.

10–6   SUMMARY

1. The arithmetic coprocessor functions in parallel with the microprocessor.
2. The data types manipulated by the arithmetic coprocessor include signed-integer, floating-point, and binary-coded decimal (BCD).
3. There are three forms of integers used with the 80X87: word (16 bits), short (32 bits), and long (64 bits). Each integer contains a signed number in true magnitude for positive numbers and two's complement form for negative numbers.
4. A BCD number is stored as an 18-digit number in 10 bytes of memory. The most significant byte contains the sign-bit, and the remaining 9 bytes contain an 18-digit packed BCD number.
5. The 80X87 supports three types of floating-point numbers: single-precision (32 bits), double-precision (64 bits), and extended-precision (80 bits). A floating-point number is formed of three parts: the sign, biased exponent, and significand. In the 80X87, the exponent is biased with a constant and the integer bit of the normalized number is not stored in the significand except in the extended-precision form.
6. Decimal numbers are converted to floating-point numbers by converting the number to binary, normalizing the binary number, adding the bias to the exponent, and storing the number in floating-point form.
7. Floating-point numbers are converted to decimal by subtracting the bias from the exponent, unnormalizing the number, and then converting it to decimal.
8. The 80X87 contains a status register that indicates busy, what conditions follow a compare or test, the location of the top of the stack, and the state of the error bits.

9. The control register of the 80X87 contains control bits that select infinity, rounding and precision control, and error masks.

10. The following directives are often used with the arithmetic coprocessor for storing data: DW (define word), DD (define double word), DQ (define quad word) and DT (define 10 bytes).

11. The 80X87 uses a stack to transfer data between itself and the memory system. Generally data are loaded to the top of the stack or removed from the top of the stack for storage.

12. All internal 80X87 data is always in the 80-bit extended-precision form. The only time that data are in any other form is when they are stored or loaded from the memory.

13. The 80X87 addressing modes include the classic stack mode, register, register with a pop, and memory. Stack addressing is implied, and the data at ST becomes the source, ST(1) the destination, and the result is found in ST after a pop. The other addressing modes are self-explanatory.

14. The 80X87 arithmetic operations include addition, subtraction, multiplication, division, and square root.

15. There are transcendental functions in the 80X87 instruction set. These functions find the partial tangent or arctangent; the sine and cosine; and $2^X - 1, Y \log_2 X$, and $Y \log_2 (X + 1)$.

16. Constants are stored inside the 80X87 that provide: $+0.0$, $+1.0$, π, $\log_2 10$, $\log_2 e$, $\log_{10} 2$, and $\log_e 2$.

10–7 GLOSSARY

Bias A number added to the exponent of a floating-point number. The bias increases performance of the floating-point arithmetic operation of the coprocessor.

Coprocessor A device that works concurrently with the microprocessor. The microprocessor and the coprocessor both execute instructions simultaneously.

Double-precision An 8-byte (64-bit) floating-point number that is often stored in memory using the DQ directive.

Exponent Part of a floating-point number that stores the binary power of 2. The exponent is usually a biased exponent that is modified by adding a bias.

Extended-precision A 10-byte (80-bit) floating-point number that is often stored in the memory using the DT directive.

Floating-point A binary number that represents an integer, fraction, or mixed number. We often call a floating-point number binary scientific notation.

Implied one-bit The whole-number portion of the floating point number (1.) that is not stored with the number, but is implied to be a part of the number by definition and operation.

Significand The fraction portion of a floating-point number, which is less than 1, but not zero except in the cases of zero and infinity.

Single-precision   A 4-byte (32-bit) floating-point number that is often stored in memory using the DD directive.

10–8   QUESTIONS

1. List the three types of data that are loaded or stored in memory by the arithmetic coprocessor.
2. List the three integer data types, the range of the integers stored in them, and the number of bits allotted to each.
3. Explain how a BCD number is stored in memory by the 80X87.
4. List the three types of floating-point numbers used with the 80X87 and the number of binary bits assigned to each.
5. Convert the following decimal numbers into single-precision floating-point numbers:
   a. 28.75 b. 624 c. −0.615 d. +0.0 e. −1000.5
6. Convert the following single-precision floating-point numbers into decimal:
   a. 11000000 11110000 00000000 00000000
   b. 00111111 00010000 00000000 00000000
   c. 01000011 10011001 00000000 00000000
   d. 01000000 00000000 00000000 00000000
   e. 01000001 00100000 00000000 00000000
   f. 00000000 00000000 00000000 00000000
7. Explain what the arithmetic coprocessor does when a normal microprocessor instruction executes.
8. Explain what the microprocessor does when an arithmetic coprocessor instruction executes.
9. What is the purpose of the C_3–C_0 bits in the status register?
10. How is the rounding mode selected in the 80X87?
11. What 80X87 instruction uses the 80286 AX register?
12. How are data stored inside the 80X87?
13. Whenever the 80X87 is reset, the top of the stack register is register number _____.
14. What does the term *chop* mean in the rounding control bits of the control register?
15. What is the difference between affine and projective infinity control?
16. What microprocessor instruction forms the opcodes for the 80X87?
17. Using assembler pseudo-opcodes, form statements that accomplish the following:
    a. Store a 23.44 into a double-precision floating-point memory location named FROG.
    b. Store a −123 into a 32-bit signed integer location named DATA3.
    c. Store a −23.8 into a single-precision floating-point memory location named DATA1.
    d. Reserve a double-precision memory location named DATA2.

FIGURE 10–11

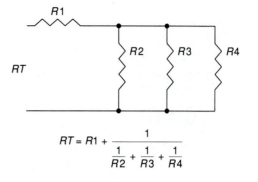

$$RT = R1 + \cfrac{1}{\cfrac{1}{R2} + \cfrac{1}{R3} + \cfrac{1}{R4}}$$

18. Describe how the FST DATA instruction functions. Assume that DATA is defined as a 64-bit memory location.
19. What does the FILD DATA instruction accomplish?
20. Form an instruction that adds the contents of register 3 to the top of the stack.
21. Describe the operation of the FADD instruction.
22. Choose an instruction that subtracts the contents of register 2 from the top of the stack and stores the result in register 2.
23. What is the function of the FBSTP DATA instruction?
24. What is the difference between a forward and a reverse division?
25. What is the difference between the FTST instruction and FXAM?
26. Explain what the F2XM1 instruction calculates.
27. What instruction pushes π onto the top of the stack?
28. What will FFREE ST(2) accomplish when executed?
29. What instruction stores the coprocessor environment?
30. What does the FSAVE instruction save?
31. Develop a procedure that finds the area of a rectangle $(A = L \times W)$. Memory locations for this procedure are single-precision A, L, and W.
32. Write a procedure that finds the inductive reactance $(XL = 2\pi FL)$. Memory locations for this procedure are single-precision XL, F, and L.
33. Develop a procedure that generates a table of square roots for the integers 2 through 10.
34. When is the FWAIT instruction used in a program?
35. Given the series/parallel circuit and equation illustrated in Figure 10–11, develop a program using single-precision values for $R1$, $R2$, $R3$, and $R4$ that finds the total resistance and stores the result at single-precision location RT.

APPENDIX A

The Assembler, Disk Operating System, and Basic Input/Output System

This appendix details the use of the assembler and also shows the DOS (*disk operating system*) and BIOS (*basic I/O system*) function calls that are used by assembly language to control the IBM-PC or its clone. The function calls control everything from reading and writing disk data to managing the keyboard and displays. The assembler represented in this text is at one time the Microsoft ML (version 6.0), at others the MASM (version 5.10) macro assembler program.

ASSEMBLER USAGE

The assembler program requires a symbolic program—written with a word processor, text editor, or the workbench program—as its input. The editor provided with version 5.10 is M.EXE, and it is strictly a full-screen editor. The editor provided with version 6.0 is PWB.EXE, a fully integrated development system that contains extensive help. Refer to the documentation that accompanies your assembler package for details on the operation of the editor program. If at all possible use version 6.0 of the assembler because it contains a detailed help file that guides the user through assembly language statements, directives, and even the DOS and BIOS interrupt function calls.

If you are using a word processor to develop your software, make sure that it is initialized to generate a pure ASCII file. The source file that you generate must use the extension .ASM, which is required for the assembler to properly identify your source program.

Once your source file is prepared, it is assembled with the assembler program. If you are using the workbench provided with version 6.0, assembly is accomplished by selecting the compile feature (or build feature) with your mouse. If you are using a word processor and DOS command lines with version 5.10, then see Example A–1 for the dialog required for version 5.10 to assemble a file called FROG.ASM. Note that this example shows in italics the portions typed by the user.

Once a program is assembled, it must be linked before it can be executed. The linker converts the object file into an executable file (.EXE). Example A–2 shows the dialog required for the linker using an MASM version 5.10 object file. If the ML version 6.0 assembler is in use, it automatically assembles and links a program using the COMPILE or BUILD command from the workbench. After compiling with ML, workbench allows the program to be debugged with a debugging tool called code view. Code view is also available with MASM, but CV must be typed at the DOS command line to access it.

ASSEMBLER MEMORY MODELS

Memory models and the .MODEL statement were introduced in Chapter 6.

EXAMPLE A–1

A>*MASM*

Microsoft (R) Macro Assembler Version 5.10
Copyright (C) Microsoft Corp 1981, 1989. All rights reserved.

Source filename [.ASM]:*FILE*
Object filename [FILE.OBJ]:*FILE*
Source filename [NUL.LST]:*FILE*
Cross reference [NUL.CRF]:*FILE*

EXAMPLE A–2

A>*LINK*

Microsoft (R) Overlay Linker Version 3.64
Copyright (C) Microsoft Corp 1983-1988. All rights reserved.

Object Modules [.OBJ]: *TEST*
Run File [TEST.EXE]: *TEST*
List File [NUL.MAP]: *TEST*
Libraries [.LIB]: *SUBR*

Here we completely define the memory models available for software development. Each model defines the way that a program is stored in the memory system. Table A–1 describes the different models available with MASM and ML.

Note that the tiny model is used to create a .COM file instead of an execute file. The .COM file is different because all data and code fits into one code segment. A .COM file must have the program originated to start at offset address 0100H. A .COM file loads from the disk and executes faster than the normal execute (.EXE) file. For most applications we normally use the execute file (.EXE) and the small memory model.

TABLE A–1 Memory models for the assembler

Type	Description
Tiny	All data and code fit into one segment, the code segment. Tiny model programs are written in the .COM file format, which means that the program must be originated at memory location 0100H. This model is most often used with small programs.
Small	All data fit into a single 64K byte data segment and all code fits into another single 64K byte code segment. This allows all code to be accessed with near jumps and calls.
Medium	All data fits into a single 64K byte data segment, and code fits into more than one code segment. This allows code to exist in multiple segments.
Compact	All code fits into a single 64K byte code segment, and data fits into more than one data segment.
Large	Both code and data fit into multiple code and data segments.
Huge	Same as large, but allows data segments that are larger than 64K bytes.
Flat	Not available in MASM version 5.10. The flat memory model uses one segment with a maximum length of 512M bytes to store data and code.

When models are used to create a program, certain defaults apply, as illustrated in Table A–2. The directives in this table are used to start a particular type of segment for the models listed in the table. If the .CODE directive is placed in a program, it indicates the beginning of the code segment. Likewise, .DATA indicates the start of a data segment. The name column indicates the name of the segment. Align indicates whether the segment is aligned on a word, double word, or a 16-byte paragraph. Combine indicates the type of segment created. The class indicates the class of the segment, such as 'CODE' or 'DATA'. The group indicates the group type of the segment.

Example A–3 shows a program that uses the small model. The small model is used for programs that contain one DATA and one CODE segment. This applies to many programs that are developed. Notice that not only is the program listed, but so is all the information generated by the assembler. Here the .DATA directive and .CODE directive indicate the start of segments. Also notice how the DS register is loaded in this program.

EXAMPLE A–3

Microsoft (R) Macro Assembler Version 6.00

```
                        .MODEL SMALL
                        .STACK 100H
          0000          .DATA

          0000 0A        FROG    DB      10
          0001 0064 [    DATA1   DB      100 DUP (2)
                  02
                  ]

          0000          .CODE

          0000 B8 —— R   BEGIN:  MOV     AX,DGROUP          ;set up DS
```

TABLE A–2 Defaults for the .MODEL directive

Model	Directive	Name	Align	Combine	Class	Group
Tiny	.CODE	_TEXT	Word	PUBLIC	'CODE'	DGROUP
	.FARDATA	FAR_DATA	Para	Private	'FAR_DATA'	
	.FARDATA?	FAR_BSS	Para	Private	'FAR_BSS'	
	.DATA	_DATA	Word	PUBLIC	'DATA'	DGROUP
	.CONST	CONST	Word	PUBLIC	'CONST'	DGROUP
	.DATA?	_BSS	Word	PUBLIC	'BSS'	DGROUP
Small	.CODE	_TEXT	Word	PUBLIC	'CODE'	
	.FARDATA	FAR_DATA	Para	Private	'FAR_DATA'	
	.FARDATA?	FAR_BSS	Para	Private	'FAR_BSS'	
	.DATA	_DATA	Word	PUBLIC	'DATA'	DGROUP
	.CONST	CONST	Word	PUBLIC	'CONST'	DGROUP
	.DATA?	_BSS	Word	PUBLIC	'BSS'	DGROUP
	.STACK	STACK	Para	STACK	'STACK'	DGROUP
Medium	.CODE	name_TEXT	Word	PUBLIC	'CODE'	
	.FARDATA	FAR_DATA	Para	Private	'FAR_DATA'	
	.FARDATA?	FAR_BSS	Para	Private	'FAR_BSS'	
	.DATA	_DATA	Word	PUBLIC	'DATA'	DGROUP
	.CONST	CONST	Word	PUBLIC	'CONST'	DGROUP
	.DATA?	_BSS	Word	PUBLIC	'BSS'	DGROUP
	.STACK	STACK	Para	STACK	'STACK'	DGROUP
Compact	.CODE	_TEXT	Word	PUBLIC	'CODE'	
	.FARDATA	FAR_DATA	Para	Private	'FAR_DATA'	
	.FARDATA?	FAR_BSS	Para	Private	'FAR_BSS'	
	.DATA	_DATA	Word	PUBLIC	'DATA'	DGROUP
	.CONST	CONST	Word	PUBLIC	'CONST'	DGROUP
	.DATA?	_BSS	Word	PUBLIC	'BSS'	DGROUP
	.STACK	STACK	Para	STACK	'STACK'	DGROUP
Large	.CODE	name_TEXT	Word	PUBLIC	'CODE'	
or	.FARDATA	FAR_DATA	Para	Private	'FAR_DATA'	
huge	.FARDATA?	FAR_BSS	Para	Private	'FAR_BSS'	
	.DATA	_DATA	Word	PUBLIC	'DATA'	DGROUP
	.CONST	CONST	Word	PUBLIC	'CONST'	DGROUP
	.DATA?	_BSS	Word	PUBLIC	'BSS'	DGROUP
	.STACK	STACK	Para	STACK	'STACK'	DGROUP
Flat	.CODE	_TEXT	Dword	PUBLIC	'CODE'	
	.FARDATA	_DATA	Dword	PUBLIC	'DATA'	
	.FARDATA?	_BSS	Dword	PUBLIC	'BSS'	
	.DATA	_DATA	Dword	PUBLIC	'DATA'	
	.CONST	CONST	Dword	PUBLIC	'CONST'	
	.DATA?	_BSS	Dword	PUBLIC	'BSS'	
	.STACK	STACK	Dword	PUBLIC	'STACK'	

```
0003  8E D8                    MOV     DS,AX
                                .
                                .
                                .
                               END     BEGIN
```

Segments and Groups:

N a m e	Size	Length	Align	Combine	Class
DGROUP	GROUP				
_DATA	16 Bit	0065	Word	Public	'DATA'
STACK	16 Bit	0100	Para	Stack	'STACK'
_TEXT	16 Bit	0005	Word	Public	'CODE'

Symbols:

N a m e	Type	Value	Attr
@CodeSize	Number	0000h	
@DataSize	Number	0000h	
@Interface	Number	0000h	
@Model	Number	0002h	
@code	Text		_TEXT
@data	Text		DGROUP
@fardata?	Text		FAR_BSS
@fardata	Text		FAR_DATA
@stack	Text		DGROUP
BEGIN	L Near	0000	_TEXT
DATA1	Byte	0001	_DATA
FROG	Byte	0000	_DATA

```
          0 Warnings
          0 Errors
```

Example A–4 lists a program that uses the large model. Notice how it differs from the small model program of Example A–3. Models can be very useful in developing software, but often we use full segment descriptions as depicted in Chapter 6.

EXAMPLE A–4

Microsoft (R) Macro Assembler Version 6.00

```
                        .MODEL LARGE
                        .STACK 1000H
0000                    .FARDATA?

0000 00                 FROG    DB      ?
0001 0064 [             DATA1   DW      100 DUP (?)
          0000
      ]

0000                    .CONST

0000 54 68 69 73 20 69  MES1    DB      'This is a character string'
```

```
                73 20 61 20 63 68
                61 72 61 63 74 65
                72 20 73 74 72 69
                6E 67
001A 53 6F 20 69 73 20 MES2      DB        'So is this!'
                74 68 69 73 21

0000                      .DATA

0000 000C        DATA2    DW        12
0002 00C8 [      DATA3    DB        200 DUP (1)
         01
       ]

0000                      .CODE

0000             FUNC     PROC      FAR
                           .
                           .
                           .
0000 CB                   RET

0001             FUNC     ENDP

                 END      FUNC
```

Segments and Groups:

N a m e	Size	Length	Align	Combine	Class	
DGROUP . . . . GROUP						
_DATA . . . . 16 Bit	16 Bit	00CA	Word	Public	'DATA'	
STACK . . . . 16 Bit	16 Bit	1000	Para	Stack	'STACK'	
CONST . . . . 16 Bit	16 Bit	0025	Word	Public	'CONST'	ReadOnly
EXA_TEXT . . . . 16 Bit	16 Bit	0001	Word	Public	'CODE'	
FAR_BSS . . . . 16 Bit	16 Bit	00C9	Para	Private	'FAR_BSS'	
_TEXT . . . . 16 Bit	16 Bit	0000	Word	Public	'CODE'	

Procedures, parameters and locals:

N a m e	Type	Value	Attr	
FUNC . . . . P Far	P Far	0000	EXA_TEXT	Length= 0001 Public

Symbols:

N a m e	Type	Value	Attr
@CodeSize . . . . Number	Number	0001h	
@DataSize . . . . Number	Number	0001h	
@Interface . . . . Number	Number	0000h	
@Model . . . . Number	Number	0005h	
@code . . . . Text	Text		EXA_TEXT
@data . . . . Text	Text		DGROUP
@fardata? . . . . Text	Text		FAR_BSS
@fardata . . . . Text	Text		FAR_DATA
@stack . . . . Text	Text		DGROUP
DATA1 . . . . Word	Word	0001	FAR_BSS

DATA2	Word	0000	_DATA
DATA3	Byte	0002	_DATA
FROG	Byte	0000	FAR_BSS
MES1	Byte	0000	CONST
MES2	Byte	001A	CONST

0 Warnings
0 Errors

DOS FUNCTION CALLS

In order to use DOS function calls, place the function number into register AH and load all other pertinent information into registers as described in the table. Once this is accomplished, follow with an INT 21H to execute the DOS function. Example A–5 shows how to display an ASCII A on the CRT screen at the current cursor position with a DOS function call. Following that is a complete listing of the DOS function calls. Note that some function calls require a segment and offset address indicated as DS:DI, for example. This means the data segment is the segment address and DI is the offset address. All of the function calls use INT 21H, and AH contains the function call number. Note that functions marked with an @ should not be used unless DOS version 2.XX is in use. As a rule, DOS function calls save all registers not used as exit data, but in certain cases some registers may change. In order to prevent problems, it is advisable to save registers where problems occur.

EXAMPLE A–5

```
0000 B4 06            MOV     AH,6
0002 B2 41            MOV     DL,'A'
0004 CD 21            INT     21H
```

00H	TERMINATE A PROGRAM
Entry	AH = 00H CS = program segment prefix address
Exit	DOS is entered
01H	READ THE KEYBOARD
Entry	AH = 01H
Exit	AL = ASCII character
Notes	If AL = 00H the function call must be invoked again to read an extended ASCII character. Refer to Chapter 6, Table 6-1, for a listing of the extended ASCII keyboard codes. This function call automatically echoes whatever is typed to the video screen.
02H	WRITE TO STANDARD OUTPUT DEVICE
Entry	AH = 02H AL = ASCII character to be displayed

Notes	This function call normally displays data on the video display.
03H	READ CHARACTER FROM COM1
Entry	AH = 03H
Exit	AL = ASCII character read from the communications port
Notes	This function call reads data from the serial communications port.
04H	WRITE TO COM1
Entry	AH = 04H DL = character to be sent out of COM1
Notes	This function transmits data through the serial communcations port.
05H	WRITE TO LPT1
Entry	AH = 05H DL = ASCII character to be printed
Notes	Prints DL on the line printer attached to LPT1
06H	DIRECT CONSOLE READ/WRITE
Entry	AH = 06H DL = 0FFH or DL = ASCII character
Exit	AL = ASCII character
Notes	If DL = 0FFH on entry, then this function reads the console. If DL = ASCII character, then this function displays the ASCII character on the console video screen. If a character is read from the console keyboard, the zero flag (ZF) indicates whether a character was typed. A zero condition indicates no key is typed and a not-zero condition indicates that AL contains the ASCII code of the key or a 00H. If AL = 00H, the function must again be invoked to read an extended ASCII character from the keyboard. Note that the key does not echo to the video screen.
07H	DIRECT CONSOLE INPUT WITHOUT ECHO
Entry	AH = 07H
Exit	AL = ASCII character
Notes	This functions exactly as function number 06H with DL = 0FFH, but it will not return from the function until the key is typed.
08H	READ STANDARD INPUT WITHOUT ECHO
Entry	AH = 08H
Exit	AL = ASCII character
Notes	Performs as function 07H, except it reads the standard input device. The standard input device can be assigned as either the keyboard or the COM port. This function also responds to a control-break, where function 06H and 07H do not. A control-break causes INT 23H to execute.
09H	DISPLAY A CHARACTER STRING
Entry	AH = 09H DS:DX = address of the character string

Notes	The character string must end with an ASCII $ (24H). The character string can be of any length and may contains control characters such as carriage return (0DH) and line feed (0AH).
0AH	BUFFERED KEYBOARD INPUT
Entry	AH = 0AH DS:DX = address of keyboard input buffer
Notes	The first byte of the buffer contains the size of the buffer (up to 255). The second byte is filled with the number of characters typed upon return. The third byte through the end of the buffer contains the character string typed followed by a cariage return (0DH). This function continues to read the keyboard (displaying data as typed) until either the specified number of characters are typed or until a carriage return (enter) key is typed.
0BH	TEST STATUS OF THE STANDARD INPUT DEVICE
Entry	AH = 0BH
Exit	AL = status of the input device
Notes	This function tests the standard input device to determine if data are available. If AL = 00, no data are available. If AL = 0FFH, then data are available that must be input using function number 08H.
0CH	CLEAR KEYBOARD BUFFER AND INVOKE KEYBOARD FUNCTION
Entry	AH = 0CH AL = 01H, 06H, 07H, or 0AH
Exit	see exit for functions 01H, 06H, 07H, or 0AH
Notes	The keyboard buffer holds keystrokes while programs execute other tasks. This function empties or clears the buffer and then invokes the keyboard function located in register AL.
0DH	FLUSH DISK BUFFERS
Entry	AH = 0DH
Notes	Erases all file names stored in disk buffers. This function does not close the files specified by the disk buffers, so care must be exercised in its usage.
0EH	SELECT DEFAULT DISK DRIVE
Entry	AH = 0DH DL = desired default disk drive number
Exit	AL = the total number of drives present in the system
Notes	Drive A = 00H, drive B = 01H, drive C = 02H, and so forth.
0FH	@OPEN FILE WITH FCB
Entry	AH = 0FH DS:DX = address of the unopened file control block (FCB)
Exit	AL = 00H if file found AL = 0FFH if file not found
Notes	The file control block (FCB) is only used with early DOS software and should never be used with new programs. File control blocks do not allow path names as do newer file access function codes presented later. Figure A-1 illustrates the structure of the FCB. To

| | open a file, the file must either be present on the disk or be created with funtion call 16H. |

FIGURE A–1 Contents of the file-control block (FCB).

Offset   Contents

Offset	Contents
00H	Drive
01H	8-character filename
09H	3-character file extension
0CH	Current block number
0EH	Record size
10H	File size
14H	Creation date
16H	Reserved space
20H	Current record number
21H	Relative record number

10H	**@CLOSE FILE WITH FCB**
Entry	AH = 10H DS:DX = address of the opened file control block (FCB)
Exit	AL = 00H if file closed AL = 0FFH if error found
Notes	Errors that occur usually indicate that either the disk is full or the media is bad.
11H	**@SEARCH FOR FIRST MATCH (FCB)**
Entry	AH = 11H DS:DX = address of the file control block to be searched
Exit	AL = 00H if file found AL = 0FFH if file not found
Notes	Wild card characters (? or *) may be used to search for a file name. The ? wild card character matches any character and the * matches any name or extension.
12H	**@SEARCH FOR NEXT MATCH (FCB)**
Entry	AH = 12H DS:DX = address of the file control block to be searched
Exit	AL = 00H if file found AL = 0FFH if file not found
Notes	This function is used after function 11H first finds a matching file name.
13H	**@DELETE FILE USING FCB**

Entry	AH = 13H DS:DX = address of the file control block to be deleted
Exit	AL = 00H if file deleted AL = 0FFH if error occured
Notes	Errors that most often occur are defective media errors.

14H	@SEQUENTIAL READ (FCB)
Entry	AH = 14H DS:DX = address of the file control block to be read
Exit	AL = 00H if read successful AL = 01H if end of file reached AL = 02H if DTA had a segment wrap AL = 03H if less than 128 bytes were read

15H	@SEQUENTIAL WRITE (FCB)
Entry	AH = 15H DS:DX = address of the file control block to be written
Exit	AL = 00H if write successful AL = 01H if disk is full AL = 02H if DTA had a segment wrap

16H	@CREATE A FILE (FCB)
Entry	AH = 16H DS:DX = address of an unopened file control block
Exit	AL = 00H if file created AL = 01H if disk is full

17H	@RENAME A FILE (FCB)
Entry	AH = 17H DS:DX = address of a modified file control block
Exit	AL = 00H if file renamed AL = 01H if error occured
Notes	Refer to Figure A-2 for the modified FCB used to rename a file.

FIGURE A–2 Contents of the modified file-control block (FCB).

Offset	Content
00H	Drive
01H	8-character filename
09H	3-character extension
0CH	Current block number
0EH	Record size
10H	File size
14H	Creation date
16H	Second file name

18H	NOT ASSIGNED

19H	RETURN CURRENT DRIVE
Entry	AH = 19H
Exit	AL = current drive
Notes	AL = 00H for drive A, 01H for drive B, and so forth.

1AH	SET DISK TRANSFER AREA
Entry	AH = 1AH DS:DX = address of new DTA
Notes	The disk transfer area is normally located within the program segment prefix at offset address 80H. The DTA is used by DOS for all disk data transfers using file control blocks.

1BH	GET DEFAULT DRIVE FILE ALLOCATION TABLE (FAT)
Entry	AH = 1BH
Exit	AL = number of sectors per cluster DS:BX = address of the media-descriptor CX = size of a sector in bytes DX = number of clusters on drive
Notes	Refer to Figure A-3 for the format of the media-descriptor byte. The DS register is changed by this function so make sure to save it before using this function.

FIGURE A–3   Contents of the media-descriptor byte.

7	6	5	4	3	2	1	0
?	?	?	?	?	?	?	?

Bit 0 = 0 if not two-sided
 = 1 if two-sided

Bit 1 = 0 if not eight sectors per track
 = 1 if eight sectors per track

Bit 2 = 0 if nonremovable
 = 1 if removable

1CH	GET ANY DRIVE FILE ALLOCATION TABLE (FAT)
Entry	AH = 1CH DL = disk drive number
Exit	AL = number of sectors per cluster DS:BX = address of the media-descriptor CX = size of a sector in bytes DX = number of clusters on drive
1DH	NOT ASSIGNED

1EH	NOT ASSIGNED
1FH	NOT ASSIGNED
20H	NOT ASSIGNED
21H	@RANDOM READ USING FCB
Entry	AH = 21H DS:DX = address of opened FCB
Exit	AL = 00H if read successful AL = 01H if end of file reached AL = 02H if the segment wrapped AL = 03H if less than 128 bytes read
22H	@RANDOM WRITE USING FCB
Entry	AH = 22H DS:DX = address of opened FCB
Exit	AL = 00H if write successful AL = 01H if disk full AL = 02H if the segment wrapped
23H	@RETURN NUMBER OF RECORDS (FCB)
Entry	AH = 23H DS:DX = address of FCB
Exit	AL = 00H number of records AL = 0FFH if file not found
24H	@SET RELATIVE RECORD SIZE (FCB)
Entry	AH = 24H DS:DX = address of FCB
Notes	Sets the record field to the value contained in the FCB.
25H	SET INTERRUPT VECTOR
Entry	AH = 25H AL = interrupt vector number DS:DX = address of new interrupt procedure
Notes	Before changing the interrupt vector, it is suggested that the current interrupt vector is first saved using DOS function 35H. This allows a back-link so that the original vector can be later restored.
26H	CREATE NEW PROGRAM SEGMENT PREFIX
Entry	AH = 26H DX = segment address of new PSP
Notes	Figure A-4 illustrates the structure of the program segment prefix.

FIGURE A–4 Contents of the program-segment prefix (PSP).

Offset	Content
00H	INT 20H
02H	Top of memory
04H	Reserved
05H	Opcode
06H	Number of bytes in segment
0AH	Terminate address (offset)
0CH	Terminate address (segment)
0EH	Control break address (offset)
10H	Control break address (segment)
12H	Critical error address (offset)
14H	Critical error address (segment)
16H	Reserved
2CH	Environment address (segment)
2EH	Reserved
50H	DOS call
52H	Reserved
5CH	File control block 1
6CH	File control block 2
80H	Command line length
81H	Command line

27H	@RANDOM FILE BLOCK READ (FCB)
Entry	AH = 27H CX = the number of records DS:DX = address of opened FCB
Exit	AL = 00H if read successful AL = 01H if end of file reached AL = 02H if the segment wrapped AL = 03H if less than 128 bytes read CX = the number of records read

28H	@RANDOM FILE BLOCK WRITE (FCB)
Entry	AH = 28H CX = the number of records DS:DX = address of opened FCB
Exit	AL = 00H if write successful AL = 01H if disk full AL = 02H if the segment wrapped CX = the number of records written

29H	@PARSE COMMAND LINE (FCB)
Entry	AH = 29H AL = parse mask DS:SI = address of FCB DS:DI = address of command line
Exit	AL = 00H if no file name characters found AL = 01H if file name characters found AL = 0FFH if drive specifier incorrect DS:SI = address of character after name DS:DI = address first byte of FCB

2AH	READ SYSTEM DATE
Entry	AH = 2AH
Exit	AL = day of the week CX = the year (1980—2099) DH = the month DL = day of the month
Notes	The day of the week is encoded as Sunday = 00H through Saturday = 06H. The year is a binary number equal to 1980 through 2099.

2BH	SET SYSTEM DATE
Entry	AH = 2BH CX = the year (1980—2099) DH = the month DL = day of the month

2CH	READ SYSTEM TIME
Entry	AH = 2CH
Exit	CH = hours (0—23) CL = minutes DH = seconds DL = hundredths of seconds

2DH	SET SYSTEM TIME

Entry	AH = 2DH CH = hours CL = minutes DH = seconds DL = hundreths of seconds
2EH	DISK VERIFY WRITE
Entry	AH = 2EH AL = 00H to disable verify on write AL = 01H to enable verify on write
2FH	READ DISK TRANSFER AREA
Entry	AH = 2FH
Exit	ES:BX = contains DTA address
30H	READ DOS VERSION NUMBER
Entry	AH = 30H
Exit	AH = fractional version number AL = whole number version number
Notes	For example, DOS version number 3.2 is returned as a 3 in AL and a 2 in AH.
31H	TERMINATE AND STAY RESIDENT (TSR)
Entry	AH = 31H AL = the DOS return code DX = number of paragraphs to reserve
Notes	A paragraph is 16 bytes and the DOS return code is read at the batch file level with ERRORCODE.
32H	NOT ASSIGNED
33H	TEST CONTROL-BREAK
Entry	AH = 33H AL = 00H to request current control-break AL = 01H to change control-break DL = 00H to disable control-break DL = 01H to enable control-break
Exit	DL = current control-break state
34H	GET ADDRESS OF InDOS FLAG
Entry	AH = 34H
Exit	ES:BX = address of InDOS flag
Notes	The InDOS flag is available in DOS versions 3.2 or newer and indicates DOS activity. If InDOS = 00H, DOS is inactive or 0FFH if DOS is active.
35H	READ INTERRUPT VECTOR
Entry	AH = 35H AL = interrupt vector number

Exit	ES:BX = address stored at vector
Notes	This DOS function is used with function 25H to install/remove interrupt handlers.
36H	DETERMINE FREE DISK SPACE
Entry	AH = 36H DL = drive number
Exit	AX = FFFFH if drive invalid AX = number of sectors per cluster BX = number of free clusters CX = bytes per sector DX = number of clusters on drive
Notes	The default disk drive is DL = 00H, drive A = 01H, drive B = 02H, and so forth.
37H	NOT ASSIGNED
38H	RETURN COUNTRY CODE
Entry	AH = 38H AL = 00H for current country code BX = 16-bit country code DS:DX = data buffer address
Exit	AX = error code if carry set BX = counter code DS:DX = data buffer address
39H	CREATE SUBDIRECTORY
Entry	AH = 39H DS:DX = address of ASCII-Z string subdirectory name
Exit	AX = error code if carry set
Notes	The ASCII-Z string is the name of the subdirectory in ASCII code ended with a 00H instead of a carriage return/line feed.
3AH	ERASE SUBDIRECTORY
Entry	AH = 3AH DS:DX = address of ASCII-Z string subdirectory name
Exit	AX = error code if carry set
3BH	CHANGE SUBDIRECTORY
Entry	AH = 3BH DS:DX = address of new ASCII-Z string subdirectory name
Exit	AX = error code if carry set
3CH	CREATE A NEW FILE
Entry	AH = 3CH CX = attribute word DS:DX = address of ASCII-Z string file name
Exit	AX = error code if carry set AX = file handle if carry cleared
Notes	The attribute word can contain any of the following (added together): 01H read-only access, 02H = hidden file or directory,

		04H = system file, 08H = volume label, 10H = subdirectory, and 20H = archive bit. In most cases a file is created with 0000H.
3DH	OPEN A FILE	
Entry	AH = 3DH AL = access code DS:DX = address of ASCII-Z string file name	
Exit	AX = error code if carry set AX = file handle if carry cleared	
Notes	The access code in AL = 00H for a read-only access, AL = 01H for a write-only access, and AL = 02H for a read/write access. For shared files in a network environment, bit 4 of AL = 1 will deny read/write access, bit 5 of AL = 1 will deny a write access, bits 4 and 5 of AL = 1 will deny read access, bit 6 of AL = 1 denies none, bit 7 of AL = 0 causes the file to be inherited by child, and if bit 7 of AL = 1 file is restricted to current process.	
3EH	CLOSE A FILE	
Entry	AH = 3EH BX = file handle	
Exit	AX = error code if carry set	
3FH	READ A FILE	
Entry	AH = 3FH BX = file handle CX = number of bytes to be read DS:DX = address of file buffer to hold data read	
Exit	AX = error code if carry set AX = number of bytes read if carry cleared	
40H	WRITE A FILE	
Entry	AH = 40H BX = file handle CX = number of bytes to write DS:DX = address of file buffer that holds write data	
Exit	AX = error code if carry set AX = number of bytes written if carry cleared	
41H	DELETE A FILE	
Entry	AH = 41H DS:DX = address of ASCII-Z string file name	
Exit	AX = error code if carry set	
42H	MOVE FILE POINTER	
Entry	AH = 42H AL = move technique BX = file handle CX:DX = number of bytes pointer moved	
Exit	AX = error code if carry set AX:DX = bytes pointer moved	
Notes	The move technique causes the pointer to move from the start of the file if AL = 00H, from the current location if AL = 01H and from the end of the file if AL = 02H. The count is stored so DX contains the least significant 16-bits and either CX or AX	

		contains the most significant 16 bits.
43H		READ/WRITE FILE ATTRIBUTES
	Entry	AH = 43H AL = 00H to read attributes AL = 01H to write attributes CX = attribute word (see function 3CH) DS:DX = address of ASCII-Z string file name
	Exit	AX = error code if carry set CX = attribute word if carry cleared
44H		I/O DEVICE CONTROL (IOTCL)
	Entry	AH = 44H AL = code (see notes) AL = 01H to write attributes BX = file handle or device number CX = number of bytes DS:DX = data or address
	Exit	AX = error code if carry set AX and DX = parameters
	Notes	The codes found in AL are as follows: 00H = read device status (DX = status) 01H = write device status (DX = status written) 02H = read data from device (DS:DX = buffer address) 03H = write data to device (DS:DX = buffer address) 04H = read data from disk drive 05H = write data to disk drive 06H = read input status (AL = 00H ready or 0FH not ready) 07H = read output status (AL = 00H ready or 0FH not ready) 08H = removable media? (AL = 00H removable, 01H fixed) 09H = local or remote device? (bit 12 of DX set for remote) 0AH = local or remote handle? (bit 15 of DX set for remote) 0BH = change entry count 0CH = generic I/O control for character devices 0DH = generic I/O control for block devices 0EH = return number of logical devices (AL = number) 0FH = change number of logical devices
45H		DUPLICATE FILE HANDLE
	Entry	AH = 45H BX = current file handle
	Exit	AX = error code if carry set AX = duplicate file handle
46H		FORCE DUPLICATE FILE HANDLE
	Entry	AH = 46H BX = current file handle CX = new file handle
	Exit	AX = error code if carry set
	Notes	This function works like function 45H except function 45H allows DOS to select the new handle while this function allows the user to select the new handle.
47H		READ CURRENT DIRECTORY
	Entry	AH = 47H DL = drive number DS:SI = address of a 64 byte buffer for directory name
	Exit	DS:SI addresses current directory name if carry cleared
48H		ALLOCATE MEMORY BLOCK

Entry	AH = 48H BX = number of paragraphs to allocate CX = new file handle
Exit	BX = largest block available if carry cleared
49H	RELEASE ALLOCATED MEMORY BLOCK
Entry	AH = 49H ES = segment address of block to be released CX = new file handle
Exit	Carry indicates an error if set
4AH	MODIFY ALLOCATED MEMORY BLOCK
Entry	AH = 4AH BX = new block size in paragraphs ES = segment address of block to be modified
Exit	BX = largest block available if carry cleared
4BH	LOAD OR EXECUTE A PROGRAM
Entry	AH = 4BH AL = function code ES:BX = address of parameter block DS:DX = address ASCII-Z string command
Exit	Carry indicates an error if set
Notes	The function codes are: AL = 00H to load and execute a program and AL = 03H to load a program but not execute it. Figure A-5 shows the parameter block used with this function.

FIGURE A–5 The parameter blocks used with function 4BH (EXEC). (a) For function code 00H. (b) For function code 03H.

(a)

Offset	Contents
00H	Environment address (segment)
02H	Command line address (offset)
04H	Command line address (segment)
06H	File control block 1 address (offset)
08H	File control block 1 address (segment)
0AH	File control block 2 address (segment)
0CH	File control block 2 address (offset)

(b)

Offset	Contents
00H	Overlay destination segment address
02H	Relocation factor

4CH	TERMINATE A PROCESS
Entry	AH = 4CH AL = error code
Exit	Returns control to DOS
Notes	This function returns control to DOS with the error code saved so it can be obtained using DOS ERROR LEVEL batch processing system. We normally use this function with an error code of 00H to return to DOS.
4DH	READ RETURN CODE
Entry	AH = 4DH
Exit	AX = return error code
Notes	This function is used to obtain the return status code created by executing a program with DOS function 4BH. The return codes are: AX = 0000H for a normal—no error—termination, AX = 0001H for a control-break termination, AX = 0002H for a critical device error, and AX = 0003H for a termination by an INT 31H.
4EH	FIND FIRST MATCHING FILE
Entry	AH = 4EH CX = file attributes DS:DX = address ASCII-Z string file name
Exit	Carry is set for file not found
Notes	This function searches the current or named directory for the first matching file. Upon exit the DTA contains the file information. See Figure A-6 for the disk transfer area (DTA).

FIGURE A–6 Data transfer area (DTA) used to find a file.

Offset	Contents
15H	Attributes
16H	Creation time
18H	Creation date
1AH	Low word file size
1CH	High word file size
1EH	Search file name

4FH	FIND NEXT MATCHING FILE
Entry	AH = 4FH
Exit	Carry is set for file not found
Notes	This function is used after the first file is found with function 4EH

50H	SET PROGRAM SEGMENT PREFIX (PSP) ADDRESS
Entry	AH = 50H BX = offset address of the new PSP
Notes	Extreme care must be used with this function because no error recovery is possible.

51H	GET PSP ADDRESS
Entry	AH = 51H
Exit	BX = current PSP segment address

52H	NOT ASSIGNED

53H	NOT ASSIGNED

54H	READ DISK VERIFY STATUS
Entry	AH = 54H
Exit	AL = 00H if verify off AL = 01H if verify on

55H	NOT ASSIGNED

56H	RENAME FILE
Entry	AH = 56H ES:DI = address of ASCII-Z string containing new file name DS:DX = address of ASCII-Z string containing file to be renamed
Exit	Carry is set for error condition

57H	READ FILE'S DATE AND TIME STAMP
Entry	AH = 57H AL = function code BX = file handle CX = new time DX = new date
Exit	Carry is set for error condition CX = time if carry cleared DX = date if carry cleared
Notes	AL = 00H to read date and time or 01H to write date and time.

58H	NOT ASSIGNED

59H	GET EXTENDED ERROR INFORMATION
Entry	AH = 59H BX = 0000H for DOS version 3.X
Exit	AX = extended error code BH = error class BL = reccomended action CH = locus

Notes	Following are the error codes found in AX:
	0001H = invalid function number 0002H = file not found 0003H = path not found 0004H = no file handles available 0005H = access denied 0006H = file handle invalid 0007H = memory control block failure 0008H = insufficient memory 0009H = memory block address invalid 000AH = environmenr failure 000BH = format invalid 000CH = access code invalid 000DH = data invalid 000EH = unknown unit 000FH = disk drive invalid 0010H = attempted to remove current directory 0011H = not same device 0012H = no more files 0013H = disk write-protected 0014H = unknown unit 0015H = drive not ready 0016H = unknown command 0017H = data error (CRC check error) 0018H = bad request structure length 0019H = seek error 001AH = unknown media type 001BH = sector not found 001CH = printer out of paper 001DH = write fault 001EH = read fault 001FH = general failure 0020H = sharing violation 0021H = lock violation 0022H = disk change invalid 0023H = FCB unavailable 0024H = sharing buffer exceeded 0025H = code page mismatch 0026H = handle end of file operation not completed 0027H = disk full 0028H — 0031H reserved 0032H = unsupported network request 0033H = remote machine not listed 0034H = duplicate name on network 0035H = network name not found 0036H = network busy 0037H = device no longer exists on network 0038H = netBIOS command limit exceeded 0039H = error in network adapter hardware 003AH = incorrect reponse from network 003BH = unexpected network error 003CH = remote adapter is incompatible 003DH = print queue is full 003EH = not enough room for print file 003FH = print file was deleted 0040H = network name deleted 0041H = network access denied 0042H = incorrect network device type 0043H = network name not found 0044H = network name exceeded limit 0045H = netBIOS session limit exceeded 0046H = temporary pause 0047H = network request not accepted 0048H = printer or disk redirection pause 0049H — 004FH reserved 0050H = file already exists 0051H = duplicate FCB 0052H = cannot make directory 0053H = failure in INT 24H (critical error) 0054H = too many redirections 0055H = duplicate redirection 0056H = invalid password 0057H = invalid parameter 0058H = network write failure 0059H = function not supported by network 005AH = required system coomponent not installed 0065H = device not selected

Following are the error class codes as found in BH:

01H = no resources available
02H = temporary error
03H = authorization error
04H = internal software error
05H = hardware error
06H = system failure
07H = application software error
08H = item not found
09H = invalid format
0AH = item blocked
0BH = media error
0CH = item already exists
0DH = unknown error

Following is the recommended action as found in BL:

01H = retry operation
02H = delay and retry operation
03H = user retry
04H = abort processing
05H = immediate exit
06H = ignore error
07H = retry with user intervention

Following is a list of loci in CH:

01H = unknown source
02H = block device error
03H = network area
04H = serial device error
05H = memory error

5AH	CREATE UNIQUE FILE NAME
Entry	AH = 5AH CX = attribute code DS:DX = address of the ASCII-Z string directory path
Exit	Carry is set for error condition AX = file handle if carry cleared DS:DX = address of the appended directory name
Notes	The ASCII-Z file directory path must end with a backslash (\). On exit the directory name is appended with a unique file name.
5BH	CREATE A DOS FILE
Entry	AH = 5BH CX = attribute code DS:DX = address of the ASCII-Z string containing the file name
Exit	Carry is set for error condition AX = file handle if carry cleared
Notes	The function only works in DOS version 3.X or higher.
5CH	LOCK/UNLOCK FILE CONTENTS
Entry	AH = 5CH BX = file handle CX:DX = offset address of locked/unlocked area SI:DI = number of bytes to lock or unlock beginning at offset
Exit	Carry is set for error condition
5DH	SET EXTENDED ERROR INFORMATION
Entry	AH = 5DH

	AL = 0AH DS:DX = address of the extended error data structure
Notes	This function is used by DOS version 3.1 or higher to store extended error information.

5EH	NETWORK/PRINTER
Entry	AH = 5EH AL = 00H (get network name) DS:DX = address of the ASCII-Z string containing network name
Exit	Carry is set for error condition CL = netBIOS number if carry cleared
Entry	AH = 5EH AL = 02H (define network printer) BX = redirection list CX = length of setup string DS:DX = address of printer setup buffer
Exit	Carry is set for error condition
Entry	AH = 5EH AL = 03H (read network printer setup string) BX = redirection list DS:DX = address of printer setup buffer
Exit	Carry is set for error condition CX = length of setup string if carry cleared ES:DI = address of printer setup buffer

62H	GET PSP ADDRESS
Entry	AH = 62H
Exit	BX = segment address of the current program
Notes	The function only works in DOS version 3.0 or higher.

65H	GET EXTENDED COUNTER INFORMATION
Entry	AH = 65H AL = function code ES:DI = address of buffer to receive information
Exit	Carry is set for error condition CX = length of country information
Notes	The function only works in DOS version 3.3 or higher.

66H	GET/SET CODE PAGE
Entry	AH = 66H AL = function code BX = code page number
Exit	Carry is set for error condition BX = active code page number DX = default code page number
Notes	A function code in AL of 01H gets the code page number and a code of 02H sets the code page number

67H	SET HANDLE COUNT
Entry	AH = 67H BX = number of handles desired
Exit	Carry is set for error condition

Notes	This function is available for DOS version 3.3 or higher
68H	COMMIT FILE
Entry	AH = 68H BX = handle number
Exit	Carry is set for error condition Else, the date and time stamp is written to directory
Notes	This function is available for DOS version 3.3 or higher
6CH	EXTENDED OPEN FILE
Entry	AH = 6CH AL = 00H BX = open mode CX = attributes DX = open flag DS:SI = address of ASCII-Z string file name
Exit	AX = error code if carry is set AX = handle if carry is cleared CX = 0001H file existed and was opened CX = 0002H file did not exist and was created
Notes	This function is available for DOS version 4.0 or higher

BIOS FUNCTION CALLS

In addition to DOS function call INT 21H, some other BIOS function calls prove useful in controlling the I/O environment of the computer. Unlike INT 21H, which exists in the DOS program, the BIOS function calls are found stored in the BIOS ROM. These BIOS functions directly control the I/O devices with or without DOS loaded into a system.

INT 10H

The INT 10H BIOS interrupt is often called the video services interrupt because it directly controls the video display in a system. The INT 10H instruction uses register AH to select the video service provided by this interrupt.

Video Mode Selection. The mode of operation for the video display is selected by placing a 00H into AH followed by one of many mode numbers in AL. Table A–3 lists the modes of operation found in video display systems using standard video modes. The VGA can use any mode listed, while the other displays are more restrictive in use. Additional, higher-resolution modes are explained later in this section.

Example A–6 lists a short sequence of instructions that places the video display in mode 03H. This mode is available on CGA, EGA, and VGA displays. This mode allows the display to draw a test with 16 colors at various resolutions dependent upon the display adapter.

TABLE A–3   Standard video display modes

Mode	Type	Columns	Rows	Resolution	Standard	Colors	Memory
00H	Text	40	25	320×200	CGA	2	B8000H
00H	Text	40	25	320×350	EGA	2	B8000H
00H	Text	40	25	360×400	VGA	2	B8000H
01H	Text	40	25	320×200	CGA	16	B8000H
01H	Text	40	25	320×350	EGA	16	B8000H
01H	Text	40	25	360×400	VGA	16	B8000H
02H	Text	80	25	640×200	CGA	2	B8000H
02H	Text	80	25	640×350	EGA	2	B8000H
02H	Text	80	25	720×400	VGA	2	B8000H
03H	Text	80	25	640×200	CGA	16	B8000H
03H	Text	80	25	640×350	EGA	16	B8000H
03H	Text	80	25	720×400	VGA	16	B8000H
04H	Gra	40	25	320×200	CGA	4	B8000H
05H	Gra	40	25	320×200	CGA	2	B8000H
06H	Gra	80	25	640×200	CGA	2	B8000H
07H	Text	80	25	720×350	EGA	4	B0000H
07H	Text	80	25	720×400	VGA	4	B0000H
0DH	Gra	80	25	320×200	CGA	16	A0000H
0EH	Gra	80	25	640×200	CGA	16	A0000H
0FH	Gra	80	25	640×350	EGA	4	A0000H
10H	Gra	80	25	640×350	EGA	16	A0000H
11H	Gra	80	30	640×480	VGA	2	A0000H
12H	Gra	80	30	640×480	VGA	16	A0000H
13H	Gra	40	25	320×200	VGA	256	A0000H

Note: Gra = graphics mode.

EXAMPLE A–6

```
0000 B4 00              MOV     AH,0            ;select mode
0002 B0 03              MOV     AL,3            ;mode is 03H
0004 CD 10              INT     10H
```

Cursor Control.   Table A–4 shows many of the function codes used to control the cursor, display characters, change colors, and so forth on the video display using the video BIOS function call INT 10H. These cursor control functions will work on any video display from the CGA display to the latest super VGA display.

TABLE A–4   Video BIOS (INT 10H) functions

00H	SELECT VIDEO MODE
Entry	AH = 00H AL = mode number
Exit	Mode changed and screen cleared

TABLE A–4 *(continued)*

01H	SELECT CURSOR TYPE
Entry	AH = 01H CH = starting line number CL = ending line number
Exit	Cursor size changed
02H	SELECT CURSOR POSITION
Entry	AH = 02H BH = page number (usuallly 0) DH = row number (beginning with 0) DL = column number (beginning with 0)
Exit	Changes cursor to new position
03H	READ CURSOR POSITION
Entry	AH = 03H BH = page number
Exit	CH = starting line (cursor size) CL = ending line (cursor size) DH = current row DL = current column
04H	READ LIGHT PEN
Entry	AH = 04H (not supported in VGA)
Exit	AH = 0, light pen triggered BX = pixel column CX = pixel row DH = character row DL = character column
05H	SELECT DISPLAY PAGE
Entry	AH = 05H AL = page number
Exit	Page number selected. Following are the valid page numbers. Mode 0 and 1 support pages 0—7 Mode 2 and 3 support pages 0—7 Mode 4, 5, and 6 support page 0 Mode 7 and D support pages 0—7 Mode E supports pages 0—3 Mode F and 10 support pages 0—1 Mode 11, 12, and 13 support page 0
06H	SCROLL PAGE UP
Entry	AH = 06H AL = number of lines to scroll (0 clears window) BH = character attribute for new lines CH = top row of scroll window CL = left column of scroll window DH = bottom row of scroll window DL = right column of scroll window
Exit	Scrolls window from the bottom toward the top of the screen. Blank lines fill the bottom using the character attribute in BH.

TABLE A–4 *(continued)*

07H	SCROLL PAGE DOWN
Entry	AH = 07H AL = number of lines to scroll (0 clears window) BH = character attribute for new lines CH = top row of scroll window CL = left column of scroll window DH = bottom row of scroll window DL = right column of scroll window
Exit	Scrolls window from the top toward the bottom of the screen. Blank lines fill from the top using the character attribute in BH.
08H	READ ATTRIBUTE/CHARACTER AT CURRENT CURSOR POSITION
Entry	AH = 08H BH = page number
Exit	AL = ASCII character code AH = character attribute Note: this function does not advance the cursor.
09H	WRITE ATTRIBUTE/CHARACTER AT CURRENT CURSOR POSITION
Entry	AH = 09H AL = ASCII character code BH = page number BL = character attribute CX = number of characters to write
Exit	Note: this function does not advance the cursor.
0AH	WRITE CHARACTER AT CURRENT CURSOR POSITION
Entry	AH = 0AH AL = ASCII character code BH = page number CX = number of characters to write
Exit	Note: this function does not advance the cursor.
0FH	READ VIDEO MODE
Entry	AH = 0FH
Exit	AL = current video mode AH = number of character columns BH = page number
10H	SET VGA PALETTE REGISTER
Entry	AH = 10H AL = 10H BX = color number (0—255) CH = green (0—63) CL = blue (0—63) DH = red (0—63)
Exit	Palette register color is changed. Note: the first 16 colors (0—15) are used in the 16 color, VGA text mode and other modes.
10H	READ VGA PALETTE REGISTER
Entry	AH = 10H

TABLE A–4 *(continued)*

	AL = 15H BX = color number (0—255)
Exit	CH = green CL = blue DH = red
11H	GET ROM CHARACTER SET
Entry	AH = 11H AL = 30H BH = 2 = ROM 8 x 14 character set BH = 3 = ROM 8 x 8 character set BH = 4 = ROM 8 x 8 extended character set BH = 5 = ROM 9 x 14 character set BH = 6 = ROM 8 x 16 character set BH = 7 = ROM 9 x 16 character set
Exit	CX = bytes per character DL = rows per character ES:BP = address of character set

If an SVGA (super VGA) EVGA (extended VGA) or XVGA (also extended VGA) adapter is available, the super VGA mode is set by using INT 10H function call AX = 4F02H with BX = to the VGA mode for these advanced display adapters. This conforms to the VESA* standard for VGA adapters. Table A–5 shows the modes selected by register BX for this INT 10H function call.

INT 11H

This function is used to determine the type of equipment installed in the system. To use this call, the AX register is loaded with an FFFFH and then the INT 11H

TABLE A–5 Extended VGA functions as defined by VESA

BX	Function
100H	640 × 400 with 256 colors
101H	640 × 480 with 256 colors
102H	800 × 600 with 16 colors
103H	800 × 600 with 256 colors
104H	1,024 × 768 with 16 colors
105H	1,024 × 768 with 256 colors
106H	1,280 × 1,024 with 16 colors
107H	1,280 × 1,024 with 256 colors
108H	80 × 60 in text mode
109H	132 × 25 in text mode
10AH	132 × 43 in text mode
10BH	132 × 50 in text mode
10CH	132 × 60 in text mode

*Video Electronics Standard Association

FIGURE A–7   The contents of AX as it indicates the equipment attached to the computer.

15	14	13	12	11	10	9	8	7	6	5	4	3	2	1	0
P1	P0			G	S2	S1	S0	D2	D1						

P1, P0 = number of parallel ports
G = 1 if game I/O attached
S2, S1, S0 = number of serial ports
D2, D1 = number of disk drives

instruction is executed. In return, an INT 11H provides information as listed in Figure A–7.

INT 12H

The memory size is returned by the INT 12H instruction. After executing the INT 12H instruction, the AX register contains the number of 1K byte blocks of memory (conventional memory in the first 1M bytes of address space) installed in the computer.

INT 13H

This call controls the diskettes (5 ¼″ or 3 ½″) and also fixed or hard disk drives attached to the system. Table A–6 lists the functions available to this interrupt via register AH. The direct control of a floppy disk or hard disk can lead to problems. Therefore we only provide a listing of the functions without detail on their usage. Before using these functions refer to the BIOS literature available from the company that produced your version of the BIOS ROM.

INT 14H

Interrupt 14H controls the serial COM (communications) ports attached to the computer. The computer system contains two COM ports, COM1 and COM2, unless you have a newer AT-style machine in which the number of communications ports are extended to COM3 and COM4. Communications ports are normally controlled with software packages that allow data transfer through a modem and the telephone lines. The INT 14H instruction controls these ports as illustrated in Table A–7.

INT 15H

The INT 15H instruction controls many of the various I/O devices interfaced to the computer. It also allows access to protected mode operation and the extended memory system on an 80286, 80386, or 80486 system. Table A–8 lists the functions supported by INT 15H.

INT 16H

The INT 16H instruction is used as a keyboard interrupt. This interrupt is accessed by DOS interrupt INT 21H, but can be accessed directly. Table A–9 shows the functions performed by INT 16H.

TABLE A–6 Disk I/O
function via INT 13H

AH	Function
00H	Reset disk system
01H	Get disk system status into AL
02H	Read sector
03H	Write sector
04H	Verify sector
05H	Format track
06H	Format bad track
07H	Format drive
08H	Get drive parameters
09H	Initialize fixed disk characteristics
0AH	Read long sector
0BH	Write long sector
0CH	Seek
0DH	Reset fixed disk system
0EH	Read sector buffer
0FH	Write sector buffer
10H	Get drive status
11H	Recalibrate drive
12H	Controller RAM diagnostics
13H	Controller drive diagnostics
14H	Controller internal diagnostics
15H	Get disk type
16H	Get disk change status
17H	Set disk type
18H	Set media type
19H	Park heads
1AH	Format ESDI drive

INT 17H

The INT 17H instruction accesses the parallel printer port usually labeled LPT1 in most systems. Table A–10 lists the three functions available for the INT 17H instruction.

TABLE A–7 COM port
interrupt INT 14H

AH	Function
00H	Initialize communications port
01H	Send character
02H	Receive character
03H	Get COM port status
04H	Extended initialize communications port
05H	Extended communications port control

TABLE A–8 The I/O sub-system interrupt INT 15H

AH	Function
00H	Cassette motor on
01H	Cassette motor off
02H	Read cassette
03H	Write cassette
0FH	Format ESDI drive periodic interrupt
21H	Keyboard intercept
80H	Device open
81H	Device closed
82H	Process termination
83H	Event wait
84H	Read joystick
85H	System request key
86H	Delay
87H	Move extended block of memory
88H	Get extended memory size
89H	Enter protected mode
90H	Device wait
91H	Device power on self test (POST)
C0H	Get system environment
C1H	Get address of extended BIOS data area
C2H	Mouse pointer
C3H	Set watch-dog timer
C4H	Programmable option select

TABLE A–9 Keyboard interrupt INT 16H

AH	Function
00H	Read keyboard character
01H	Get keyboard status
02H	Get keyboard flags
03H	Set repeat rate
04H	Set keyclick (PCjr only)
05H	Push character and scan code

TABLE A–10 Parallel printer interrupt INT 17H

AH	Function
00H	Print character
01H	Initialize printer
02H	Get printer status

DOS SYSTEM MEMORY MAP

Figure A–8 illustrates the memory map used by a DOS computer system. The first 1M byte of memory is listed with all the areas containing different devices and programs used with DOS. The transient program area (TPA) is where DOS applications programs are loaded and executed. The size of the TPA is usually between 500K and 628K bytes, unless many TSR programs and drivers fill memory before the TPA.

DOS Low Memory Assignments

Table A–11 shows the low memory assignments (00000H–005FFH) for the DOS-based microprocessor system. This area of memory contains the interrupt vectors, BIOS data area, and the DOS/BIOS data area illustrated in Figure A–8.

DOS Version 5.0 Memory Map

Microsoft DOS version 5.0 has a slightly different memory map than earlier versions of DOS because of its ability to load drivers and programs in the system area. If the microprocessor is an 80386 or 80486, memory between the ROM memory located between addresses A0000H and FFFFFH can be backfilled with extended memory for drivers and programs. In many systems, memory area D0000H–DFFFFH is unused, as is E0000H–EFFFFH. These areas can be filled with extended memory through memory paging found in the 80386 and 80486 microprocessors. This new memory area can then be filled with and addressed by normal real-mode memory programs, extending the memory available to DOS applications.

The drivers HIMEM.SYS and EMM386.SYS are used to accomplish the backfilling. If you want to use memory area E0000H–EFFFFH, you must load EMM386.SYS as EMM386.SYS I = E000–EFFF. Using these drivers increases the

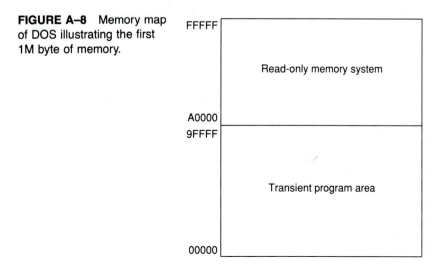

FIGURE A–8 Memory map of DOS illustrating the first 1M byte of memory.

TABLE A–11   DOS low memory assignments

Location	Purpose
00000H-002FFH	System interrupt vectors
00300H-003FFH	System interrupt vectors, power on, and bootstrap area
00400H-00407H	COM1-COM4 I/O port base addresses
00408H-0040FH	LPT1-LPT4 I/O port base addresses
00410H-00411H	Equipment flag word, returned in AX by an INT 11H

Bit	Purpose
15–14	Number of parallel printers (LPT1-LPT4)
13	Internal MODEM installed
12	Joystick installed
11–9	Number of serial ports (COM1-COM4)
8	Unused
7–6	Number of disk drives
5–4	Video mode
3–2	Unused
1	Math coprocessor installed
0	Disk installed

Location	Purpose
00412H	Reserved
00413H–00414H	Memory size in Kbytes (0–640K)
00415H–00416H	Reserved
00417H	Keyboard control byte

Bit	Purpose
7	Insert locked
6	Caps locked
5	Numbers locked
4	Scroll locked
3	Alternate key pressed
2	Control key pressed
1	Left shift key pressed
0	Right shift key pressed

Location	Purpose
00418H	Keyboard control byte

Bit	Purpose
7	Insert key pressed
6	Caps lock key pressed
5	Numbers lock key pressed
4	Scroll lock key pressed
3	Pause locked
2	System request key pressed

Location	Purpose
1	Left alternate key pressed
0	Right control key pressed
00419H	Alternate keyboard entry
0041AH–0041BH	Keyboard buffer header pointer
0041CH–0041DH	Keyboard buffer tail pointer
0041EH–0043DH	32 byte keyboard buffer area
0043EH–00448H	Disk drive control area
00449H–00466H	Video control data area
00467H–0046BH	Reserved
0046CH–0046FH	Timer counter
00470H	Timer overflow
00471H	Break key state
00472H–00473H	Reset flag
00474H–00477H	Hard disk drive data area
00478H–0047BH	LPT1–LPT4 timeout area
0047CH–0047FH	COM1–COM4 timeout area
00480H–00481H	Keyboard buffer start offset pointer
00482H–00483H	Keyboard buffer end offset pointer
00484H–0048AH	Video control data area
0048BH–00495H	Hard drive control area
00496H	Keyboard mode, state, and type flag
00497H	Keyboard LED flags
00498H–00499H	Offset address of user wait complete flag
0049AH–0049BH	Segment address of user wait complete flag
0049CH–0049DH	User wait count (low word)
0049EH–0049FH	User wait count (high word)
004A0H	Wait active flag
004A1H–004A7H	Reserved
004A8H–004ABH	Pointer to video parameters
004ACH–004EFH	Reserved
004F0H–004FFH	Applications program communications area
00500H	Print screen status
00501H–00503H	Reserved
00504H	Single drive mode status
00505H–0050FH	Reserved
00510H–00521H	Used by ROM BASIC
00522H–0052FH	Used by DOS for disk initialization
00530H–00533H	Used by MODE command
00534H–005FFH	Reserved

DOS TPA to more than 600K bytes. A typical CONFIG.SYS file for DOS 5.0 appears in Example A–7. Notice that drivers after EMM386.SYS are loaded in high memory with the DEVICEHIGH directive instead of the DEVICE directive. Programs are loaded using the LOADHIGH or LH directive in front of the program name.

EXAMPLE A–7

(CONFIG.SYS file)

```
FILES=30
BUFFERS=30
STACKS=64,128
FCBS=48
SHELL=C:\DOS\COMMAND.COM C:\DOS\ /E:256 /P
DEVICE=C:\DOS\HIMEM.SYS
DOS=HIGH,UMB
DEVICE=C:\DOS\EMM386.EXE I=C800-EFFF NOEMS
DEVICEHIGH SIZE=1EB0 C:\LASERLIB\SONY_CDU.SYS /D:SONY_001 /B:340 /Q:* /T:* /M:H
DEVICEHIGH SIZE=0190 C:\DOS\SETVER.EXE
DEVICEHIGH SIZE=3150 C:\MOUSE1\MOUSE.SYS
LASTDRIVE = F
```

(AUTOEXEC.BAT file)

```
PATH C:\DOS;C:\;C:\MASM\BIN;C:\MASM\BINB\;C:\UTILITY;C:\WS;C:\LASERLIB
SET BLASTER=A220 I7 D1 T3
SET INCLUDE=C:\MASM\INCLUDE\
SET HELPFILES=C:\MASM\HELP\*.HLP
SET INIT=C:\MASM\INIT\
SET ASMEX=C:\MASM\SAMPLES\
SET TMP=C:\MASM\TMP
SET SOUND=C:\SB
LOADHIGH C:\LASERLIB\MSCDEX.EXE /D:SONY_001 /L:F /M:8
LOADHIGH C:\LASERLIB\LLTSR.EXE ALT-Q
LOADHIGH C:\DOS\FASTOPEN C:=256
LOADHIGH C:\DOS\DOSKEY /BUFSIZE=1024
LOADHIGH C:\LASERLIB\PRINTF.COM
DOSKEY GO=DOSSHELL
DOSSHELL
```

APPENDIX B

Instruction Set Summary

The instruction set summary, which follows this introduction, contains a complete listing of all instructions for the 8086, 8088, 80286, 80386, and 80486 microprocessors. Note that numeric coprocessor instructions for the 80486 microprocessor and the 8087, 80287, and 80387 numeric coprocessors appear in Chapter 10.

Each instruction entry lists the mnemonic opcode plus a brief description of the purpose of the instruction. Also listed is the binary machine language coding for each instruction plus any other data required to form the instruction such as displacement or immediate data. Next to the binary machine language version of the instruction appears the flag register bits and any change that might occur for a given instruction. In this listing a blank indicates no change, a ? indicates a change with an unpredictable outcome, a * indicates a predictable change, a 1 indicates the flag is set, and a 0 indicates the flag is cleared.

Before the instruction listing begins, some information about the bit settings in the binary machine language versions of the instructions is required. Table B–1 shows the modifier bits, coded as oo in the instruction listings, so instructions can be formed with a register, displacement, or no displacement.

Table B–2 lists the memory-addressing modes available with the register/ memory field, coded as mmm. This table applies to all versions of the microprocessor.

TABLE B–1 The modifier bits, coded as oo in the instruction listing

oo	Function
00	If mmm = 110, then a displacement follows the opcode; otherwise, no displacement is used
01	An 8-bit signed displacement follows the opcode
10	A 16-bit signed displacement follows the opcode
11	mmm specifies a register, instead of an addressing mode

TABLE B–2
Register/memory field
(mmm) description

mmm	Function
000	DS:[BX + SI]
001	DS:[BX + DI]
010	SS:[BP + SI]
011	SS:[BP + DI]
100	DS:[SI]
101	DS:[DI]
110	SS:[BP]
111	DS:[BX]

Table B–3 lists the register options (rrr) when encoded for either an 8-bit or a 16-bit register. This table also lists the 32-bit registers used with the 80386 and 80486 microprocessors.

Table B–4 lists the segment register bit assignments (rrr) for the MOV, PUSH, and POP instructions, which use these segment registers.

TABLE B–3   Register field
(rrr) options

rrr	W = 0	W = 1	reg32
000	AL	AX	EAX
001	CL	CX	ECX
010	DL	DX	EDX
011	BL	BX	EBX
100	AH	SP	ESP
101	CH	BP	EBP
110	DH	SI	ESI
111	BH	DI	EDI

TABLE B–4   Register field
assignments (rrr) that are
used to represent the segment registers

rrr	Register
000	ES
001	CS
010	SS
011	DS
100	FS
101	GS

When the 80386 and 80486 microprocessors are used, some of the definitions provided in the prior tables will change. Refer to Tables B–5 and B–6 for these changes as they apply to the 80386 and 80486 microprocessors.

The instruction set summary that follows lists all of the instructions, with examples, for the 8086, 8088, 80286, 80386, and 80486 microprocessors. Missing are the segment override prefixes: CS (2EH), SS (36H), DS (3EH), ES (26H), FS (64H), and GS (65H). These prefixes are one byte in length and placed in memory before the instruction that is prefixed.

The D-bit, in the code segment descriptor, indicates the default size of the operand and the addresses for the 80386 and 80486 microprocessors. If D = 1, then all addresses and operands are 32 bits, and if D = 0, all addresses and operands are 16 bits. In the real mode, the D-bit is set to zero by the 80386 and 80486 microprocessors so operands and addresses are 16 bits.

The address-size prefix (67H) must be placed before instructions in the 80386 and 80486 to change the default size as selected by the D-bit. For example, the MOV AX,[ECX] instruction must have the address-size prefix placed before it in machine code if the default size is 16 bits. If the default size is 32 bits, the address prefix is not needed with this instruction. The operand-override prefix (66H) functions in much the same manner as the address-size prefix. In the previous example, the operand size is 16 bits. If the D-bit selects 32-bit operands and addresses, this instruction requires the operand-size prefix.

TABLE B–5 Index registers are specified with rrr in the 80386 and 80486 microprocessor

rrr	Index register
000	EAX
001	ECX
010	EDX
011	EBX
100	No index
101	EBP
110	ESI
111	EDI

TABLE B–6 Possible combinations of oo, mmm, and rrr for the 80386 and 80486 instruction set using the 32-bit addressing mode

oo	mmm	rrr	Function
00	000	—	DS:[EAX]
00	001	—	DS:[ECX]
00	010	—	DS:[EDX]
00	011	—	DS:[EBX]
00	100	000	DS:[EAX + scaled-index]
00	100	001	DS:[ECX + scaled-index]
00	100	010	DS:[EDX + scaled-index]
00	100	011	DS:[EBX + scaled-index]
00	100	100	SS:[ESP + scaled-index]
00	100	101	DS:[disp32 + scaled-index]
00	100	110	DS:[ESI + scaled-index]
00	100	111	DS:[EDI + scaled-index]
00	101	—	DS:disp32
00	110	—	DS:[ESI]
00	111	—	DS:[EDI]
01	000	—	DS:[EAX + disp8]
01	001	—	DS:[ECX + disp8]
01	010	—	DS:[EDX + disp8]
01	011	—	DS:[EBX + disp8]
01	100	000	DS:[EAX + scaled-index + disp8]
01	100	001	DS:[ECX + scaled-index + disp8]
01	100	010	DS:[EDX + scaled-index + disp8]
01	100	011	DS:[EBX + scaled-index + disp8]
01	100	100	SS:[ESP + scaled-index + disp8]
01	100	101	SS:[EBP + scaled-index + disp8]
01	100	110	DS:[ESI + scaled-index + disp8]
01	100	111	DS:[EDI + scaled-index + disp8]
01	101	—	SS:[EBP + disp8]
01	110	—	DS:[ESI + disp8]
01	111	—	DS:[EDI + disp8]
10	000	—	DS:[EAX + disp32]
10	001	—	DS:[ECX + disp32]
10	010	—	DS:[EDX + disp32]
10	011	—	DS:[EBX + disp32]
10	100	000	DS:[EAX + scaled-index + disp32]
10	100	001	DS:[ECX + scaled-index + disp32]
10	100	010	DS:[EDX + scaled-index + disp32]
10	100	011	DS:[EBX + scaled-index + disp32]
10	100	100	SS:[ESP + scaled-index + disp32]
10	100	101	SS:[EBP + scaled-index + disp32]
10	100	110	DS:[ESI + scaled-index + disp32]
10	100	111	DS:[EDI + scaled-index + disp32]
01	101	—	SS:[EBP + disp32]
01	110	—	DS:[ESI + disp32]
01	111	—	DS:[EDI + disp32]

Notes: disp8 = 8-bit displacement, disp32 = 32-bit displacement

INSTRUCTION SET SUMMARY

AAA	ASCII adjust after addition	(AL)

00110111	O D I T S Z A P C ? ? ? * ? *
Example	
AAA	

AAD	ASCII adjust before division	(AX)

11010101 00001010	O D I T S Z A P C ? * * ? * ?
Example	
AAD	

AAM	ASCII adjust after multiplication	(AX)

11010100 00001010	O D I T S Z A P C ? * * ? * ?
Example	
AAM	

AAS	ASCII adjust after subtraction	(AL)

00111111	O D I T S Z A P C ? ? ? * ? *
Example	
AAS	

ADC	Add with carry	

000100dw oorrrmmm disp	O D I T S Z A P C * * * * * *

Format	Examples
ADC reg,reg	ADC AX,BX ADC AL,BL ADC EAX,EBX
ADC mem,reg	ADC DATA,AL ADC LIST,SI ADC DATA[DI],CL ADC [EAX],AL
ADC reg,mem	ADC BL,DATA ADC SI,LIST ADC CL,DATA[DI] ADC EDX,[EBX+100H]

100000sw oo010mmm disp data

Format	Examples
ADC reg,imm	ADC CX,3 ADC DL,1AH ADC EAX,12345
ADC mem,imm	ADC DATA,33 ADC LIST,'A' ADC DATA[DI],2 ADC BYTE PTR [EAX],3

0001010w data	
Format	Examples
ADC acc,imm	ADC AX,3 ADC AL,1AH ADC EAX,3

ADD   Add

000000dw oorrrmmm disp		O DI T S Z A P C * * * * *
Format	Examples	

ADD reg,reg	ADD AX,BX
	ADD AL,BL
	ADD ESI,EDI	

ADD mem,reg	ADD DATA,AL
	ADD LIST,SI
	ADD DATA[DI],CL
	ADD [EAX],AL	

ADD reg,mem	ADD BL,DATA
	ADD SI,LIST
	ADD CL,DATA[DI]
	ADD EDX,[EBX+100H]	

100000sw oo000mmm disp data	
Format	Examples

ADD reg,imm	ADD CX,3
	ADD DL,34H
	ADD EAX,12345

ADD mem,imm	ADD DATA,33
	ADD LIST,'A'
	ADD DATA[DI],2
	ADD WORD PTR[DI],669H

0000010w data	
Format	Examples

ADD acc,imm	ADD AX,3
	ADD AL,1AH
	ADD EAX,3

AND   Logical AND

001000dw oorrrmmm disp		O DI T S Z A P C 0 * * ? * 0
Format	Examples	
AND reg,reg	AND CX,BX AND DL,BL AND ECX,EBX	
AND mem,reg	AND BIT,CH AND LIST,DI AND DATA[BX],CL AND [EDX+4*ECX],EDI	
AND reg,mem	AND BL,DATA AND SI,LIST AND CL,DATA[DI] AND CL,[EAX]	

100000sw oo100mmm disp data	
Format	Examples
AND reg,imm	AND BP,1 AND DL,34H AND EBP,12345

AND mem,imm	AND DATA,33 AND DATA[SI],2 AND DWORD PTR[EAX],3 AND WORD PTR[DI],669H
0010010w data	
Format	Examples
AND acc,imm	AND AX,15 AND AL,1FH AND EAX,3

ARPL   Adjust requested privilege level   (80286/80386/80486)

01100011 oorrrmmm disp	O D I T S Z A P C *
Format	Examples
ARPL reg,reg	ARPL AX,BX ARPL BX,SI
ARPL mem,reg	ARPL NUMB,AX ARPL 0EDX+4*ECX],DI

BOUND   Check array bounds   (80286/80386/80486)

01100010 oorrrmmm disp	O D I T S Z A P C
Format	Examples
BOUND reg,mem	BOUND AX,BETS BOUND BX,[DI]

BSF   Bit scan forward   (80386/80486)

00001111 10111100 oorrmmm disp	O D I T S Z A P C *
Format	Examples
BSF reg,reg	BSF AX,BX BSF ECX,EBX
BSF reg,mem	BSF AX,DATA BSF EAX,DATA6

BSR   Bit scan reverse   (80386/80486)

00001111 10111101 oorrmmm disp	O D I T S Z A P C *
Format	Examples
BSR reg,reg	BSR AX,BX BSR ECX,EBX
BSR reg,mem	BSR AX,DATA BSR ECX,MEMORY

BSWAP   Byte swap   (80486)

00001111 11001rrr	O D I T S Z A P C
Format	Examples
BSWAP reg32	BSWAP EAX BSWAP EBX

BT   Bit test   (80386/80486)

00001111 10111010 oo100mmm disp data		O DI T S ZA P C *
Format	Examples	
BT reg,imm8	BT AX,2 BT CX,4 BT EBX,2	
BT mem,imm8	BT DATA1,2 BT LIST,2	

00001111 10100011 disp		
Format	Examples	
BT reg,reg	BT AX,CX BT SI,CX	
BT mem,reg	BT DATA1,AX BT DATA[DI],AX	

BTC   Bit test and complement   (80386/80486)

00001111 10111010 oo111mmm disp data		O DI T S ZA P C *
Format	Examples	
BTC reg,imm8	BTC AX,2 BTC CX,4	
mem,imm8	BTC DATA1,2 BTC [BX],1	

00001111 10111011 disp		
Format	Examples	
reg,reg	BTC AX,CX BTC DI,CX	
BTC mem,reg	BTC DATA1,AX BTC DATA9,BX	

BTR   Bit test and reset   (80386/80486)

00001111 10111010 oo110mmm disp data		O DI T S ZA P C *
Format	Examples	
BTR reg,imm8	BTR AX,2 BTR CX,8	
mem,imm8	BTR DATA1,2 BTR FROG,3	

00001111 10110011 disp		
Format	Examples	
BTR reg,reg	BTR AX,CX BTR CX,BP	
BTR mem,reg	BTR DATA1,AX BTR DATA[DI+6],AX	

BTS   Bit test and set   (80386/80486)

00001111 10111010 oo101mmm disp data		O DI T S ZA P C *
Format	Examples	
BTS reg,imm8	BTS AX,2 BTS BP,10H	

mem,imm8	BTS DATA1,2 BTS LIST,2

00001111 10101011 disp	
Format	Examples
BTS reg,reg	BTS AX,CX BTS DI,AX
BTS mem,reg	BTS DATA1,AX BTS WORD PTR[DI+2],BX

CALL — Call procedure (subroutine)

11101000 disp	O DI T S Z A P C
Format	Examples
CALL label (near)	CALL FOR_FUN CALL HOME CALL ET

10011010 disp	
Format	Examples
CALL label (far)	CALL FAR PTR DATES CALL WHAT CALL WHERE

11111111 oo010mmm	
Format	Examples
CALL reg (near)	CALL AX CALL BX CALL CX
CALL mem (near)	CALL ADDRESS CALL [DI] CALL HERO

11111111 oo011mmm	
Format	Examples
CALL mem (far)	CALL FAR_LIST[SI] CALL FROM_HERE CALL TO_THERE

CBW — Convert byte to word (AX = AL)

10011000	O DI T S Z A P C
Example	
CBW	

CDQ — Convert doubleword to quadword (EDX,EAX = EAX)

10011001	O DI T S Z A P C
Example	
CDQ	

CLC — Clear carry flag

11111000	O DI T S Z A P C 0

Example
CLC

CLD   Clear direction flag

11111100	O D I T S Z A P C 0
Example	
CLD	

CLI   Clear interrupt flag

11111010	O D I T S Z A P C 0
Example	
CLI	

CLTS   Clear task switched flag   (80286/80386/80486)

00001111 00000110	O D I T S Z A P C
Example	
CLTS	

CMC   Complement carry flag

10011000	O D I T S Z A P C *
Example	
CMC	

CMP   Compare operands

001110dw oorrrmmm disp	O D I T S Z A P C * * * * *

Format	Examples

CMP reg,reg	CMP AX,BX
	CMP AL,BL
	CMP EAX,EBX

CMP mem,reg	CMP DATA,AL
	CMP LIST,SI
	CMP [EAX],AL

CMP reg,mem	CMP BL,DATA
	CMP CX,DATA[DI]
	CMP EDX,[EBX+100H]

100000sw oo111mmm disp data

Format	Examples		

CMP reg,imm	CMP CX,3
	CMP DL,34H
	CMP EBX,12345		

CMP mem,imm	CMP DATA,33
	CMP LIST,'A'
	CMP DATA[DI],2
	CMP BYTE PTR [EAX],3
	CMP WORD PTR[DI],669H
		80286	6

		80386	5
		80486	2

0011110w data
CMP acc,imm	CMP AX,3 CMP AL,1AH

	CMP EAX,3

CMPS — Compare strings

	O D I T S Z A P C
1010011w	* * * * *
Examples	
CMPSB CMPSW CMPSD	

CMPXCHG — Compare and exchange (80486)

	O D I T S Z A P C
00001111 1011000w 11rrrrrr	* * * * * *

Format	Examples
CMPXCHG reg,reg	CMPXCHG EAX,EBX CMPXCHG ECX,EDX

00001111 1011000w oorrrmmm

Format	Examples
CMPXCHG mem,reg	CMPXCHG DATA,EAX CMPXCHG DATA2,EBX

CWD — Convert word to doubleword (DX,AX = AX)

	O D I T S Z A P C
10011001	
Example	
CWD	

CWDE — Convert word to extended doubleword (EAX = AX)

	O D I T S Z A P C
10011000	
Example	
CWDE	

DAA — Decimal adjust after addition (AL)

	O D I T S Z A P C
00100111	? * * * *
Example	
DAA	

DAS — Decimal adjust after subtraction (AL)

	O D I T S Z A P C
00101111	? * * * *
Example	
DAS	

DEC — Decrement

	O D I T S Z A P C
1111111w oo001mmm disp	* * * * *

Format	Examples
DEC reg8	DEC BL

	DEC DH
DEC mem	DEC DATA DEC BYTE PTR [EAX]

01001rrr	
Format	Examples
DEC reg16 DEC reg32	DEC AX DEC EAX

DIV   Unsigned division

1111011w oo110mmm disp	O D I T S Z A P C ? ? ? ? ? ?
Format	Examples
DIV reg	DIV BL DIV BX DIV ECX
DIV mem	DIV DATA DIV LIST DIV DWORD PTR [EAX]

ENTER   Create a stack frame   (80286/80386/80486)

11001000 data	O D I T S Z A P C
Format	Examples
ENTER imm,0	ENTER 4,0 ENTER 8,0
ENTER imm,1	ENTER 4,1 ENTER 10,1
imm,imm	ENTER 3,6 ENTER 100,3

ESC   Escape

11011nnn oonnnmmm	O D I T S Z A P C
nnnnnn = opcode for coprocessor	
Format	Examples
ESC imm,reg	ESC 5,AL ESC 5,BH ESC 6,CH
ESC imm,mem	ESC 2,DATA ESC 3,FROG FMUL FROG

HLT   Halt

11110100	O D I T S Z A P C
Example	
HLT	

IDIV   Signed division

1111011w oo111mmm disp	O D I T S Z A P C ? ? ? ? ? ?
Format	Examples
IDIV reg	IDIV BL

	IDIV DI IDIV ECX	
IDIV mem	IDIV DATA IDIV BYTE PTR [EAX] IDIV WORD PTR [DI]	

IMUL   Signed multiplication

1111011w oo101mmm disp		O D I T S ZA P C * ? ? ? ? *
Format	Examples	
IMUL reg	IMUL BL IMUL SI IMUL ECX	
IMUL mem	IMUL DATA IMUL BYTE PTR [EAX] IMUL DWORD PTR [DI]	
011010sl oorrrmmm disp data		(80286/80386/80486)
Format	Examples	
IMUL reg,imm	IMUL CX,16 IMUL DX,100 IMUL EAX,20	
reg,reg,imm	IMUL DX,AX,2 IMUL CX,DX,3 IMUL BX,AX,33	
reg,mem,imm	IMUL CX,DATA,4	
00001111 10101111 oorrrmmm disp		(80386/80486)
Format	Examples	
IMUL reg,reg	IMUL CX,DX IMUL DX,BX IMUL EAX,ECX	
IMUL reg,mem	IMUL DX,DATA IMUL CX,FROG	

IN   Input data from port

1110010w port number		O DI T S ZA P C
Format	Examples	
IN acc,pt	IN AL,12H IN AX,12H IN EAX,10H	
1110110w		
Format	Examples	
IN acc,DX	IN AL,DX IN AX,DX IN EAX,DX	

INC   Increment

1111111w oo000mmm disp		O DI T S ZA P C * * * * *
Format	Examples	
INC reg8	INC BL INC CH	
INC mem	INC DATA INC LIST INC WORD PTR [DI]	

01000rrr	
Format	Examples
INC reg16 INC reg32	INC AX INC ESI

INS   Input string from port   (80286/80386/80486)

0110110w	O D I T S Z A P C
Examples	
INSB INSW INSD	

INT   Interrupt

11001101 type	O D I T S Z A P C 0 0
Format	Examples
INT type	INT 10H INT 255 INT 21H

11001100	
Example	
INT 3	

INTO   Interrupt on overflow

11001110	O D I T S Z A P C * *
Example	
INTO	

INVD   Invalidate data cache   (80486)

00001111 00001000	O D I T S Z A P C
Example	
INVD	

INVLPG   Invalidate TLB entry   (80486)

00001111 00000001 oo111mmm	O D I T S Z A P C
Format	Examples
INVLPG mem	INVLPG DATA INVLPG LIST

IRET   Interrupt return

11001101 data	O D I T S Z A P C * * * * * * * * *
Format	Examples
IRET IRETD	IRET IRETD IRET 10H

Jconditional Conditional jump

0111cccc disp	O DI T S Z A P C

Format	Examples
Jcc label	JA BELOW JB ABOVE JG GREATER

00001111 1000cccc disp	(80386/80486)

Format	Examples
Jcc label	JNE NOT_MORE JLE LESS_THAN

Condition Codes	Mnemonic	Flag	Description
0000	JO	O = 1	Jump if overflow
0001	JNO	O = 0	Jump if no overflow
0010	JC/JB/JNAE	C = 1	Jump if carry/below
0011	JNC/JAE/JNB	C = 0	Jump if no carry/above or equal
0100	JE/JZ	Z = 1	Jump if equal/zero
0101	JNE/JNZ	Z = 0	Jump if not equal/not zero
0110	JBE/JNA	C = 1 + Z = 1	Jump if below or equal
0111	JA/JNBE	C = 0 $\cdot$ Z = 0	Jump if above
1000	JS	S = 1	Jump if sign
1001	JNS	S = 0	Jump if no sign
1010	JP/JPE	P = 1	Jump if parity even
1011	JNP/JPO	P = 0	Jump if parity odd
1100	JL/JNGE	S <> O	Jump if less than
1101	JGE/JNL	S = O	Jump if greater or equal
1110	JLE/JNG	Z = 1 + S <> O	Jump if less than or equal
1111	JG/JNLE	Z = 0 $\cdot$ S = O	Jump if greater

JCXZ/JECXZ Jump if CX (ECX) equals zero

11100011	O DI T S Z A P C

Format	Examples
JCXZ label JECXZ label	JCXZ LOTSA JCXZ OVER JECXZ UPPER

JMP Unconditional jump

11101011 disp	O DI T S Z A P C

Format	Examples
JMP label (short)	JMP SHORT UP JMP SHORT DOWN JMP SHORT OVER

11101001 disp	

Format	Examples
JMP label (near)	JMP VER JMP FROG JMP UNDER

11101010 disp	

Format	Examples
JMP label (far)	JMP VER JMP FROG JMP FAR PTR THERE

11111111 oo100mmm

Format	Examples
JMP reg (near)	JMP AX JMP EAX JMP CX
JMP mem (near)	JMP DATA JMP LIST JMP DATA[DI]

11111111 oo101mmm

Format	Examples
JMP mem (far)	JMP WAYOFF JMP TABLE JMP UP

LAHF Load AH from flags

10011111	O D I T S Z A P C

Example

LAHF

LAR Load access rights (80286/80386/80486)

00001111 00000010 oorrrmmm disp	O D I T S Z A P C *

Format	Examples
LAR reg,reg	LAR AX,BX LAR CX,DX LAR EAX,ECX
LAR reg,mem	LAR CX,DATA LAR AX,LIST LAR ECX,FROG

LDS Load far pointer

11000101 oorrrmmm	O D I T S Z A P C

Format	Examples
LDS reg,mem	LDS DI,DATA LDS SI,LIST

LES Load far pointer

11000100 oorrrmmm	O D I T S Z A P C

Format	Examples
LES reg,mem	LES DI,DATA LES SI,LIST

LFS Load far pointer (80386/80486)

00001111 10110100 oorrrmmm disp	O D I T S Z A P C

Format	Examples
LFS reg,mem	LFS DI,DATA LFS SI,LIST

LGS — Load far pointer (80386/80486)

00001111 10110101 oorrrmmm disp	O D I T S Z A P C
Format	Examples
LGS reg,mem	LGS DI,DATA LGS SI,LIST

LSS — Load far pointer (80386/80486)

00001111 10110010 oorrrmmm disp	O D I T S Z A P C
Format	Examples
LSS reg,mem	LSS DI,DATA LSS SI,LIST

LEA — Load effective address

10001101 oorrrmmm disp	O D I T S Z A P C
Format	Examples
LEA reg,mem	LEA DI,DATA LEA BX,ARRAY

LEAVE — Leave high-level procedure (80286/80386/80486)

11001001	O D I T S Z A P C
Example	
LEAVE	

LGDT — Load global descriptor table (80286/80386/80486)

00001111 00000001 oo010mmm disp	O D I T S Z A P C
Format	Examples
LGDT mem64	LGDT DESCRIP LGDT TABLE

LIDT — Load interrupt descriptor table (80286/80386/80486)

00001111 00000001 oo011mmm disp	O D I T S Z A P C
Format	Examples
LIGT mem64	LIDT DATA LIDT DESCRIPT

LLDT — Load local descriptor table (80286/80386/80486)

00001111 00000000 oo010mmm disp	O D I T S Z A P C
Format	Examples
LLDT reg	LLDT AX LLDT CX
LLDT mem	LLDT DATA LLDT LIST

LMSW   Load machine status word

00001111 00000001 oo110mmm disp	O D I T S Z A P C

should only be used with the 80286

Format	Examples
LMSW reg	LMSW AX LMSW CX
LMSW mem	LMSW DATA LMSW LIST

LOCK   Lock the bus

11110000	O D I T S Z A P C

Format	Examples
LOCK inst	LOCK:XCHG AX,BX LOCK:MOV AL,AH

LODS   Load string operand

1010110w	O D I T S Z A P C

Examples

LODSB
LODSW
LODSD

LOOP   Loop until CX = 0
LOOPD   Loop until ECX = 0

11100010 disp	O D I T S Z A P C

Format	Examples
LOOP label LOOPD label	LOOP DATA LOOPD BACK

LOOPE   Loop while equal (CX)
LOOPED   Loop while equal (ECX)

11100001 disp	O D I T S Z A P C

Format	Examples
LOOPE label LOOPZ label LOOPED label LOOPZD label	LOOPE NEXT LOOPZ AGAIN LOOPED REPEAT LOOPZD FALL

LOOPNE   Loop while not equal (CX)
LOOPNED   Loop while not equal (ECX)

11100000 disp	O D I T S Z A P C

Format	Examples
LOOPNE label LOOPNZ label LOOPNED label LOOPNZD label	LOOPNE AGAIN LOOPNZ BACK LOOPNED REPL LOOPNZD LEFT

LSL   Load segment limit   (80286/80386/80486)

00001111 00000011 oorrrmmm disp		O D I T S Z A P C
		*
Format	Examples	
LSL reg,reg	LSL AX,BX LSL CX,BX	
LSL reg,mem	LSL AX,LIMIT LSL EAX,NUMB	

LTR   Load task register   (80286/80386/80486)

00001111 00000000 oo011mmm disp		O D I T S Z A P C
Format	Examples	
LTR reg	LTR AX LTR CX	
LTR mem	LTR TASK LTR EDGE	

MOV   Move data

100010dw oorrrmmm disp		O D I T S Z A P C
Format	Examples	

MOV reg,reg	MOV CL,CH
	MOV CX,DX
	MOV EBP,ESI	

MOV mem,reg	MOV DATA,DL
	MOV NUMB,CX
	MOV TEMP,EBX	

MOV reg,mem	MOV DL,DATA
	MOV DX,NUMB
	MOV EBX,TEMP	

1100011w oo000mmm disp data	
Format	Examples

MOV mem,imm	MOV DATA,23H
	MOV LIST,12H
	MOV BYTE PTR [DI],2

1011wrrr data	
Format	Examples

MOV reg,imm	MOV BX,23H
	MOV CL,2
	MOV ECX,123423H

101000dw disp	
Format	Examples

MOV mem,acc	MOV DATA,AL
	MOV NUMB,AX
	MOV NUMB1,EAX

MOV acc,mem	MOV AL,DATA
	MOV AX,NUMB
	MOV EAX,TEMP

100011d0 oosssmmm disp	
Format	Examples

MOV seg,reg	MOV SS,AX
	MOV DS,DX
	MOV ES,CX

MOV seg,mem	MOV SS,DATA
	MOV DS,NUMB
	MOV ES,TEMP1

MOV reg,seg	MOV AX,DS
	MOV DX,ES
	MOV CX,CS

MOV mem,seg	MOV DATA,SS
	MOV NUMB,ES
	MOV TEMP1,DS

00001111 001000d0 11rrrmmm		(80386/80486)
Format	Examples	

MOV reg,cr	MOV EAX,CR0
	MOV EBX,CR2
	MOV ECX,CR3	

MOV cr,reg	MOV CR0,EAX
	MOV CR2,EBX
	MOV CR3,ECX	

00001111 001000d1 11rrrmmm		(80386/80486)
Format	Examples	
MOV reg,dr	MOV EBX,DR6 MOV EAX,DR6	
MOV dr,reg	MOV DR1,ECX MOV DR2,ESI	

00001111 001001d0 11rrrmmm		(80386/80486)
Format	Examples	
MOV reg,tr	MOV EAX,TR6 MOV EDX,TR7	
MOV tr,seg	MOV TR6,EDX MOV TR7,ESI	

MOVS   Move string data

1010010w	O D I T S Z A P C
Examples	
MOVSB MOVSW MOVSD	

MOVSX   Move with sign extend   (80386/80486)

00001111 1011111w oorrrmmm disp	O D I T S Z A P C
Format	Examples
MOVSX reg,reg	MOVSX BX,AL MOVSX EAX,DX
reg,mem	MOVSX AX,DATA MOVSX EAX,NUMB

MOVZX   Move with zero extend   (80386/80486)

00001111 1011011w oorrrmmm disp	O D I T S Z A P C

Format	Examples
MOVZX reg,reg	MOVZX BX,AL MOVZX EAX,DX
reg,mem	MOVZX AX,DATA MOVZX EAX,NUMB

MUL   Unsigned multiplication

1111011w oo100mmm disp	O D I T S Z A P C * ? ? ? ? *
Format	Examples
MUL reg	MUL BL MUL CX MUL ECX
MUL mem	MUL DATA MUL BYTE PTR [SI] MUL DWORD PTR [ECX]

NEG   Negate

1111011w oo011mmm disp	O D I T S Z A P C * * * * * *
Format	Examples
NEG reg	NEG AX NEG CX NEG EDX
NEG mem	NEG DATA NEG NUMB NEG WORD PTR [DI]

NOP   No operation

10010000	O D I T S Z A P C
Example	
NOP	

NOT   One's complement

1111011w oo010mmm disp	O D I T S Z A P C
Format	Examples
NOT reg	NOT AX NOT CX NOT EDX
NOT mem	NOT DATA NOT NUMB NOT DWORD PTR [DI]

OR   Inclusive-OR

000010dw oorrrmmm disp	O D I T S Z A P C 0 * * ? * 0
Format	Examples
OR reg,reg	OR CL,BL OR CX,DX OR ECX,EBX
OR mem,reg	OR DATA,CL OR NUMB,CX OR [DI],CX

OR reg,mem	OR CL,DATA OR CX,NUMB OR CX,[SI]

100000sw oo001mmm disp data	
Format	Examples

OR reg,imm	OR CL,3 OR DX,1000H OR EBX,100000H

OR mem,imm	OR DATA,33 OR NUMB,4AH OR NUMS,123498H

0000110w data	
Format	Examples

OR acc,imm	OR AL,3 OR AX,1000H OR EAX,100000H

OUT   Output data to port

1110011w port number		O DI T S ZA P C
Format	Examples	

OUT pt,acc	OUT 12H,AL OUT 12H,AX OUT 10H,EAX

1110111w	
Format	Examples

OUT DX,acc	OUT DX,AL OUT DX,AX OUT DX,EAX

OUTS   Output string data to port   (80286/80386/80486)

0110111w port number	O DI T S ZA P C
Examples	

OUTSB
OUTSW
OUTSD

POP   Pop data from stack

01011rrr		O DI T S ZA P C
Format	Examples	

POP reg	POP CX POP AX POP EBX

10001111 oo000mmm disp	
Format	Examples

POP mem	POP DATA POP LISTS POP NUMBS

00sss111	
Format	Examples

POP seg	POP DS

	POP ES POP SS
00001111 10sss001	
Format	Examples
POP seg	POP FS POP GS

POPA/POPAD   Pop all registers from stack

01100001	O D I T S Z A P C
Examples	
POPA POPAD	

POPF/POPFD   Pop flags from stack

10011101	O D I T S Z A P C * * * * * * * * *
Examples	
POPF POPFD	

PUSH   Push data onto stack

01010rrr	O D I T S Z A P C
Format	Examples
PUSH reg	PUSH CX PUSH AX PUSH ECX
11111111 oo110mmm disp	
Format	Examples
PUSH mem	PUSH DATA PUSH LISTS PUSHD NUMB
00sss110	
Format	Examples
PUSH seg	PUSH DS PUSH CS
00001111 10sss000	
Format	Examples
PUSH seg	PUSH FS PUSH GS
011010s0 data	
Format	Examples
PUSH imm	PUSH 2000H PUSHD 5322H

PUSHA/PUSHAD   Push all registers

01100000	O D I T S Z A P C
Examples	
PUSHA	

PUSHAD

PUSHF/PUSHFD   Push flags onto stack

10011100	O D I T S Z A P C

Examples

PUSHA PUSHFD

RCL/RCR/ROL/ROR   Rotate

1101000w ooTTTmmm disp	O D I T S Z A P C
	* *

TTT = 000 = ROL
TTT = 001 = ROR
TTT = 010 = RCL
TTT = 011 = RCR

Format	Examples
ROL reg,1 ROR reg,1	ROL CL,1 ROL DX,1 ROR EDX,1
RCL reg,1 RCR reg,1	RCL CL,1 RCL SI,1 RCR EAX,1
ROL mem,1 ROR mem,1	ROL DATA,1 ROL BYTE PTR [DI],1 ROR WORD PTR [ECX],1
RCL mem,1 RCR mem,1	RCL DATA,1 RCL BYTE PTR [DI],1 RCR NUMB,1

1101001w ooTTTmmm disp

Format	Examples
ROL reg,CL ROR reg,CL	ROL CH,CL ROL DX,CL ROR EBX,CL
RCL reg,CL RCR reg,CL	RCL DL,CL RCL SI,CL RCR EDI,CL
ROL mem,CL ROR mem,CL	ROL DATA,CL ROL BYTE PTR [DI],CL
RCL mem,CL RCR mem,CL	RCL DATA,CL RCL BYTE PTR [DI],CL

1100000w ooTTTmmm disp data

Format	Examples
ROL reg,imm ROR reg,imm	ROL CL,4 ROL DX,5 ROR ESI,12
RCL reg,imm RCR reg,imm	RCL CL,2 RCL SI,3 RCR EAX,5
ROL mem,imm ROR mem,imm	ROL DATA,4 ROL BYTE PTR [DI],2
RCL mem,imm RCR mem,imm	RCL DATA,6 RCL BYTE PTR [DI],7

REP   Repeat prefix

11110010 1010010w	O DI T S Z A P C
Format	**Examples**
REP MOVS	REP MOVSB REP MOVSW REP MOVSD

11110010 1010101w	
Format	**Examples**
REP STOS	REP STOSB REP STOSW REP STOSD

11110010 0110110w	
Format	**Examples**
REP INS	REP INSB REP INSW REP INSD

11110010 0110111w	
Format	**Examples**
REP OUTS	REP OUTSB REP OUTSW REP OUTSD

REPE/REPNE   Repeat conditional

11110011 1010011w	O DI T S Z A P C *
Format	**Examples**
REPE CMPS	REPE CMPSB REPE CMPSW REPE CMPSD

11110011 1010111w	
Format	**Examples**
REPE SCAS	REPE SCASB REPE SCASW REPE SCASD

11110010 1010011w	
Format	**Examples**
REPNE CMPS	REPNE CMPSB REPNE CMPSW REPNE CMPSD

11110010 1010111w	
Format	**Examples**
REPNE SCAS	REPNE SCASB REPNE SCASW REPNE SCASD

RET   Return from procedure

11000011	O DI T S Z A P C
Example	
RET (near)	
11000010 data	

Format	Examples
RET imm (near)	RET 4 RET 100H

11001011
Example

RET (far)

11001010 data

Format	Examples
RET imm (far)	RET 4 RET 100H

SAHF   Store AH into flags

10011110	O DI T S Z A P C * * * *
Example	

SAHF

SAL/SAR/SHL/SHR   Shift

1101000w ooTTTmmm disp	O DI T S Z A P C * * * ? * *

TTT = 100 = SHL/SAL
TTT = 101 = SHR
TTT = 111 = SAR

Format	Examples
SAL reg,1 SHL reg,1 SHR reg,1	SAL CL,1 SHL DX,1 SHR EDX,1
SAL mem,1 SHL mem,1	SAL DATA,1 SHL BYTE PTR [DI],1

1101001w ooTTTmmm disp

Format	Examples
SAL reg,CL SHL reg,CL SHR reg,CL	SAL CH,CL SHL DX,CL SHR EDI,CL
SAL mem,CL SHL mem,CL	SAL DATA,CL SHL BYTE PTR [DI],CL

1100000w ooTTTmmm disp data

Format	Examples
SAL reg,imm SHL reg,imm SHR reg,imm	SAL CL,4 SHL DX,5 SHR EAX,12
SAL mem,imm SHL mem,imm	SAL DATA,6 SHL BYTE PTR [DI],7

SBB   Subtract with borrow

000110dw oorrrmmm disp	O DI T S Z A P C * * * * *
Format	Examples
SBB reg,reg	SBB CL,DL SBB AX,DX

	SBB EAX,EBX
SBB mem,reg	SBB DATA,CL SBB BYTES,CX SBB NUMBS,ECX
SBB reg,mem	SBB CL,DATA SBB CX,BYTES SBB ECX,NUMBS

100000sw oo011mmm disp data	
Format	Examples
SBB reg,imm	SBB CL,4 SBB DX,5 SBB EDX,12
SBB mem,imm	SBB DATA,6 SBB BYTE PTR [DI],7

0001110w data	
Format	Examples
SBB acc,imm	SBB AL,4 SBB AX,5 SBB EAX,9

SCAS   Scan string

1010111w	O DI T S ZAP C
	* * * * *
Examples	
SCASB SCASW SCASD	

SET   Set on condition   (80386/80486)

00001111 1001cccc oo000mmm	O DI T S ZAP C
Format	Examples
SETcd reg8	SETA BL SETB CH SETG DL
SETcd mem8	SETE DATA SETLE BYTES

Condition Codes	Mnemonic	Flag	Description
0000	SETO	O = 1	Set if overflow
0001	SETNO	O = 0	Set if no overflow
0010	SETB/SETNAE	C = 1	Set if below
0011	SETAE/SETNB	C = 0	Set if above or equal
0100	SETE/SETZ	Z = 1	Set if equal/zero
0101	SETNE/SETNZ	Z = 0	Set if not equal/not zero
0110	SETBE/SETNA	C = 1 + Z = 1	Set if below or equal
0111	SETA/SETNBE	C = 0 • Z = 0	Set if above
1000	SETS	S = 1	Set if sign
1001	SETNS	S = 0	Set if no sign
1010	SETP/SETPE	P = 1	Set if parity even
1011	SETNP/SETPO	P = 0	Set if parity odd
1100	SETL/SETNGE	S = <> O	Set if less than
1101	SETGE/SETNL	S = O	Set if greater or equal
1110	SETLE/SETNG	Z = 1 + S <> O	Set if less than or equal
1111	SETG/SETNLE	Z = 0 + S = O	Set if greater

SGDT/SIDT/SLDT   Store descriptor table   (80286/80386/80486)

00001111 00000001 oo000mmm disp	O DI T S ZAP C

Format	Examples
SGDT mem	SGDT MEMORY SGDT GLOBAL

00001111 00000001 oo001mmm disp

Format	Examples
SIDT mem	SIDT DATAS SIDT INTERRUPT

00001111 00000000 oo000mmm disp

Format	Examples
SLDT reg	SLDT CX SLDT DX
SLDT mem	SLDT NUMBS SLDT LOCALS

SHLD/SHRD   Double precision shift   (80386/80486)

00001111 10100100 oorrrmmm disp data

	O	D	I	T	S	Z	A	P	C
	?				*	*	?	*	*

Format	Examples
SHLD reg,reg,imm	SHLD AX,CX,10 SHLD DX,BX,8 SHLD CX,DX,2
SHLD mem,reg,imm	SHLD DATA,CX,8 SHLD NUMB,DX,2

00001111 10101100 oorrrmmm disp data

Format	Examples
SHRD reg,reg,imm	SHRD CX,DX,2
SHRD mem,reg,imm	SHRD DATA,CX,3

00001111 10100101 oorrrmmm disp

Format	Examples
SHLD reg,reg,CL	SHLD DX,BX,CL
SHLD mem,reg,CL	SHLD DATA,AX,CL

00001111 10101101 oorrrmmm disp

Format	Examples
SHRD reg,reg,CL	SHRD DX,BX,CL
SHRD mem,reg,CL	SHRD DATA,AX,CL

SMSW   Store machine status word

00001111 00000001 oo100mmm disp   O D I T S Z A P C

should only be used with the 80286

Format	Examples
SMSW reg	SMSW AX SMSW DX

SMSW mem	SMSW DATA

STC   Set carry flag

11111001	O DI T S ZA P C
	1
Example	
STC	

STD   Set direction flag

11111101	O DI T S ZA P C
	1
Example	
STD	

STI   Set interrupt flag

11111011	O DI T S ZA P C
	1
Example	
STI	

STOS   Store string data

1010101w	O DI T S ZA P C
Examples	
STOSB STOSW STOSD	

STR   Store task register   (80286/80386/80486)

00001111 00000000 oo001mmm disp	O DI T S ZA P C
Format	Examples
STR reg	STR DX STR CX
STR mem	STR DATA

SUB   Subtract

001010dw oorrrmmm disp	O DI T S ZA P C
	* * * * *
Format	Examples
SUB reg,reg	SUB CL,DL SUB AX,DX SUB EAX,EBX
SUB mem,reg	SUB DATA,CL SUB BYTES,CX SUB NUMBS,ECX
SUB reg,mem	SUB CL,DATA SUB CX,BYTES SUB ECX,NUMBS

100000sw oo101mmm disp data	
Format	Examples

SUB reg,imm	SUB CL,4
	SUB DX,5
	SUB ESI,9

SUB mem,imm	SUB DATA,6 SUB BYTE PTR [DI],7 SUB WORD PTR [ECX],5

0010110w data

Format	Examples

SUB acc,imm	SUB AL,4
	SUB AX,5
	SUB EAX,9

TEST   Test operands (logical compare)

		O D I T S Z A P C
1000011w oorrrmmm disp		0 * * ? * 0
Format	Examples	
TEST reg,reg	TEST CL,DL TEST CX,DX TEST ECX,EBX	
TEST mem,reg reg,mem	TEST DATA,CL TEST CL,DATA	

1111011w oo000mmm disp data

Format	Examples
TEST reg,imm	TEST CL,4 TEST DX,5 TEST ESI,256
mem,imm	TEST DATA,6

1010100w data

Format	Examples
TEST acc,imm	TEST AL,4 TEST AX,5 TEST EAX,9

VERR/VERW   Verify read or write   (80286/80386/80486)

		O DI T S Z A P C
00001111 00000000 oo100mmm disp		*
Format	Examples	
VERR reg16	VERR BX VERR CX	
VERR mem16	VERR DATA	

00001111 00000000 oo101mmm disp

Format	Examples
VERW reg16	VERW AX VERW CX
VERW mem16	VERW DATA

WAIT   Wait for coprocessor

	O DI T S Z A P C
10011011	
Examples	
WAIT	

FWAIT			

WBINVD — Write back and invalidate data cache (80486)

00001111 00001001	O D I T S Z A P C
Example	
WBINVD	

XADD — Exchange and add (80486)

00001111 1100000w 11rrrrr	O D I T S Z A P C
	* * * * *

Format	Examples
XADD reg,reg	XADD EBX,ECX
	XADD EDX,EAX

00001111 1100000w oorrrmmm disp	
Format	Examples
XADD mem,reg	XADD DATA,EAX
	XADD [DI],EAX

XCHG — Exchange

1000011w oorrrmmm disp	O D I T S Z A P C

Format	Examples
XCHG reg,reg	XCHG BH,CH
	XCHG CX,DX
	XCHG ECX,EBX
reg,mem	XCHG CL,DATA2
mem,reg	XCHG DATA3,EBX

10010rrr	
Format	Examples
XCHG acc,reg	XCHG AX,CX
XCHG reg,acc	XCHG ECX,EAX

XLAT — Translate

11010111	O D I T S Z A P C
Example	
XLAT	

XOR — Exclusive-OR

001100dw oorrrmmm disp	O D I T S Z A P C
	0 * * ? * 0

Format	Examples
XOR reg,reg	XOR BL,CL
	XOR CX,DX
	XOR EAX,EBX
XOR mem,reg	XOR DATA,CL
	XOR BYTES,CX
	XOR [EAX],CX
XOR reg,mem	XOR CL,DATA
	XOR CX,BYTES
	XOR ECX,NUMBS

	XOR ECX,NUMBS
100000sw oo110mmm disp data	
Format	Examples
XOR reg,imm	XOR CL,4 XOR DX,5 XOR ESI,9
XOR mem,imm	XOR DATA,6 XOR BYTE PTR [DI],7 XOR WORD PTR [ECX],5
0011010w data	
Format	Examples
XOR acc,imm	XOR AL,4 XOR AX,5 XOR EAX,9

APPENDIX C

Answers to Even-Numbered Problems

Chapter 1

2. Herman Hollerith
4. Konrad Zuse
6. ENIAC
8. The 8080 microprocessor
10. The 8086 microprocessor
12. 4G bytes
14. 1,024 bytes
16. 1,000,000 typewritten pages
18. 640K bytes
20. 1M bytes
22. The 80386 and 80486 microprocessors
24. Basic Input/Output System
26. The XT computer contains an 8-bit peripheral bus and the AT contains a 16-bit peripheral bus.
28. A terminate and stay resident program stays in the memory until it is activated by a hot-key or other event.
30. The CONFIG.SYS file configures the computer system by loading drivers and setting buffer sizes.
32. The AUTOEXEC.BAT file
34. These statements are found in the AUTOEXEC.BAT file.
36. See Figure C–1
38. Address, data, and control buses
40. The $\overline{MRDC}$ signal
42. A byte is an 8-bit wide number, a word is a 16-bit wide number, and a double-word is 32-bits wide.
44. The standard ASCII character ranges in value from 00H–7FH, while the extended ASCII character ranges in value from 80H–FFH.
46. (a) 0000 0011 1110 1000, (b) 1111 1111 1000 1000, (c) 1111 1100 1110 0000, and (d) 1111 0011 0111 1100.

Buses

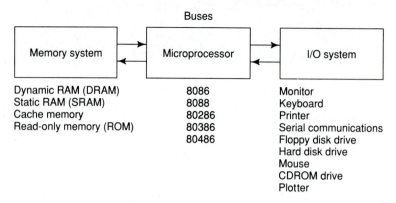

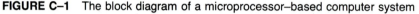

Memory system	Microprocessor	I/O system
Dynamic RAM (DRAM)	8086	Monitor
Static RAM (SRAM)	8088	Keyboard
Cache memory	80286	Printer
Read-only memory (ROM)	80386	Serial communications
	80486	Floppy disk drive
		Hard disk drive
		Mouse
		CDROM drive
		Plotter

FIGURE C–1   The block diagram of a microprocessor–based computer system

48. (a) 00000001 00000010 (packed) and 00000001 00000000 00000010 (unpacked)
    (b) 01000100 (packed) and 00000100 00000100 (unpacked)
    (c) 00000011 00000001 (packed) and 00000011 00000000 00000001 (un-packed)
    (d) 00010000 00000000 (packed) and 00000001 00000000 00000000 00000000 (unpacked)
50. (a) 89, (b) 09, (c) 32, and (d) 01.
52. (a) +3.5, (b) −1.0, (c) +12.5, and (d) −0.875

Chapter 2

2. 16
4. The extended 32-bit registers are found in the 80386 and 80486 microprocessor.
6. The instruction pointer is used by the microprocessor to locate the next instruction in a program.
8. No
10. The interrupt flag (I)
12. The segment register addresses the start of a 64K byte memory segment.
14. (a) 12000H, (b) 21000H, (c) 23A00H, (d) 25000H, and (e) 3F12DH.
16. DI
18. Stack segment (SS) plus stack pointer (SP/ESP) offset address
20. (a) 23000H, (b) 1C000H, (c) CA000H, (d) 89000H, and (e) 1CC90H.
22. The segment register selects a descriptor from a descriptor register that describes the starting address, length, and access to a memory segment.
24. A00000H–A01000H
26. 00280000H–00290000H
28. 11_2, or 3.
30. 64K bytes
32. FF 2F 00 00 00 92 80 03
34. The local descriptor is addressed through an entry in the global descriptor table.

Chapter 3

2. AH, AL, BH, BL, CH, CL, DH, and DL.
4. EAX, EBX, ECX, EDX, ESP, EBP, EDI, and ESI.
6. Mixed register sizes are not allowed.
8. (a) MOV EDX,EBX (b) MOV CL,BL (c) MOV BX,SI (d) MOV AX,DS and (e) MOV AH,AL
10. The # symbol is used by some assemblers to denote immediate data.
12. The [] symbols indicate indirect addressing through a register or registers.
14. A memory to memory transfer is not allowed with the MOV instruction.
16. MOV WORD PTR[SI],2
18. The MOV BX,DATA instruction transfers a copy of the word-sized memory location DATA into BX. The MOV BX,OFFSET DATA instruction loads the address of DATA into BX.
20. (a) 12100H, (b) 12350H, and (c) 12200H.
22. (a) 15500H, (b) 15500H, and (c) 13000H.
24. (a) 03100H, (b) 05100H, and (c) 07100H.
26. The far direct jump instruction requires 5 bytes of memory. The first byte contains the opcode, the next 2 bytes contain the IP value (offset address), and the last 2 bytes contain the CS value (segment address).
28. Because a segment is cyclic in nature, the displacement of ±32K bytes allows the near jump to jump to any location within the current code segment.
30. SHORT
32. JMP BX
34. In the 8086–80286, 2 bytes are stored on the stack by a PUSH. In the 80386/80486, either 2 bytes or 4 bytes are stored by a PUSH.
36. AX, BX, CX, DX, SP, BP, DI, and SI.

Chapter 4

2. The D-bit selects the direction from REG to R/M or from R/M to REG and the W-bit selects a byte or a word/doubleword register.
4. DL
6. DS:[BX+DI]
8. MOV AX,[BX]
10. 8B7702
12. The CS register may not be changed with a MOV instruction.
14. CS
16. EAX, EBX, ECX, EDX, ESP, EBP, EDI, and ESI.
18. The PUSH BX instruction places BH into memory location 011FFH and BL into location 011FEH. After storing BX, the stack pointer changes to 01FEH.
20. 2
22. The MOV DI,NUMB instruction copies the word-sized contents of memory location NUMB into DI. The LEA DI,NUMB copies the address of NUMB into DI.
24. The MOV with an OFFSET is the more efficient.

26. Both instructions are the same except that LDS loads DS, while LSS loads SS.
28. The direction flag is used with the string instructions to select the autoincrement function (D = 0) or the autodecrement function (D = 1) for DI/SI.
30. The DI register addresses data in the extra segment and the SI register addresses data in the data segment.
32. The STOSW instruction copies the word-sized contents of AX into the extra segment memory location addressed by DI. After the transfer, DI is either incremented (D = 0) or decremented (D = 1) by two.
34. The REP prefix is used with the string instructions to repeat them the number of times found in register CX.
36. DX
38. In software that translates 8085 code into 8086 code.
40.

EXAMPLE C–1

```
0000                    CODE     SEGMENT 'CODE'
                                 ASSUME     CS:CODE


0000                    CONVERT PROC  FAR

0000 8C C8                       MOV     AX,CS               ;load DS
0002 8E D8                       MOV     DS,AX
0004 BB 0009 R                   MOV     BX,OFFSET TABLE     ;address table
0007 D7                          XLAT
0008 CB                          RET

0009                    CONVERT ENDP

0009 30 31 32 33        TABLE    DB      30H,31H,32H,33H
000D 34 35 36 37                 DB      34H,35H,36H,37H
0011 38 39                       DB      38H,39H

0013                    CODE     ENDS

                                 END     CONVERT
```

42. The OUT DX,AX instruction copies the 16-bit contents of AX to the output device addressed by register DX.
44. MOV AH,ES:[BX]
46. An assembly language directive is a command to the assembler program.
48. LIST1 DB 30 DUP (?)
50. The .386 directive informs the assembler to generate code for the 80386 microprocessor.
52. Memory models

54. The INT 21H instruction using 4CH in the AH register terminates a program and returns control to DOS.
56.

EXAMPLE C–2

```
0000                    CODE    SEGMENT 'CODE'
                        ASSUME      CS:CODE

0000                    COPY    PROC  NEAR

0000 2E: A1 000D R              MOV   AX,CS:DATA1
0004 8B D8                      MOV   BX,AX
0006 8B C8                      MOV   CX,AX
0008 8B D0                      MOV   DX,AX
000A 8B F0                      MOV   SI,AX
000C C3                         RET

000D                    COPY    ENDP

000D 000                DATA1   DW    ?

000F                    CODE    ENDS

                                END   COPY
```

Chapter 5

2. You cannot mix the sizes of operands. In this case a 16-bit operand is added to a 32-bit operand, which is not allowed.
4. The sum is 3100H found in AX. The flags change as follows: carry = 0, overflow = 0, sign = 0, parity = 0, zero = 0, and auxiliary carry = 1.
6.

EXAMPLE C–3

```
0000  03 C3                     ADD   AX,BX
0002  03 C1                     ADD   AX,CX
0004  03 C2                     ADD   AX,DX
0006  03 C4                     ADD   AX,SP
0008  8B F8                     MOV   DI,AX
```

8. ADC DX,BX
10. The assembler cannot determine if the data stored in memory is a byte, word, or doubleword.

12. The difference is 81H. The flags change as follows: carry = 0, auxiliary carry = 0, parity = 1, zero = 0, sign = 1, and overflow = 0.
14. DEC EBX
16. Both instructions subtract and change the flag bits. The difference is that the SUB instruction leaves the difference in the destination operand, while the CMP instruction does not.
18. The product is in DX (most-significant) and AX (least-significant).
20. The product is in EDX–EAX.
22.

EXAMPLE C–4

```
0000 8A C2                MOV   AL,DL
0002 F6 E2                MUL   DL
0004 F6 E2                MUL   DL
```

24. The dividend is in AX.
26. Overflow and divide-by-zero
28. The remainder is in AH.
30. The DAA and DAS instructions
32. The AAM instruction divides the contents of AX by a 10. This results in a 2-digit, unpacked BCD number in AH and AL.
34. (a) AND BX,DX, (b) AND DH,0EAH, (c) AND DI,BP, (d) AND EAX,1122H, (e) AND [BP],CX, (f) AND DX,[SI-8], and (g) AND WHAT,AL
36. (a) OR AH,BL, (b) OR ECX,88H, (c) OR SI,DX, (d) OR BP,1122H, (e) OR [BX],CX, (f) OR AL,[BP+40], and (g) OR WHEN,AH.
38. (a) XOR AH,BH, (b) XOR CL,99H, (c) XOR DX,DI, (d) XOR ESP,1A23H, (e) XOR [EBX],DX, (f) XOR DI,[BP+30], and (g) XOR DI,WELL.
40. The AND and TEST instruction both AND two numbers together and change the flags. The difference is that AND moves the logical product into the destination, while the TEST instruction does not.
42. The NOT instruction one's complements a number, while the NEG instruction two's complements a number.
44. The SCASB (scan a byte) instruction compares AL with the contents of the extra segment memory location addressed by DI. After the comparison, the contents of DI are incremented by 1 if D = 0, or decremented by 1 if D = 1.
46. The D flag bit selects increment (D = 0) or decrement (D = 1) for the string instruction pointers.
48. When CX reaches a zero or when an equal condition is detected.
50.

EXAMPLE C–5

```
0000 BF 0300 R            MOV   DI,OFFSET LIST
0003 B9 0300              MOV   CX,300H
0006 B0 66                MOV   AL,66H
```

```
0008 F2/ AE                    REPNE SCASB
000A 75 03 E9 01F1             JE      FOUND_66H
```

Chapter 6

2. The near jump instruction allows a branch to any location within a segment.
4. The FAR jump instruction is 5 bytes in length.
6. A colon is used to denote a label that functions as a near or short address, for use only with the jump or call instructions.
8. The far jump instruction changes the contents of CS and IP to effect the jump.
10. The JMP DI instruction jumps to the near address indexed by the contents of the DI register. The JMP [DI] instruction jumps to the memory location stored in the memory location indexed by the DI register.
12. The carry (C), overflow (O), sign (S), parity (P), and zero (Z) flag bits are tested by conditional jump instructions.
14. The JO (jump if overflow is set) jumps to the operand address if an overflow condition is indicated by the overflow flag bit.
16. The JA, JAE, JB, or JBE instructions follow comparison of unsigned numbers.
18. The JCXZ instruction jumps to the operand address if the contents of CX is zero.
20. CX
22. The LOOPE instruction loops (jumps) to the operand address if CX is not a zero and as long as an equal condition exists. The loop terminates whenever a not-equal condition exists or whenever the counter (CX) reaches a zero.
24.

EXAMPLE C–6

```
0000 BF 0500 R             MOV    DI,OFFSET BLOCK      ;address data
0003 B9 0100               MOV    CX,100H              ;load count
0006 32 C0                 XOR    AL,AL                ;clear UP and DOWN
0008 2E: A2 0600 R         MOV    UP,AL
000C 2E: A2 0601 R         MOV    DOWN,AL
0010 B0 42                 MOV    AL,42H               ;set AL to 42H
0012             LOOP1:
0012 AE                    SCASB                       ;test against 42H
0013 74 0E                 JE     LOOP3
0015 72 07                 JB     LOOP2
0017 2E: FE 06 0600 R      INC    UP                   ;if above 42H
001C EB 05                 JMP    LOOP3
001E            LOOP2:
001E 2E: FE 06 0601 R      INC            DOWN         ;if below 42H
0023            LOOP3:
0023 E2 ED                 LOOP   LOOP1
```

26. The near CALL instruction pushes the contents of IP onto the stack and then changes IP to call a procedure. The far CALL instruction pushes both IP and CS onto the stack and then changes both registers to effect the call.

28. RET (return)
30. The near and far procedure are usually identified with the PROC statement and the keyword NEAR or FAR following PROC.
32.

EXAMPLE C–7

```
0000                    CUBES   PROC  NEAR

0000 50                         PUSH  AX
0001 52                         PUSH  DX
0002 8B C1                      MOV   AX,CX
0004 F7 E1                      MUL   CX
0006 F7 E1                      MUL   CX
0008 8B C8                      MOV   CX,AX
000A 5A                         POP   DX
000B 58                         POP   AX
000C C3                         RET

000D                    CUBES   ENDP
```

34. An interrupt is a hardware initiated procedure.
36. 256
38. The interrupt vector requires 4 bytes of memory. The first 2 bytes contain the new instruction pointer (IP) value and the last 2 bytes contain the new code segment (CS) value.
40. The IRETD instruction is used to return from an interrupt in the 80386/80486 system.
42. AT location 40H times 4, or 100H
44. 17H
46. The WAIT instruction.
48. 16
50. ESC (escape)

Chapter 7

2. MASM will normally generate an object (.OBJ) and list (.LST) file from the source.
4. The .EXE file can be of any length and the .COM is limited to a maximum of 64K bytes in length. Another difference is that the .EXE file is slightly larger than the .COM file due to internal overhead.
6. The EXTRN directive indicates that a label is external to the module.
8. The linker removes only the procedures used from the library file during linking.
10. A macro sequence is a group of instructions that are placed into your program each time the macro is invoked.

12.

EXAMPLE C-8

```
ADD32     MACRO

          ADD     CX,AX
          ADC     DX,BX

          ENDM
```

14. The LOCAL directive indicates that a label is local to the macro.
16.

EXAMPLE C-9

```
ADDM      MACRO   LIST,LENGTH
          LOCAL   ADDM1,ADDM2

          MOV     SI,OFFSET LIST
          MOV     CX,LENGTH
          XOR     AX,AX
ADDM1:
          ADD     AL,[SI]
          JNC     ADDM2
          INC     AH
ADDM2:
          LOOP    ADDM1

          ENDM
```

18.

EXAMPLE C-10

```
0000                  RANDOM PROC FAR

0000 52                      PUSH  DX
0001 B4 06                   MOV   AH,6
0003 B2 FF                   MOV   DL,0FFH
0005                  RANDOM1:
0005 FE C1                   INC   CL        ;increment RANDOM
0007 CD 21                   INT   21H       ;test for keystroke
0009 74 FA                   JZ    RANDOM1   ;if no keystroke
000B 5A                      POP   DX
000C CB                      RET

000D                  RANDOM ENDP
```

20.

EXAMPLE C–11

```
STRING    MACRO
          LOCAL  STRING1,STRING2

          PUSH   AX
          PUSH   DX
          PUSH   SI
          MOV    SI,DX
          MOV    AH,9
STRING1:
          LODSB
          OR     AL,AL
          JZ     STRING2
          MOV    DL,AL
          INT    21H
          JMP    STRING1
STRING2:
          POP    SI
          POP    DX
          POP    AX

          ENDM
```

22.

EXAMPLE C–12

```
0000                    NEW    PROC NEAR

0000 B6 03                     MOV    DH,3
0002 B2 06                     MOV    DL,6
0004 B3 00                     MOV    BL,0
0006 CD 10                     INT    10H
0008 C3                        RET

0009                    NEW    ENDP
```

24. This conversion is accomplished by division by 10.
26. 30H
28. First, each digit has a 30H subtracted, and then they are combined by using addition and multiplication by 10.

30.

EXAMPLE C–13

```
0009                    LOOK    PROC  FAR

0009 1E                         PUSH  DS
000A 50                         PUSH  AX
000B 8C C8                      MOV   AX,CS
000D 8E D8                      MOV   DS,AX
000F 58                         POP   AX
0010 BB 0018 R                  MOV   BX,OFFSET TABLE
0013 24 0F                      AND   AL,0FH
0015 D7                         XLAT
0016 1F                         POP   DS
0017 CB                         RET

0018                    LOOK    ENDP

0018 30 31 32 33        TABLE   DB    30H,31H,32H,33H
001C 34 35 36 37                DB    34H,35H,36H,37H
0020 38 39 41 42                DB    38H,39H,41H,42H
0024 43 44 45 46                DB    43H,44H,45H,46H
```

32. The instruction XLAT SS:LOOK used the stack segment and table LOOK.
34.

EXAMPLE C–14

```
0000                    STACK   SEGMENT STACK
0000 0080 [                     DW    128 DUP (?)
        0000
              ]
0100                    STACK   ENDS

0000                    DATA    SEGMENT

0000 32 20 20 3D 20     MES             '2 = $'
     24
0006                    DATA    ENDS

0000                    CODE    SEGMENT 'CODE'
                                ASSUME     CS:CODE,DS:DATA,SS:STACK

0000                    MAIN    PROC  FAR

0000 B8 ----                    MOV   AX,DATA          ;load DS
0003 8E D8                      MOV   DS,AX
```

```
0005 E8 000B              CALL   CRLF            ;display CR and LF

0008 B9 0008              MOV    CX,8            ;set power count
000B E8 001D              CALL   POWER           ;display powers

000E B8 4C00              MOV    AX,4C00H        ;end program
0011 CD 21                INT    21H

0013            MAIN      ENDP

0013            CRLF      PROC   NEAR

0013 B0 0D                MOV    AL,13           ;display CR
0015 E8 0006              CALL   OUTPUT
0018 B0 0A                MOV    AL,10           ;display LF
001A E8 0001              CALL   OUTPUT
001D C3                   RET

001E            CRLF      ENDP

001E            OUTPUT    PROC   NEAR

001E B4 06                MOV    AH,6            ;display AL
0020 8A D0                MOV    DL,AL
0022 CD 21                INT    21H
0024 C3                   RET

0025            OUTPUT    ENDP

0025            SPACE     PROC   NEAR

0025 B0 20                MOV    AL,' '          ;display space
0027 E8 FFF4              CALL   OUTPUT
002A C3                   RET

002B            SPACE     ENDP

002B            POWER     PROC   NEAR

002B E8 FFE5              CALL   CRLF
002E E8 FFF4              CALL   SPACE
0031 B0 08                MOV    AL,8            ;display power
0033 2A C1                SUB    AL,CL
0035 04 30                ADD    AL,'0'
0037 E8 FFE4              CALL   OUTPUT
003A E8 FFD6              CALL   CRLF
003D BA 0000 R            MOV    DX,OFFSET MES   ;display MES
0040 B4 09                MOV    AH,9
0042 CD 21                INT    21H
```

```
0044 51                          PUSH   CX                      ;find value
0045 B0 08                       MOV    AL,8
0047 2A C1                       SUB    AL,CL
0049 8A C8                       MOV    CL,AL
004B B0 01                       MOV    AL,1
004D 74 02                       JE     POWER1
004F D2 E0                       SHL    AL,CL
0051                 POWER1:
0051 59                          POP    CX
0052 E8 0003                     CALL   NUMB                    ;display value

0055 E2 D4                       LOOP   POWER                   ;repeat for all powers
0057 C3                          RET
0058                 POWER  ENDP

0058                 NUMB   PROC   NEAR

0058 3C 64                       CMP    AL,100                  ;display hundred's
005A 72 07                       JB     NUMB1
005C 2C 64                       SUB    AL,100
005E 50                          PUSH   AX
005F B0 31                       MOV    AL,'1'
0061 EB 03                       JMP    NUMB2
0063                 NUMB1:
0063 50                          PUSH   AX
0064 B0 20                       MOV    AL,' '
0066                 NUMB2:
0066 E8 FFB5                     CALL   OUTPUT
0069 58                          POP    AX
006A 32 E4                       XOR    AH,AH
006C D4 0A                       AAM                            ;convert to BCD
006E 05 3030                     ADD    AX,3030H                ;convert to ASCII
0071 80 FC 30                    CMP    AH,30H                  ;suppress leading zero
0074 75 03                       JNE    NUMB3
0076 80 EC 10                    SUB    AH,10H
0079                 NUMB3:
0079 50                          PUSH   AX
007A 8A C4                       MOV    AL,AH
007C E8 FF9F                     CALL   OUTPUT
007F 58                          POP    AX
0080 E8 FF9B                     CALL   OUTPUT
0083 C3                          RET

0084                 NUMB   ENDP

0084                 CODE   ENDS

                                 END    MAIN
```

36.

EXAMPLE C–15

```
0000                    STACK    SEGMENT STACK
0000 0080 [                      DW      128 DUP (?)
         0000
              ]
0100                    STACK    ENDS

0000                    DATA     SEGMENT

0000 45 6E 74 65 72     MES1     DB      'Enter address = $'
     20 61 64 64 72
     65 73 73 20 3D
     20 24

0011                    DATA     ENDS

0000                    CODE     SEGMENT 'CODE'
                                 ASSUME      CS:CODE,DS:DATA,SS:STACK
                        .286
0000                    MAIN     PROC FAR

0000 B8 —— R                     MOV     AX,DATA          ;load DS
0003 8E D8                       MOV     DS,AX
0005 FC                          CLD                      ;select increment

0006 BA 0000 R                   MOV     DX,OFFSET MES1   ;display MES1
0009 B4 09                       MOV     AH,9
000B CD 21                       INT     21H

000D E8 000                      CALL    GET_ADR          ;get address

0010 E8 0088                     CALL    BLOCK            ;display data

0013 B8 4C00                     MOV     AX,4C00H         ;exit to DOS
0016 CD 21                       INT     21H

0018                    MAIN     ENDP

0018                    GET_ADR PRO     NEAR

0018 33 FF                       XOR     DI,DI            ;clear address
001A 33 DB                       XOR     BX,BX
001C B9 0005                     MOV     CX,5             ;count 5 digits
001F                    GET_ADR1:
001F E8 0055                     CALL    READ             ;read keyboard
0022 3C 08                       CMP     AL,8
0024 74 3C                       JE      BS               ;if a backspace key
```

```
0026 3C 61              CMP     AL,'a'
0028 72 06              JB      GET_ADR2        ;if below a
002A 3C 7A              CMP     AL,'z'
002C 77 02              JA      GET_ADR2        ;if above z
002E 2C 20              SUB     AL,20H          ;make upper case
0030            GET_ADR2:
0030 3C 30              CMP     AL,'0'
0032 72 EB              JB      GET_ADR1        ;if below 0
0034 3C 39              CMP     AL,'9'
0036 76 08              JBE     GET_ADR3        ;if good number 0—9
0038 3C 41              CMP     AL,'A'
003A 72 E3              JB      GET_ADR1        ;if below A
003C 3C 46              CMP     AL,'F'
003E 77 DF              JA      GET_ADR1        ;if above F
0040            GET_ADR3:
0040 E8 0045            CALL    OUTPUT          ;echo to video display
0043 2C 30              SUB     AL,30H          ;convert from ASCII
0045 3C 09              CMP     AL,9
0047 76 02              JBE     GET_ADR4
0049 2C 07              SUB     AL,7
004B            GET_ADR4:
004B D1 E7              SHL     DI,1            ;shift BX:DI left 4 places
004D D1 D3              RCL     BX,1
004F D1 E7              SHL     DI,1
0051 D1 D3              RCL     BX,1
0053 D1 E7              SHL     DI,1
0055 D1 D3              RCL     BX,1
0057 D1 E7              SHL     DI,1
0059 D1 D3              RCL     BX,1
005B 32 E4              XOR     AH,AH
005D 03 F8              ADD     DI,AX
005F E2 BE              LOOP    GET_ADR1
0061 C3                 RET
0062            BS:
0062 83 F9 05           CMP     CX,5            ;backspace
0065 74 B8              JE      GET_ADR1
0067 41                 INC     CX
0068 E8 001D            CALL    OUTPUT
006B B0 20              MOV     AL,' '
006D E8 0018            CALL    OUTPUT
0070 B0 08              MOV     AL,8
0072 E8 0013            CALL    OUTPUT
0075 EB A8              JMP     GET_ADR1

0077            GET_ADR ENDP

0077            READ    PROC NEAR

0077 B4 06              MOV     AH,6            ;read a key
```

```
0079 B2 FF              MOV     DL,0FFH
007B CD 21              INT     21H
007D 74 F8              JZ      READ            ;if no key
007F 0A C0              OR      AL,AL
0081 75 04              JNZ     READ1           ;if good key
0083 CD 21              INT     21H             ;if extended key
0085 EB F0              JMP     READ
0087            READ1:
0087 C3                 RET

0088            READ    ENDP

0088            OUTPUT  PROC  NEAR

0088 B4 06              MOV     AH,6            ;display character
008A 8A D0              MOV     DL,AL
008C CD 21              INT     21H
008E C3                 RET

008F            OUTPUT  ENDP
008F            CRLF    PROC  NEAR

008F B0 0D              MOV     AL,13           ;display carriage return
0091 E8 FFF4            CALL    OUTPUT
0094 B0 0A              MOV     AL,10           ;display line feed
0096 E8 FFEF            CALL    OUTPUT
0099 C3                 RET

009A            CRLF    ENDP

009A            SPACE   PROC  NEAR

009A B0 20              MOV     AL,' '          ;display space
009C E8 FFE9            CALL    OUTPUT
009F C3                 RET

00A0            SPACE   ENDP

00A0            BLOCK   PROC  NEAR

00A0 81 E7 FF00         AND     DI,0FF00H       ;mask right digits
00A4 D1 EB              SHR     BX,1            ;get segment
00A6 D1 DF              RCR     DI,1
00A8 D1 EB              SHR     BX,1
00AA D1 DF              RCR     DI,1
00AC D1 EB              SHR     BX,1
00AE D1 DF              RCR     DI,1
00B0 D1 EB              SHR     BX,1
00B2 D1 DF              RCR     DI,1
```

```
00B4 8E DF              MOV   DS,DI              ;load DS
00B6 33 F6              XOR   SI,SI              ;clear SI
00B8 B9 0010            MOV   CX,16              ;get byte count
00BB          BLOCK1:
00BB E8 0030            CALL  ADR                ;display address
00BE E8 FFD9            CALL  SPACE              ;display space
00C1          BLOCK2:
00C1 E8 000C            CALL  BYTES              ;display memory byte
00C4 E2 FB              LOOP  BLOCK2             ;repeat 16 times
00C6 B9 0010            MOV   CX,16              ;reload byte count
00C9 81 E6 00FF         AND   SI,00FFH           ;test for block end
00CD 75 EC              JNZ   BLOCK1
00CF C3                 RET

00D0          BLOCK  ENDP

00D0          BYTES  PROC  NEAR

00D0 AC                 LODSB                    ;get memory byte
00D1 50                 PUSH  AX
00D2 C1 E8 04           SHR   AX,4
00D5 E8 0008            CALL  CON                ;display digit
00D8 58                 POP   AX
00D9 E8 0004            CALL  CON                ;display digit
00DC E8 FFBB            CALL  SPACE              ;display space
00DF C3                 RET

00E0          BYTES  ENDP

00E0          CON    PROC  NEAR

00E0 24 0F              AND   AL,15              ;mask byte
00E2 04 30              ADD   AL,30H             ;convert to ASCII
00E4 3C 39              CMP   AL,39H
00E6 76 02              JBE   CON1
00E8 04 07              ADD   AL,7
00EA          CON1:
00EA E8 FF9B            CALL  OUTPUT             ;display digit
00ED C3                 RET

00EE          CON    ENDP

00EE          ADR    PROC  NEAR

00EE E8 FF9E            CALL  CRLF               ;get new line
00F1 8B C7              MOV   AX,DI              ;get 3 digits
00F3 C1 E8 04           SHR   AX,4
00F6 50                 PUSH  AX
00F7 C1 E8 04           SHR   AX,4
```

```
00FA 50                      PUSH   AX
00FB C1 E8 04                SHR    AX,4
00FE E8 FFDF                 CALL   CON
0101 58                      POP    AX
0102 E8 FFDB                 CALL   CON
0105 58                      POP    AX
0106 E8 FFD7                 CALL   CON
0109 8B C6                   MOV    AX,SI
010B C1 E8 04                SHR    AX,4
010E E8 FFCF                 CALL   CON
0111 8B C6                   MOV    AX,SI
0113 E8 FFCA                 CALL   CON
0116 E8 FF81                 CALL   SPACE
0119 C3                      RET


011A             ADR         ENDP


011A             CODE        ENDS

                             END    MAIN
```

38. 64,000 bytes of memory
40. The bit mask register selects which bits change.
42.

EXAMPLE C–16

```
0000             CODE        SEGMENT 'CODE'
                             ASSUME     CS:CODE


0000             MAIN        PROC FAR


0000 B8 0013                 MOV    AX,13H          ;select mode 13H
0003 CD 10                   INT    10H


0005 B8 A000                 MOV    AX,0A000H       ;address PEL memory
0008 8E C0                   MOV    ES,AX
000A FC                      CLD


000B BF 0000                 MOV    DI,0            ;address upper left PEL
000E B0 02                   MOV    AL,2            ;get green


0010 B9 00A0                 MOV    CX,160          ;count
0013             MAIN1:
0013 AA                      STOSB                  ;PEL to green
0014 AA                      STOSB                  ;PEL to green
0015 81 C7 0140              ADD    DI,320          ;skip to next line
```

```
0019 E2 F8                       LOOP  MAIN1              ;repeat 160 times
001B                   MAIN2:
001B B4 06                       MOV   AH,6               ;pause for a key
001D B2 FF                       MOV   DL,0FFH
001F CD 21                       INT   21H
0021 74 F8                       JE    MAIN2

0023 B8 4C00                     MOV   AX,4C00H           ;exit to DOS
0026 CD 21                       INT   21H

0028                   MAIN      ENDP

0028                   CODE      ENDS

                                 END   MAIN
```

44.

EXAMPLE C–17

```
0000                   CODE      SEGMENT 'CODE'
                                 ASSUME       CS:CODE

0000                   MAIN      PROC FAR

0000 B8 0012                     MOV   AX,12H             ;select mode 12H
0003 CD 10                       INT   10H
0005 B8 A000                     MOV   AX,0A000H          ;address video RAM
0008 8E D8                       MOV   DS,AX
000A 8E C0                       MOV   ES,AX
000C 33 F6                       XOR   SI,SI              ;address location A0000H
000E B9 01E0                     MOV   CX,480             ;set line count
0011                   MAIN1:
0011 E8 000F                     CALL  LINE               ;display a line
0014 E2 FB                       LOOP  MAIN1
0016                   MAIN2:
0016 B4 06                       MOV   AH,6               ;wait for any key
0018 B2 FF                       MOV   DL,0FFH
001A CD 21                       INT   21H
001C 74 F8                       JE    MAIN2
001E B8 4C00                     MOV   AX,4C00H           ;exit to DOS
0021 CD 21                       INT   21H

0023                   MAIN      ENDP
```

```
0023                    LINE    PROC  NEAR

0023 BA 03CE                    MOV   DX,3CEH              ;address GAR
0026 B0 08                      MOV   AL,8                 ;select bit mask
0028 EE                         OUT   DX,AL

0029 BA 03CF                    MOV   DX,3CFH              ;address BMR
002C B0 FF                      MOV   AL,0FFH
002E EE                         OUT   DX,AL

002F BA 03C4                    MOV   DX,3C4H              ;select SAR
0032 B0 02                      MOV   AL,2
0034 EE                         OUT   DX,AL

0035 BA 03C5                    MOV   DX,3C5H              ;address MMR
0038 B3 0F                      MOV   BL,0FH               ;color F
003A            LINE1:
003A B7 05                      MOV   BH,5                 ;count of 5
003C            LINE2:
003C B0 0F                      MOV   AL,0FH
003E EE                         OUT   DX,AL
003F AC                         LODSB
0040 C6 44 FF 00                MOV   BYTE PTR [SI-1],0    ;color to black
0044 8A C3                      MOV   AL,BL
0046 EE                         OUT   DX,AL                ;select new color
0047 C6 44 FF FF                MOV   BYTE PTR [SI-1],0FFH ;write new color
004B FE CF                      DEC   BH
004D 75 ED                      JNZ   LINE2                ;write 40 PELs
004F FE CB                      DEC   BL                   ;write all colors
0051 79 E7                      JNS   LINE1
0053 C3                         RET

0054                    LINE    ENDP

0054                    CODE    ENDS

                                END   MAIN
```

46.

EXAMPLE C–18

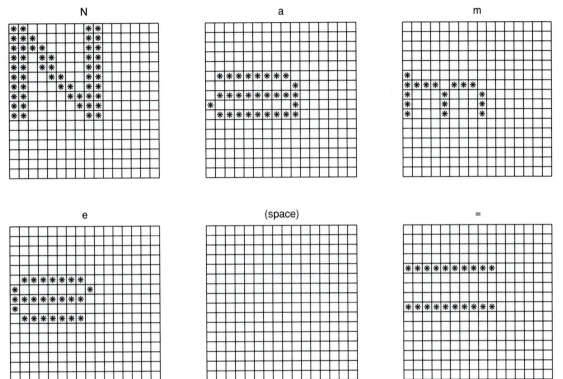

Chapter 8

2. Micro-floppy disk
4. 1.44M bytes
6. (a) a track is a complete ring of data on a side of a disk; (b) a sector is a pie-shaped wedge of data within a track; (c) a cluster is usually a grouping of 4 sectors on many hard-disk memory systems; and (d) a cylinder is the pair of tracks on both sides of a floppy disk.
8. The boot sector holds a program called a bootstrap loader that loads DOS into memory.
10. As many as needed.
12. 3
14.

EXAMPLE C–19

```
0000                    OPEN    PROC  NEAR

0000 B8 3D00                    MOV   AX,3D00H           ;open for read-only
0003 BA 234F R                  MOV   DX,OFFSET FILEN
```

```
0006 CD 21                    INT     21H
0008 C3                       RET

0009               OPEN       ENDP
```

16.

EXAMPLE C–20

```
0000               DATA       SEGMENT

0000 46 52 4F 47 2E  FILE     DB      'FROG.LST',0
     4C 53 54 00
0009 0200 [          BUFF     DB      512 DUP (0)
         00
             ]

0209               DATA       ENDS

0000               CODE       SEGMENT 'CODE'
                              ASSUME       CS:CODE,DS:DATA
                   ;note that this program is written for an error free environment.
0000               MAIN       PROC        FAR

0000 B8 ---- R               MOV     AX,DATA
0003 8E D8                   MOV     DS,AX

0005 B4 3C                   MOV     AH,3CH              ;create
0007 33 C9                   XOR     CX,CX
0009 BA 0000 R               MOV     DX,OFFSET FILE
000C CD 21                   INT     21H
000E 8B D8                   MOV     BX,AX               ;handle to BX

0010 B4 40                   MOV     AH,40H              ;write
0012 B9 0200                 MOV     CX,512
0015 BA 0009 R               MOV     DX,OFFSET BUFF
0018 CD 21                   INT     21H

001A B4 3E                   MOV     AH,3EH              ;close
001C CD 21                   INT     21H
001E B8 4C00                 MOV     AX,4C00H            ;exit to DOS
0021 CD 21                   INT     21H

0023               MAIN       ENDP

0023               CODE       ENDS

                              END     MAIN
```

18.

EXAMPLE C–21

```
APPEND   MACRO  BUFFER,COUNT,HANDLE,BUFFER
         LOCAL  APPEND1

         MOV    AX,4202H                    ;;find end of file
         XOR    CX,CX
         XOR    DX,DX
         MOV    BX,HANDLE
         INT    21H
         JC     APPEND1
         MOV    AH,40H
         MOV    CX,COUNT
         MOV    DX,OFFSET BUFFER
         INT    21H
APPEND1:

         ENDM
```

20.

EXAMPLE C–22

```
ACCESS   MACRO  BUFFER,RECORD,HANDLE
         LOCAL  ACCESS1

         MOV    AX,RECORD
         MOV    CX,100
         MUL    CX
         XCHG   DX,AX
         MOV    CX,AX
         MOV    AX,4200H
         MOV    BX,HANDLE
         INT    21H                         ;;locate record
         JC     ACCESS1
         MOV    AH,3FH                      ;;read record
         MOV    CX,100
         MOV    DX,OFFSET BUFFER
         INT    21H
ACCESS1:

         ENDM
```

Chapter 9

2. 35H
4. 25H

6. A .COM program must be organized with all of its data in a single memory segment. The first executable step of the .COM program must exist at memory location 100H.

8.

EXAMPLE C–23

```
0000                    CODE   SEGMENT 'CODE'
                               ASSUME    CS:CODE

0000                    MAIN   PROC  FAR

0000  B0 B6                    MOV   AL,0B6H           ;set timer
0002  E6 43                    OUT   43H,AL

0004  BA 0012                  MOV   DX,12H            ;find count
0007  B8 34DC                  MOV   AX,34DCH
000A  BB 0BB8                  MOV   BX,3000           ;3000 Hz
000D  F7 F3                    DIV   BX

000F  E6 42                    OUT   42H,AL            ;program count
0011  8A C4                    MOV   AL,AH
0013  E6 42                    OUT   42H,AL

0015  E4 61                    IN    AL,61H            ;speaker on
0017  0C 03                    OR    AL,3
0019  E6 61                    OUT   61H,AL

001B  33 C0                    XOR   AX,AX             ;segment 0000H
001D  8E C0                    MOV   ES,AX

001F  BA 0024                  MOV   DX,36             ;get clock tick plus 36
0022  33 C9                    XOR   CX,CX
0024  26: 03 16 046C           ADD   DX,ES:[46CH]
0029  26: 13 0E 046E           ADC   CX,ES:[46EH]

002E                    MAIN1:                         ;wait for 2 seconds

002E  26: 8B 1E 046C           MOV   BX,ES:[46CH]
0033  26: A1 046E              MOV   AX,ES:[46EH]
0037  2B DA                    SUB   BX,DX
0039  1B C1                    SBB   AX,CX
003B  72 F1                    JC    MAIN1             ;repeat for 2 seconds

003D  E4 61                    IN    AL,61H            ;speaker off
003F  34 03                    XOR   AL,3
0041  E6 61                    OUT   61H,AL

0043  B8 4C00                  MOV   AX,4C00H          ;exit to DOS
0046  CD 21                    INT   21H
```

```
0048                    MAIN    ENDP

0048                    CODE    ENDS

                                END    MAIN
```

10. The immediate data loaded into BX at memory location 1179H determine the tone generated by the speaker.
12. By loading the number of paragraphs required into the DX register before executing the DOS function 31H.
14. A special key or combination of keys that invokes execution of a TSR.
16. By using INT 15H
18. Bit position 0 of location 0000:0417H.
20. Access time through the video BIOS is very slow compared to the OUT instruction and its direct access to the PEL registers.
22.

EXAMPLE C–24

```
0000                    CODE    SEGMENT 'CODE'
                                ASSUME      CS:CODE
                        .286                                    ;80286 or newer

                                ORG     100H

0100 E9 0087            MAIN:   JMP     START

0103 00000000           VEC9    DD      ?                       ;original vector 9
0107 00000000           VEC8    DD      ?                       ;original vector 8
010B 00                 H_FLAG  DB      0                       ;hot-key flag
010C 57                 H_CODE  DB      57H                     ;F11 code
010D 08                 MASKS   DB      8                       ;alternate mask
010E 08                 HOT     DB      8                       ;alternate key

010F                    VECS9:
010F FB                         STI                             ;interrupts on
0110 50                         PUSH    AX
0111 E4 60                      IN      AL,60H                  ;get scan code
0113 2E: 3A 06 010C R           CMP     AL,H_CODE               ;test for hot-key
0118 74 09                      JE      VECS92                  ;if possible hot-key
011A                    VECS91:
011A 58                         POP     AX
011B FA                         CLI                             ;interrupts off
011C 9C                         PUSHF                           ;do original vector 9
011D 2E: FF 1E 0103 R           CALL    CS:VEC9
0122 CF                         IRET
0123                    VECS92:
0123 06                         PUSH    ES
```

```
0124 33 C0                        XOR     AX,AX
0126 8E C0                        MOV     ES,AX
0128 26: A0 041                   MOV     AL,ES:[417H]              ;get key status
012C 2E: 22 06 010D R             AND     AL,MASKS
0131 2E: 3A 06 010E R             CMP     AL,HOT
0136 07                           POP     ES
0137 75 E1                        JNE     VECS91                   ;if not alternate F11
0139 FA                           CLI                              ;interrupts off
013A E4 61                        IN      AL,61H                   ;throw away F11
013C 0C 80                        OR      AL,80H
013E E6 61                        OUT     61H,AL
0140 24 7F                        AND     AL,7FH
0142 E6 61                        OUT     61H,AL
0144 B0 20                        MOV     AL,20H                   ;reset interrupt controller
0146 E6 20                        OUT     20H,AL
0148 FB                           STI                              ;enable interrupts
0149 58                           POP     AX
014A 2E: C6 06 010B R             MOV     H_FLAG,0FFH              ;indicate hot-key
     FF
0150 CF                           IRET
0151               VECS8:
0151 2E: 80 3E 010B R             CMP     H_FLAG,0                 ;test hot-key flag
     00
0157 75 05                        JNZ     VECS81                   ;if hot-key active
0159 2E: FF 2E 0107 R             JMP     CS:VEC8                  ;do original vector 8
015E               VECS81:
015E 2E: C6 06 010B R             MOV     H_FLAG,0                 ;clear hot-key flag
     00
0164 9C                           PUSHF                            ;do original vector 8
0165 2E: FF 1E 0107 R             CALL    CS:VEC8
016A FB                           STI                              ;enable interrupts
016B 60                           PUSHA
016C 2A FF                        SUB     BH,BH                    ;get page
016E B4 08                        MOV     AH,8
0170 CD 10                        INT     10H
0172 8A DF                        MOV     BL,BH                    ;blank screen
0174 8A FC                        MOV     BH,AH
0176 2B C9                        SUB     CX,CX
0178 BA 184F                      MOV     DX,184FH
017B B8 0600                      MOV     AX,0600H
017E CD 10                        INT     10H
0180 8A FB                        MOV     BH,BL                    ;home cursor
0182 2B D2                        SUB     DX,DX
0184 B4 02                        MOV     AH,2
0186 CD 10                        INT     10H
0188 61                           POPA
0189 CF                           IRET
018A               START:
018A 8C C8                        MOV     AX,CS
018C 8E D8                        MOV     DS,AX                    ;address data segment
```

```
018E B8 3509              MOV   AX,3509H              ;get vector 9
0191 CD 21                IN    21H
0193 2E: 89 1E 0103 R     MOV   WORD PTR VEC9,BX
0198 2E: 8C 06 0105 R     MOV   WORD PTR VEC9+2,ES

019D B8 3508              MOV   AX,3508H              ;get vector 8
01A0 CD 21                INT   21H
01A2 2E: 89 1E 0107 R     MOV   WORD PTR VEC8,BX
01A7 2E: 8C 06 0109 R     MOV   WORD PTR VEC8+2,ES

01AC B8 2509              MOV   AX,2509H              ;install new vector 9
01AF BA 010F R            MOV   DX,OFFSET VECS9
01B2 CD 21                INT   21H

01B4 B8 2508              MOV   AX,2508H              ;install new vector 8
01B7 BA 0151 R            MOV   DX,OFFSET VECS8
01BA CD 21                INT   21H

01BC BA 018A R            MOV   DX,OFFSET START       ;make TSR
01BF C1 EA 04             SHR   DX,4
01C2 42                   INC   DX
01C3 B8 3100              MOV   AX,3100H
01C6 CD 21                INT   21H

01C8             CODE     ENDS

                 END      MAIN
```

Chapter 10

2. (1) Word: $\pm32K$, (2) short (32-bit): $\pm2.0 \times 10^9$, and (3) long (64-bit): $\pm9.0 \times 10^{18}$.

4. (1) Short: 32 bits, (2) long: 64 bits, and (3) temporary: 80 bits.

6. (a) -7.5, (b) $+0.5625$, (c) $+306$, (d) $+2.0$, (e) $+10$, and (f) $+0.0$.

8. The microprocessor is free to fetch and execute normal (noncoprocessor) instructions.

10. The rounding mode is selected through the control register by using the FLDCW instruction.

12. All data are stored as 80-bit floating-point numbers within the coprocessor.

14. Truncate toward zero

16. ESC (escape)

18. The 64-bit floating-point number stored at memory location DATA is loaded to the top of the stack as an 80-bit temporary floating-point number.

20. FADD ST,ST(3)

22. FSUBR ST,ST(2)

24. Reverse division divides the top of the stack into the memory operand.

26. $2^x - 1$

28. This clears register ST(2)

30. The FSAVE instruction saves the entire state of the coprocessor to the memory.
32.

EXAMPLE C–25

```
0000                    CODE    SEGMENT 'CODE'
                        ASSUME      CS:CODE

                        .387

0000 00000000   XL      DD    ?
0004 42700000   F       DD    60.0                    ;60 Hz
0008 41700000   L       DD    15.0                    ;15 H

000C            REACT   PROC FAR

000C 9B D9 E8           FLD1
000F 9B D8 C0           FADD  ST,ST(0)
0012 9B D9 EB           FLDPI
0015 9B DE C9           FMUL
0018 9B 2E: D8 0E 0004 R    FMUL  CS:F
001E 9B 2E: D8 0E 0008 R    FMUL  CS:L
0024 9B 2E: D9 1E 0000 R    FSTP  CS:XL
002A CB                 RET

002B            REACT   ENDP

002B            CODE    ENDS

                        END   REACT
```

34. Whenever the microprocessor must wait for a result from the coprocessor.

INDEX